AF335280

Aurora Torealis

*Studies in the History of Science and Ideas
in Honor of Tore Frängsmyr*

Marco Beretta
Karl Grandin
Svante Lindqvist

Editors

SCIENCE HISTORY PUBLICATIONS/USA
a division of
Watson Publishing International LLC
Sagamore Beach
2008

First published in the United States of America
by Science History Publications/USA
a division of
Watson Publishing International LLC
P. O. Box 1240, Sagamore Beach, MA 02562

www.shpusa.com

*This book has been produced thanks to the generous support of Lisbet Rausing;
The Bank of Sweden Tercentenary Foundation;
Sven och Dagmar Saléns Stiftelse; and the Nobel Foundation.*

Library of Congress Cataloging-in-Publication Data

Aurora Torealis : studies in the history of science and ideas in honor of Tore Frängsmyr /
Marco Beretta, Karl Grandin, Svante Lindqvist, editors.
     p. cm.
  Includes bibliographical references and index.
  ISBN 978-0-88135-398-3 (alk. paper)
  1. Science—History. 2. Science—Philosophy. I. Frängsmyr, Tore, 1938– II. Beretta,
Marco. III. Grandin, Karl. IV. Lindqvist, Svante.
  Q125.A83 2008
  509—dc22

                   2008005311

Endpapers: Prints of auroras from *Voyages de la Commission scientifique du Nord en Scandinavie, en Laponie, au Spitzberg et aux Feröe pendant les années 1838, 1839 et 1840 sur la corvette La Recherche, commandée par M. Fabvre* (Paris: A. Bertrand, 1842–45). These prints were given to J. J. Berzelius by Paul Gaimard 24 July 1844. (front cover): "Apparence de l'aurore boreale dans le sud à Bossekop (Finmark), le 19 Janvier 1839, à 7h 40m du soir"; (back cover): "Apparence de l'aurore boreale dans le nord-est à Bossekop (Finmark), le 16 Janvier 1839, à 10h 5m du soir". Bossekop is Sami for Whale Creek and is located in the very north of Norway near the Alta River.

Designed and typeset by Publishers' Design and Production Services, Inc.
Sagamore Beach, Massachusetts. *www.pdps.com*

Manufactured in the U.S.A.

Contents

Preface   v

*Enlightenment in Antiquity?: Evolution and Progress in
the Fifth Book of Lucretius'* De rerum natura   *1*
Marco Beretta

*Leonardo da Vinci's Concept of "Nature":
"More Cruel Stepmother than Mother"*   *13*
Paolo Galluzzi

*How Torricelli Improved on Galileo's Laws of Free Fall
and Projectile Motion*   *31*
William R. Shea and Tiziana Bascelli

*Solomon's Houses: Francesco Bianchini's
Academic Places*   *51*
John L. Heilbron

The Origin of Species from Linnaeus to Darwin   *71*
Nicolaas Rupke

*The Natural Economy of Households:
Charles Darwin's Account Books*   *87*
Janet Browne

*The Persistent Use and Abuse of Wilhelm von Humboldt
in History and Politics—noch einmal 111*
Thorsten Nybom

*Fruit Nationalism: Horticulture in the United States—
From the Revolution to the First Centennial 129*
Daniel J. Kevles

Jules Verne—The Divided Modernist 147
Jean-François Battail

Toward a History of Reason 165
Lorraine Daston

*Experience-Experiments: The Changing
Experiential Basis of Physics 181*
H. Otto Sibum

Intermediate Theoretical Physics 193
Karl Grandin

*A Nobel Prize for Scientific Revolutions?:
Hannes Alfvén and Conflicting Cosmologies 215*
Svante Lindqvist

*Re-Reading Bernal: History of Science at the Crossroads
in 20th-Century Britain 235*
Mary Jo Nye

History of Science in the Age of Policy 259
Sven Widmalm

A Note on the "Integrity" of the University 277
Sheldon Rothblatt

Old Books and E-Books 299
Robert Darnton

Notes on Contributors 305

Index 309

Preface

*T*he compass-needle points to the north. In 1747, the astronomer Anders Celsius of Uppsala University and his assistant Olof Hiorter showed the connection between the variations in the position of the compass-needle and the *Aurora Borealis*. Their findings were published in the Proceedings of the Royal Swedish Academy of Sciences; an academy founded by Linnaeus and others only a few years earlier. Celsius' and Hiorter's discovery is one of the earliest examples of Swedish science reaching out and influencing the international scientific community. An earlier example is the young Linnaeus's renowned journey in the north of Sweden in 1732; a journey which brought him laurels when he toured the continent in his *Lapland* dress in the 1730s.

Tore Frängsmyr—to whom this volume is dedicated on the occasion of his 70th birthday on July 8, 2008—has like Celsius and Linnaeus worked with both Uppsala University and the Royal Swedish Academy of Sciences as his institutional base. Like Celsius and Linneaus in their days, he, too, has been instrumental in reaching out.

Tore Frängsmyr was appointed to a personal Chair in the History of Science in 1982 by the Swedish Research Council for the Humanities and Social Sciences (HSFR); a chair which was made permanent at Uppsala University in 1984. Since then, the development of the history of science and ideas in Sweden has benefited from the prolific and generous contributions by Tore. His impressive scholarly production has been coupled by his extraordinary successful attempt to renew and expand its institutional setting. The Office for History of Science at Uppsala University and the Center for History of Science at the Royal Swedish Academy of Sciences in Stockholm testify to the enduring significance of Tore's effort.[1] In his scholarly work he has focused on the sciences in 18th-century Enlightenment (to the

---

[1] Tore Frängsmyr, *Avdelningen för vetenskapshistoria 1982–2002*, Salvia Småskrifter No. 3 (Uppsala: Avdelningen för vetenskapshistoria, 2003); *idem, Vetenskapshistoria under 25 år 1982–2007*, Salvia Småskrifter No. 9 (Uppsala: Avdelningen för vetenskapshistoria, 2007). Cf. "Leaving His Chair after Twenty-five Years," *Uppsala Newsletter History of Science*, No. 37 (Spring 2006), pp. 1–2.

extent that he admits that such a thing ever existed). In 2000, Tore published his *magnum opus*, a two-volume work on Swedish history of science, ideas and learning, *Svensk idéhistoria: Bildning och vetenskap under tusen år.*

In the early 2000s, with an impressive scholarly career which few could match to his credit—and when a younger generation already had begun to go about tattling about his possible successor—Tore surprised them all. In 2002, he was appointed to a new chair: the Hans Rausing Chair in History of Science, a new personal chair at Uppsala University donated by the historian of science Dr. Lisbet Rausing in honor of her father. Tore's scholarly output has continued to grow since then, e.g. in 2006 he published his remarkable book on the Peking man, *Pekingmänniskan: En historia utan slut.*

The compass-needle points to the north when it comes to Tore's background in Västerbotten, a mythical province in *Ultima Thule* described by him in almost Rudbeckian terms. Many times have we heard him say of any well-known Swede "You know, it's very interesting. He [*or she*] comes from the same village as I do." To our minds, this village must have been a huge metropolis, the size of a major capital. Yes, it seems as if the higher echelons of Swedish academic, cultural, ecclestical, business and political life are saturated with childhood friends, neighbours and third cousins of Tore from this modest little hamlet in Västerbotten.

Tore loves, tongue in cheek, to exagerate the harsh conditions of his northern background under the bleak Pole Star. Of this there is little when on December 10 each year he hands the Nobel medals and diplomas to the King during the Nobel Prize Award Ceremony; his Linnaeus medal jingling against the other medals on the chest of his formal attire; its lapels being embroidered with the Pole Star of the Royal Swedish Academy of Sciences.

The contributions to the history of science and ideas from antiquity to the present day published in this volume are offered as a tribute to an exceptional career, and as a gift to Tore Frängsmyr from his many friends in the international scholarly community of the history of science and ideas. The compass-needle does indeed point to the north, influenced as it is by *Aurora Torealis.*

Firenze, Uppsala, and Stockholm
Marco Beretta, Karl Grandin, and Svante Lindqvist

Enlightenment in Antiquity?

Evolution and Progress in the Fifth Book of *Lucretius'* De rerum natura[1]

MARCO BERETTA

On various occasions during the past century, historians and philologists have suggested, albeit with due prudence, that among the Greek philosophers there was a conscious awareness of progress in human knowledge, and that they saw in this process yet further evidence of their intellectual superiority to the Romans. With regard to the latter in fact, there has been an unusual consensus of opinion among historians as to the congenital resistance of Latin culture to any sort of innovation. In 1939, in his celebrated work *Roman Revolution*, Ronald Syme described this reactionary mentality in a passage that reads with undiminished authority three-quarters of a century later:

> The Romans as a people were possessed by an especial veneration for authority, precedent and tradition, by a rooted distaste of change unless change could be shown to be in harmony with ancestral custom, 'mos maiorum'—which in practice meant the sentiments of the oldest senators. *Lacking any perception of the dogma of progress—for it had not yet been invented—the Romans regarded novelty with distrust and aversion.* The word 'novus' had an evil ring.[2]

While providing a more nuanced and suggestive framework, Elisa Romano concurred that "[. . .] in Rome the very idea of novelty was accepted with the greatest reluctance"[3] and that any stimulus to recognize the need for change arose only during moments of profound political transformation, such as the gradual and inexorable mutation of the Republic into Empire. This transformation was experienced as a genuine crisis in cultural identity and therefore assumed an ambiguous, primarily negative connotation. Significantly, Elisa Romano identified in Lucretius the sole exception to the general tendency among Romans to regard the *res novae* as perilous threats to hallowed tradition and the constituted order.

Indeed, of all the classical authors who have left concrete traces of a philosophy of history guided by the idea of a progressive evolution in events, Lucretius is certainly the most explicit. His six-book philosophical poem *De rerum natura* (*DRN: On the nature of things*) not only contains numerous references, but above all utilizes terms that unequivocally express the concept of a historical progression. All the same, it remains a bitterly disputed question whether Lucretius actually sustained with conviction the idea that human history consisted of a progressive unfolding of events, and given the sheer volume of the literature that has been published on this topic over the past sixty years,[4] it will not be possible here to summarize, even in the broadest outlines, the complex and subtle arguments presented by various scholars, in which ideological interpretation has often taken precedence over a contextual reading of the poem.

While there are more than a few objectively controversial points in *DRN*, one can only arrive at a full understanding of Lucretius' position on progress after a careful reading of the entire text of Book V rather than—as is usually done—concentrating on its concluding verses. It is also necessary to take into account the significance of Book V within the overall structure and philosophical design of the poem. Only in this way, I believe, will it be possible to overcome the perception of a fundamental cosmic pessimism in the prophecy divulged to Memmius at the outset (Book V, 95–96 and 104–106) regarding the imminent collapse of the machine of the world due to a cataclysmic earthquake. Or the picture of irreversible decay painted by Lucretius when he likens the sterility of the earth to "a woman worn with the lapse of age" (V, 827), a vision that would have done little to raise his readers' hopes for a better future, prompting them instead—in accordance with time-honored literary tradition—to look back with nostalgia on the fecundity of nature in an irrevocably lost golden age.

And yet, contrary to general belief, not only are these images (which are echoed in other passages of *DRN*) consistent with the fundamental optimism of Epicureanism—they constitute one of its most profound doctrinal justifications. Indeed, man will succeed in liberating himself from the terrors of religion only when he is able to provide a scientific, naturalistic, and immanent explanation of death. The alternation of life and death, which Lucretius describes in the first four books of *DRN* as the result of the combining and the dissolution of atoms and molecules (*concilia*), leads ineluctably to a discussion at the macroscopic level of the death and dissolution of worlds, and therefore of our own earth, as inevitable consequences of this philosophical emancipation. Life and death, evolution and dissolution are closely intertwined in a union whose dynamics are revealed by the molecular structure of matter and whose outcome, manifested by the solidity of the atoms that constitute its essence, will in this endless cycle always produce the same quantity of matter.

Within this framework of dynamic equilibrium, death and the proximity of the world's end are typical *eventa*, transitory qualities of matter that, like the qualities of life, are necessarily and eternally destined to mutate. Changes in natural phenomena are in fact attributed by Lucretius, and by the atomist philosophers in general, to the perpetual motion of these atoms. Lucretius provides the elements necessary to understand these changes and anchors them in a comprehensive and rational conception of nature in the first two books of *DRN*, in his description of the forms of the atoms, their motion, their weight and their combinations. The atom therefore is the microscopic but inherent building block of matter, providing an explanation for phenomena that other philosophies of nature, *in primis* that of Aristotle, were forced to account for by returning to an ontological, metaphysical and immutable principle of nature. According to Epicurean atomism, while the atom on the individual level is immutable and eternal, its true essence is manifested only when it comes into contact with and combines with other atoms, expressing itself through its motion and specific weight, and generating constant and often unpredictable variations.

The fifth book of *DRN* is dedicated to the origin of the universe, the world, and our civilization. Here the history of nature and the history of civilization converge in a unified design where there is no longer room for belief in divine providence or an Aristotelian teleology, and the passage of time is marked by the inexorable attraction, dissolution, and regeneration of atoms. Aware of the exceptional novelty of this idea, in his proem (V, 1–17) Lucretius cites with reverence Epicurus, the philosopher whose creative genius paved the way for these new, almost divine discoveries—*repertis* that constituted a decisive step forward in terms of human progress, comparable to the early inventions of agriculture and viticulture. Going beyond mankind's achievements in the arts and technology, which he nevertheless adopted as his departure point and basis of comparison, Lucretius shows how the teachings of Epicurus unveiled the underlying cause of things (*sapientia*), conducting the human spirit out of the shadows into a safe, sunlit haven. The acquisition of knowledge, or more exactly of the tools required to attain this knowledge, was therefore a fairly recent step, and one that signified for Lucretius a crucial moment of transition, and of progress in human culture. The acceptance of the natural philosophy of Epicurus in fact denoted man's attainment of self-awareness and his consequent emancipation. For centuries to come, science would never succeed as well as Epicureanism in qualifying itself as the principal instrument of human emancipation, leaving this role to be filled first and foremost by religion, and then by philosophy.

After demonstrating the transience and mortality of the world and the necessity for the formation of new aggregations and new worlds, Lucretius links the fate of the universe to that of human history, associating the destiny of man with that of the cosmos while at the same time rejecting the classical concept of the cyclic nature of history:

> If there was no birth and beginning of the earth and sky, and they were always everlasting, why beyond the Theban war and the doom of Troy have not other poets sung of other happenings as well? Whither have so many deeds of men so often passed away? Why are they nowhere enshrined in glory in the everlasting memorials of fame? But indeed, I believe, our world is in its youth, and quite new is the nature of the firmament, nor long ago did it receive its first-beginnings. Wherefore even now certain arts are being perfected, even now are progressing (*augescunt*); much now has been added to ships, recently musicians gave birth to tuneful harmonies. (*DRN*, V, 324–337)

Evidently therefore, human history has achieved progress, such that the written record of the events marking this history and the consequent birth of the collective memory were relatively recent developments. The world itself, although destined to dissolution,[5] still exhibited—through man's progress in the arts and in the new natural philosophy of Epicurus—the dynamic characteristics of its youthful age and rapid evolution. Here the intentions of the poet and the consequentiality of the apparently pessimistic opening to Book V become clearer: born into a world already old, tired and infertile, man has been forced to develop arts that would favor the development of new ways of sustaining himself through his own labor and ingenuity. It appears that these useful arts triggered the impulse to create the other arts that were more closely linked to abstract knowledge, such as music.

Lucretius, who was proud to present himself as the first Latin advocate of the philosophy of Epicurus, associates his teacher's discoveries (*reperta*) on the nature of things with the—obviously much more recent—advances in the arts of navigation and music, making an illuminating connection between progress in the arts and the true atomist doctrine that provided the substrate for the evolution of the cosmos. The very recentness of these advances—signs of the birth of civilization—is used by Lucretius to demonstrate the relatively young age of the world and therefore its intrinsic temporality.

Typically, daily experience demonstrated—through progress—the transformation of time into historical experience, and served Lucretius to shed light on the evolution of the cosmos, which is subject to the same laws and hence the same alternations as the birth and development of civilizations.

After describing the non-teleological origin of the cosmos, the movement of the planets, the shape of the earth, and the cause of eclipses, Lucretius turns his attention, beginning with verse 772, to the origins of life on earth and how, during the youth of the world, a still prosperous and fecund Mother Earth was capable of generating not only the animal species known to us today (*mortalia saecla*), but also many prodigious creatures (*portenta*) "born with strange faces and strange limbs"

(*DRN*, V, 838). Despite the perennially mild climate of those remote times, that knew no extremes of heat or cold, nature did not allow unadapted species to reproduce and only those animals endowed with "cunning," "valor," and "swiftness" (*DRN*, V, 858) managed to secure their lineages and survive to our own day. A cruel struggle for survival seems to have animated the first stirrings of the world, which Lucretius describes unperturbedly:

> Wherefore, again and again, rightly has the earth worn and keeps the name of mother, since she herself formed the race of men, and almost at a fixed time brought forth every animal which ranges madly everywhere on the mighty mountains, and with them the fowls of the air with their diverse forms. But because she must needs come to some end of child-bearing, she ceased, like a woman worn with the lapse of age.[6] For time changes the nature of the whole world, and one state after another must needs overtake all things, nor does anything abide like itself; all things change, nature alters all things and constrains them to turn. *DRN* (V, 821–831)

This passage unmistakably evinces a world whose myriad species did not appear simultaneously in a creative act analogous to the one attributed by the metaphysical philosophers to God, but at separate moments in history and under different environmental conditions. According to Lucretius, therefore, it seems that time plays a fundamental role in biological transformations as well.

The possibility that such a temporal progression might include a process of evolution is apparently denied by Lucretius' assertion that the species "all preserve their separate marks by fixed natural laws" (*DRN*, V, 924) and that every living thing grows in accordance with its own established rhythm, never transcending its essential characteristics. And yet, at various points the poet describes the changes that different species can undergo with the passage of time and in response to changes in the environment. Indeed, *natura creatrix* herself provided a simple demonstration—in the natural grafting of plants—of how the ecosystem was susceptible to change. The combination of artifice and nature that gives rise to evolution is captured by Lucretius in a lyrical passage:

> But nature herself, creatress of things, was first a pattern for sowing and the origin of grafting (*origo insitionis*[7]), since berries and nuts fallen from the trees in due time put forth swarms of shoots beneath. (*DRN*, V, 1361–1363)

Grafting is plainly an event that produces a wholly new species and the aim of Lucretius in ascribing the original process to nature is to eliminate the traditional opposition between *physis* and *techne* by showing how man's technical manipulations are simply the result of his emulation of nature. Without attempting to read into Lucretius' words an (improbable) presage of Darwin's theory of evolution, they can be interpreted as a negation of the concept of biological fixity, and the indication of a position disposed to a less schematic vision of life than we are accustomed to think. For the rest, there were other *auctoritates* who challenged the Aristotelian notion of the fixity of species. Theophrastus, for instance, in the opening to Book IV of *De causis plantarum* admitted that some cereal plants could change species as a result of cultivation or other circumstances, and in *Historia plantarum* (II, 13.3) he cited the case of rice being transformed into wheat by artificial intervention, i.e., by man beating the grains.[8]

In a vivid exposition, Lucretius uses the prehistory and history of man to illustrate this process of biological transformism. Primitive man was equipped by nature with a robust physique that "was built up on larger and more solid bones within, fastened with strong sinews traversing the flesh; not easily to be harmed by heat or cold or strange food or any taint of the body" (*DRN*, V, 925–930). Thanks to these biological characteristics adapted to resist a hostile environment, primitive men "prolonged their lives after the roving manner of wild beasts" (V, 932) and lived without the assistance of any arts, simply following their instincts, "trusting in their wondrous strength of hand and foot" (*DRN*, V, 966). When they were tired, "like bristly boars, these woodland men would lay their limbs naked on the ground . . . wrapping themselves up around with leaves and foliage" (*DRN*, V, 970–972). The condition of these wretched (*miseris*) beings—at the mercy of the most disparate adversities and perils—was made bearable only by their ignorant state, unconscious of events and incapable of imagining how they might compete with the overwhelming forces of nature.

Everything changed once man discovered fire, because ". . . their chilly limbs could not now so well bear cold under the roof of heaven" (*DRN*, V, 1015). This discovery came about by sheer serendipity—a lightning bolt or the friction of tree branches that resulted in a burst of flames and demonstrated to our wondering ancestors how heat could be propagated. "Either of these happenings may have given fire to mortals. And then the sun taught them how to cook food and soften it by the heat of the flame" (*DRN*, V, 1101–1102). The mastery of fire brought about a gradual change in the morphology of man, who lost his primitive constitution and could no longer survive like the other animals in the wild, where he indeed now risked extinction. It was therefore not nature that dictated man's evolution in a particular direction; it was the discovery of fire and the possibility of recreating it artificially that, according to Lucretius, influenced man's biological evolution even before his cultural development. The softening of his constitution by the warmth of fire compelled man first of all to construct a hut for shelter, then to protect his body from the inclement weather with the skins of animals, and finally to recognize himself as a social animal.

> Then, too, neighbours began to form friendships one with another, longing
> neither to hurt nor to be harmed, and they commended to mercy children and
> the race of women, when with cries and gestures they taught by broken words

that 'tis right for all men to have pity on the weak. Yet not in all ways could unity be begotten, but a good part, the larger part, would keep their compacts [*foedera*] loyally; or else the human race would even then have been all destroyed, nor could breeding have prolonged the generations until now. (*DRN*, V 1019–1027)

Hence social evolution became a necessary requirement for the survival of the species, but this development was possible only when men learned to regulate their relationships within the community through language. Disputing Plato and all the other philosophers who believed in the innate origin of the word, Lucretius revived the theory of Democritus,[9] according to whom the names of things were shaped by *utilitas* (*DRN*, V, 1029) and man, like the other animals, elaborated a language that was capable, through the emission of sounds, of expressing his needs and psychological states (*DRN*, V, 1050–1061).

Fire therefore gave rise to society and, as they united around the community hearth, men of necessity learned to build social relationships regulated by sophisticated forms of communication. Once this crucial step had been made, the evolution of man was transformed into the history of civilization:

And day by day those who excelled in understanding and were strong in mind showed them more and more how to change their former life and livelihood for new habits and for fire. Kings began to build cities and to found a citadel, to be for themselves a stronghold and a refuge; and they parcelled out and gave flocks and fields to each man for his beauty or his strength or understanding. (*DRN*, V, 1105–1112)

So it appears that Lucretius assigned a role of particular significance to the native ingenuity that led the first civilized men to extend the applications of fire. By building fortresses and in general transforming themselves into the masters of new and socially useful discoveries, some were able to assume the command of communities and the title of rulers. Of course, not all of these discoveries were equally positive in their consequences and, reviving a traditional theme in the philosophical literature, Lucretius identified the discovery of gold as the source of human avarice and of the resulting contest (fuelled by envy rather than reason) between men to extend their power, influence and territory by means of war. However, even war[10]—that fearsome echo of the bestial origins of man—could be, if not completely eliminated, at least domesticated by the introduction of rules and principles, such that: "the race of men, worn out with leading a life of violence, [. . .] gave in to ordinances and the close mesh of laws" (*DRN*, V, 1145–1148).

Laws, like language and the other discoveries of man, spring from need or contingent circumstances. At this point, Lucretius discusses the origin of religions, yet another invention that contains echoes of the primitive condition of man. According to the poet, the existence of the gods—manifested by the perpetual recurrence of their simulacra in the imagination and in dreams—led man to attribute to them the creation of the world and all natural phenomena, such as the thunderbolts whose regular appearance could instill fear even in the mighty. This type of projection was, according to Lucretius, caused by a fundamental defect in perception (*DRN*, V, 1211) that would only be corrected when men embraced a new philosophy of nature unencumbered by metaphysical preconceptions and prepared to recognize the intrinsic fortuitousness and complexity of the cosmos. As John Masson acutely observed, Lucretius' critique of religion in this passage was aimed at demonstrating the necessity of a cosmos in constant evolution.[11] If one negated the divine origin of the cosmos, its teleological order and its eternal nature, then one was obliged to recognize a process of constant evolution in the cosmos, and *ergo* of perennial progress within the specific context of human history. Yet, as already noted, one cannot assume that this progress was a vector only of positive values, given that (and here Lucretius provides his highly critical analysis of progress in metallurgy) man tended not to employ his discoveries in the pursuit of knowledge, instead allowing himself to be blinded by cupidity, and often resorting to the violence that was its natural consequence.

The principle of utility, which according to Lucretius is the driving force behind progress, functions in identical fashion for the discoveries that lead mankind to vice and those that favor an increase in knowledge and happiness. All the same, the modalities by which these "benefits" accumulate can differ. In the sphere of the material world, objects are substituted out of caprice and any progress is more illusory than real. Lucretius writes in this regard:

> For what is here at hand, unless we have learnt anything sweeter before, pleases us above all, and is thought to excel, but for the most part the better thing found later on destroys it and changes our feeling for all the old things. So loathing for their acorns set in, and the old couches strewn with grass and piled with leaves were deserted. [. . .] And so the race of men toils fruitlessly and in vain for ever, and wastes its life [in] idle cares, because, we may be sure, it has not learned what are the limits of possession, nor at all how far true pleasure can increase. (*DRN*, V, 1412–1415; V, 1430–1432)

These achievements—mere ends in themselves—file by leaving no trace of their passage or their essential cause, which Lucretius declares clearly at the beginning of his reconstruction to be man's reason. Contrary to what is often retained, in this passage Lucretius does not criticize technology and technical progress *per se*, but rather the absence of a theoretical framework to govern them. He is denouncing not the process of evolution dictated by utility, but the applications of its advances and the refusal to acknowledge the philosophical and moral reasons that render this evolution all the more necessary in his eyes. The emancipatory philosophy of Epicurus is in fact not a moral philosophy, but a system of ethics founded upon a true knowledge of nature—that is, a vision of nature in

FIGURE 1. A portrayal of *Veritas filia temporis*. Veritas, shown in the guise of Nature, looks up at Chronos while at her feet two cherubs are engaged in chemical experiments. The image is accompanied by a verse from Lucretius (V, 1452–1453) praising progress realized through the powers of reason. This engraving was drawn from *Description méthodique d'une collection de minéraux, du cabinet de m. D.R.D.L* by Jean Baptiste Romé de l'Isle (Paris: Didot jeune, 1773).

constant and necessary evolution. It would therefore be in contradiction with the entire system of Epicurus to negate the existence of progress in human history. However, in order for progress to be given a positive connotation it is necessary to recognize and appreciate not only its utilitarian origin but also, and above all, its cultural valence. It is within this framework of justified optimism that Lucretius closes Book V:

> Ships and the tilling of the land, walls, laws, weapons, roads, dress, [and] all things of this kind, all the prizes, and the luxuries of life, one and all, songs and pictures, and the quaintly wrought polished statues, practice (*usus*) and therewith the inventiveness of the eager mind (*experientia mentis*) taught them little by little, as they went forward step by step. So, little by little, time brings out each thing into view, and reason raises it up into the coasts of light. For they saw one thing after another grow clear in their mind, until by their arts they reached the topmost pinnacle. (*DRN*, V, 1448–1457)

Necessity, the vestige of our primitive origins, is now guided by *experientia mentis*, a quality that was born in the moment when man became the master, and fully conscious protagonist, of his own history. In this state of self-awareness—at once biological and philosophical—the discoveries made in the arts and sciences become the lamps that will light man's way during his advance, independently of what lies at journey's end, along the road marking the emancipation of the human spirit.

Notes

1. All the citations from Lucretius' *De rerum natura* (hereafter abbreviated as *DRN*) have been taken, with a few modifications, from Volume 1 of *Titi Lucreti Cari De rerum natura libri sex*. Edited with Prolegomena, Critical Apparatus, Translation, and Commentary by Cyril Bailey (Oxford: The Clarendon Press, 1947), 3 vols.
2. Ronald Syme, *The Roman Revolution* (Oxford: Clarendon Press, 1939), p. 315. The italics are mine.
3. Elisa Romano, "L'ambiguità del nuovo: *res novae* e cultura romana" in *Laboratoire Italien. Politique et société*, 6 (2005), pp. 17–35 (p. 20).
4. The most recent is the book by Gordon Campbell, *Lucretius on Creation and Evolution: A Commentary on De Rerum Natura Book Five, Lines 772–1104* (Oxford: Oxford University Press, 2003), which includes an exhaustive bibliography on the subject. Among the most significant studies that underline the originality of Lucretius' position and his idea of progress are: Margaret Taylor, "Progress and Primitivism in Lucretius," *The American Journal of Philology*, 68 (1947), pp. 180–194; William R. Nethercut, "The Conclusion of Lucretius' Fifth Book: Further Remarks," *The Classical Journal*, 63 (1967), pp. 97–106; Jean Bayet, *Mélanges de literature latine* (Rome: Edizioni di storia e letteratura, 1967), pp. 27–84; David J. Furley, "Lucretius the Epicurean. On the History of Man," in *Lucrèce* (Geneva: Fondation Hardt, 1978), pp. 1–37; David Konstan, *Lucrezio e la psicologia epicurea* 2nd revised edition of *Some Aspects of Epicurean Psychology* (Milano: Vita e Pensiero, 2007), pp. 105–153.

5. Lucretius is somewhat vague regarding the proximity of this event, which he counts among the order of necessary things within the framework of atomism. In Book II (1150 ff.), he describes a world by now in inexorable decline and close to destruction. In Book V, he emphasizes the present sterility of the earth compared to past epochs (826–836) and announces to Memmius the impending annihilation of the *machina mundi* (95–95).

6. The biological evolution of man is a clear demonstration of what takes place in a cosmos fixed in the eternal alternation of birth and death.

7. This technical term seems to have been coined by Lucretius himself and was used, albeit infrequently, by Columella and Palladio in their agricultural treatises.

8. Pliny as well refers to the natural origins of grafting in his discussion of its role in the history of agriculture (*Naturali Historia*, XVII, 23–24).

9. Hermann Diels, *Die Fragmente der Vorsokratiker. Herausgegeben von Wilhelm Krantz*, Eighteenth Edition (Zürich, Weidmann, 1989), vol. 3, p. 654; and Salomon Luria, *Democritea* (Leningrad: Nauka, 1970), fr. 558.

10. On the verses devoted to the military arts and to a presumed polemic against Posidonius, see the very interesting considerations of Hermann Diels, "Lukrezstudien. IV," *Sitzungberichte der Preussichen Akademie der Wissenschaften*, 9 (1921): 237–244.

11. "But where religion is ignored, any theory of the world must be based on a belief in Evolution, in a slow progress upward from a primitive chaos." John Masson, *Lucretius Epicurean and Poet* (London: John Murray, 1909), vol. 2, p. 120.

Leonardo da Vinci's Concept of "Nature"

"More Cruel Stepmother than Mother"[*]

PAOLO GALLUZZI

"*H*ere nature appears with many animals to have been rather a cruel stepmother than a mother, and with others not a stepmother, but a most tender mother."[1] This is an isolated text in which the denunciation of the contradictory attitude of nature shows a style evocative of the "prophecies" that recur frequently in Leonardo's manuscripts. A judgment that is similar, even in its terminology, emerges in fact in another prophecy entitled "Of beaten donkeys":

> O indifferent Nature! Wherefore art thou so partial, being to some of thy children a tender and benignant mother, and to others a most cruel and pitiless stepmother? I see children given up to slavery to others, without any sort of advantage, and instead of remuneration for the good they do, they are paid with the severest suffering, and spend their whole life in benefiting their oppressor.[2]

The "beaten donkeys" are, obviously, the humble, obedient men subjected to the tyranny of their masters. In his prophecies Leonardo represents situations in which living beings are suffering under the indifferent gaze of nature. It is a bitter vision, often marked by an accusing finger pointed at religion and the clergy, in order to denounce the foolish *naiveté* of those who expect the wrongs suffered on earth to be repaid by divine justice. An emblematic example of Leonardo's attitude and style is offered by the prophecy entitled "Paradise for sale":

> An infinite number of men will sell publicly and unhindered things of the very highest price, without leave from the Master of them, while they never were theirs nor in their power; and human justice will not prevent it.[3]

The hostile attitude of nature is denounced again in another prophecy: "certainly this seems as though nature wished to eradicate the human race as being useless to the world and as spoiling all created things."[4] A concept reiterated in the prophecy entitled "On the cruelty of man":

Animals will be seen on the earth who will always be fighting against each other with the greatest loss and frequent deaths on each side. And there will be no end to their malice; by their strong limbs we shall see a great portion of the trees of the vast forests laid low throughout the universe; and when they are filled with food, the satisfaction of their desires will be to deal death and grief and labor and fears and flight to every living thing; and from their immoderate pride they will desire to rise towards heaven, but the excessive weight of their limbs will keep them down. Nothing will remain on earth, or under the earth, or in the waters, which will not be persecuted, disturbed, and spoiled, and those of one country removed into another. And their bodies will become the tomb and means of transit of all the living bodies they have killed. O Earth, why dost thou not open and engulf them in the fissures of thy vast abyss and caverns, and no longer display in the sight of heaven so cruel and horrible a monster?[5]

While there are several examples of how Leonardo contrasts a good and provident nature with a fierce, merciless humanity, much more numerous are the texts in which nature appears forced to obey the strict laws imposed on her; laws that do not provide for any privileged treatment of mankind. Nature is, in fact, indifferent to the fate of her own creatures, man included. The false impression that nature is hostile to man's fate, acting in his regard like a "cruel stepmother" rather than like a "loving mother," is only the consequence of his presumption of belonging to a species privileged by nature.

In Leonardo, along with admiration for mankind—"nature's highest instrument"—the awareness that nature does not side with man is thus clearly perceptible. Although Leonardo did not remain insensitive to the model of man as the center and measure of all things—one need only recall his celebrated *Vitruvian Man*—he seems to distance himself from the anthropocentrism of 15th-century culture.

An emblematic example of the early manifestation of his awareness of the human condition as suspended between chance and the drama of nature's indifference appears in a youthful text on the bitter fate of the "animated tool of nature's artifice," the lightning bolt, a metaphor for the haughtiness and presumption of mankind. Obliged to "obey the law of God and the time granted it by its progenitor, nature," the lightning bolt violently discharges its energy, sowing death and destruction, but simultaneously ending its own existence.[6]

Here the "formative nature" appears subject to precise laws imposed by God and by Time. Time is the, "consumer of things [. . .], swift despoiler of all created things [. . .]."[7] Leonardo reflects on the demolishing effects of time, exerted especially through the action of water, which, digging deep into the belly of the planet, is then locked at its center, leaving the land dry:

> In this way the fertile and fruitful earth will remain deserted, arid, and sterile from the water being shut up in its interior, and from the activity of nature it will continue a little time to increase until the cold and subtle air being gone.

The pressure exerted by water will cause the earth to swell; thus it "will be forced to end with the element of fire; and then its surface will be left burnt up to cinder and this will be the end of all terrestrial nature."[8]

The vision of the end of the human species appears central to the concept developed by Leonardo in these texts:

> *Contra*: why did nature not ordain that one animal should not live by death of another?
>
> *Pro*: nature, being inconstant and taking pleasure in creating and making constantly new lives and forms, because she knows that her terrestrial materials become thereby augmented, is more ready and more swift in her creating than time in his destruction; and so she has ordained that many animals shall be food for others. Nay, this not satisfying her desire, to the same end she frequently sends forth certain poisonous and pestilential vapors and frequent plagues upon the vast increase and congregation of animals; and most all upon men, who increase vastly because other animals do not feed upon them; and, the causes being removed, the effects would cease. This earth therefore seeks to lose its like, desiring only continual reproduction as you bring forward and demonstrate by argument; like effects always follow like causes, animals are a type of the life of the world.[9]

The concluding statement, in which the destiny of mankind is likened to that of the world, is clarified in the immediately following passage:

> Now you see that the hope and the desire of returning to the first state of chaos is like the moth to the light, and that the man who with constant longing awaits with joy each new springtime, each new summer, each new month and new year—deeming that the thing he longs for are ever too late in coming—does not perceive that he is longing for his own destruction. But this desire is the very quintessence, the spirit of the elements, which finding itself imprisoned with the soul is ever longing to return from the human body to its giver. And you must know that this same longing is that quintessence, inseparable from nature, and the man is the image of the world.[10]

In this extraordinary text, the anthropocentrism of humanistic culture is replaced by a strongly naturalist vision. Man is indeed the model of the world, but in the sense that, like all of nature, he is subject to the iron law of time. There is no trace of any expectation of an afterlife that will put an end to the ineluctable course toward dissolution. Time dominates nature, and therefore man as well. Man is subject to the common fate of a brief life progressing toward an outcome that will mark the end of "human generation" and of "terrestrial nature." Leonardo responds to this awareness by undertaking a courageous line of research, which, rejecting the promises of religion, strives to throw light on the reality of the human condition. The apologue, modeled on the Platonic myth of the cavern, inserted among these texts, clearly illustrates the young artist's state of mind:

> Unable to resist my eager desire and wanting to see the great multitude of the various and strange shapes made by formative nature [. . .], I came to the entrance of a great cavern [. . .] Bending my back into an arch, I rested my tired hand on my knee and held my right hand over my downcast and contracted eyebrows: often bending first one way and then the other to see whether I could discover anything inside, and this being forbidden by the deep darkness [. . .] two contrary emotions arose in me, fear and desire [. . .].[11]

Fear and desire, this is the dramatic polarity of the human condition, as seen by Leonardo. The desire to know the truth is innate in man, but is hindered by fear, which urges him to put his faith in those who offer soothing hopes.

We should not forget that Leonardo—the young Leonardo in particular—was not an erudite humanist, but above all an artist. It is not surprising, then, to find evident reflections of this naturalistic vision in his paintings from this period, such as the *Saint Jerome*, dating from 1479–80. Here Leonardo emphasizes the identification of man with nature, both portrayed in the aspect of sterile aridity. Moreover, the dramatic tension in the Saint's expression perfectly reflects the attitude of man, suspended between fear and the desire of hope before the mystery of the Revelation. Also, the landscape in which he is immersed appears devoid of any ornament, to underline the identical fate shared by man and nature. In the same months when he was painting the *Saint Jerome*, Leonardo was also working on the *Adoration of the Magi*, in which the same conceptual models appear. The Magi and the throng of onlookers before the Virgin are not portrayed in the attitude of adoration, but rather of suffering stupefaction, as if torn between the desire to believe in the redemption promised by the Advent and the fear of being deceived. We find this ambiguity again in the *Virgin of the Rocks*, inserted in a context that clearly shows both the proliferating force of nature (the luxuriant vegetation in the foreground) and its condition as the theater of ceaseless transformation (the geological stratifications produced by the work of time). Furthermore, the mysterious smile and finger pointing toward the sky of the much later *Saint John the Baptist* express a knowledge of the destiny of mankind, torn between a yearning for eternity and the awareness of a fleeting existence. This smile expresses Leonardo's disenchanted gaze at man and nature, like the inscrutable one of *Mona Lisa*.

Many of Leonardo's texts describe the genealogies that link nature to necessity, to time and to the soul or spirit. In a famous passage, he proclaims the full subordination of nature to necessity: "Necessity is the mistress and guide of nature. Necessity is the theme and the inventress of nature, the curb and law and theme."[12]

In Leonardo's maturity, "necessity" became identified with the principles of geometry, which govern the operations of nature. Each of nature's processes thus unfolds according to a plan of absolute rationality: "In nature there is no effect without cause; once the cause is understood there is no need to test it by experience."[13] The salient feature of nature's operations is maximum economy and absence of superfluity: "Nature operates, according to the causes, in the swiftest possible way"[14]; and again: "nothing is superfluous and nothing is lacking in any species of animals or product of nature."[15] To unveil the processes of nature, man must thus apply rigorous methods of investigation. This is the basic premise that drives Leonardo's efforts to decipher the mathematical principles of nature. Leonardo hails direct experience of natural phenomena as the only way to discover the principles governing the operations of a nature dominated by necessity.

One of the consequences of this conviction is the marked teleological character of the anatomical and geological research conducted by Leonardo in his maturity. He analyzes the human machine and that of the Earth according to the paradigm of the structural relationship between form and function. Necessity has imposed this paradigm on nature, which has rigorously applied it in creating the complex devices of man.

In confronting the "mechanisms" of the human organism, Leonardo strives to determine the relationship between dissected organs and their functions. "Why the muscles of the anus are odd in number? And if this disparity were necessary, why were there three or seven chosen rather than five?"[16] His method of investigating is fixed in strict terms: "Describe the nature of all the organs [. . .] Of nature which of necessity makes the vital instruments of suitable and necessary shapes and positions. How necessity is the companion of nature."[17] Many of Leonardo's demonstrations of the relationship between form and functions of the organs begin with the formula "It is proved by necessity,"[18] or "Necessity causes [. . .],"[19] or again, in regard to the geometric structure of the tricupid valves in the heart, "[. . .] necessarily the thing reflected from the hemicycle *aob* exerts the whole of its utmost power in shutting again the angle *acb*. Thus one has interpreted why nature made these valve cups threefold in number and not fourfold or in any other number."[20] Leonardo's admiration for the edifice of the human body reached its highest peak before the structure of the hand and of the tongue, the two organs that express to the highest degree the "marvelous works of nature"[21] which is the body of man. Dissection shows him the hand as a device capable of performing movements of infinite variety: to exert force, to carry out heavy manual work, or to perform the controlled movements required for playing musical instruments or painting.[22] The tongue, moreover, with its twenty-four muscles, can pronounce all of the names of the things "which are actually in, or in the power of nature."[23]

The iron discipline of nature excludes any possibility of prodigies or miracles. In one of his tirades, justly famous, against sorcerers and magicians, Leonardo declares that it is nature herself who punishes those who claim to counterfeit her operations:

> Mental matters which have not passed through the *senso comune* are vain and they beget nothing but prejudiced truth. And because such discourses arise from poverty of wits, such reasoners are always poor, and if they are born rich they will die poor in old age because it seems that nature revenges herself on those who want to work miracles, so that they have less than other quieter men. And those who want to grow rich in a day live for a long time in great poverty, as happens and will happen to eternity, to the alchemists, searchers after the creation of gold and silver and to those engineers who want dead water to give itself moving life with perpetual motion. And to the supreme fools, the necromancer and the enchanter.[24]

If he cannot emulate nature, man is however capable of manipulating her creations:

> [. . .] man does not differ from animals except in accidental things; in this it is that he shows himself to be a thing divine; for where nature finishes in the production of her forms of species, there man begins to make with the aid of nature an infinite number of forms [. . .].[25]

In another text, his negative judgment on alchemists seems attenuated, finding partial justification in the benefits for treating diseases that derive from the substances generated through their skillful combinations of medicinal herbs: "which function cannot be exercised by nature herself because in her there are no organic instruments with which she might be able to do the work which man performs with his hands [. . .]."[26]

Nature creates both medicinal herbs and creatures through the spiritual principle instilled in her. In Leonardo's desperate effort to subject all of the phenomena of nature to the principles of geometry and mechanics, the theme of the presence of the spiritual principle that animates the world constantly emerges. Leonardo began early to question himself about this fleeting entity. He is convinced that he can find the mechanical causes that govern the dynamic processes, by means of scrupulous observation of the effects. But there is a limit beyond which he knows that the mind of man cannot venture. That limit concerns the life-breathing force "nature's companion." Not all knowledge, then, is resolved through experience: "Nature—he states in a beautiful text—is full of infinite causes that have never occurred in experience."[27]

In a late sheet of his anatomical manuscripts, Leonardo reiterates that nature, in order to operate, requires mechanical instruments. "Why nature cannot give movement to animals without mechanical instruments is demonstrated by me in this book on the active movements made by nature in animals. And for this reason I have drawn up the rules of the four powers of nature without which nothing through her can give local motion to these animals."[28] The four powers (motion, weight, force, percussion) explain the mode of action of the spiritual energy that animates nature. Through the four powers, the quintessence, or soul, generates all of the motions of living beings: "Weight, force, and casual impulse, together with resistance, are the four external powers in which all the visible actions of mortals have their being and their end."[29]

The conviction that nature is impotent without mechanical instruments lies at the basis of Leonardo's striving, in his studies on geology and anatomy, to lay bare the operation of the devices responsible for motion in the machines of the world and of man. The soul of man endeavors to emulate the creative capacity of nature not only by combining its medicinal herbs, but also by attempting to replicate its most marvelous productions. Man's ambition to compete with nature is, however, destined to frustration:

> Though human ingenuity by various inventions with different instruments yields the same end, it will never devise an invention either more beautiful, easier, or shorter than does nature, because in her inventions nothing is lacking, and nothing superfluous; and she does not involve counter-weights when she makes organs suitable for motion in the bodies of animals, but puts there the soul, the composer of the body, that is the soul of the mother which first composes in the womb the shape of man, and in due time awakens the soul that is to be its inhabitant. This at first remains asleep, under the guardianship of the soul of the mother, who nourishes and vitalizes it through the umbilical vein, with all its spiritual organs. And so it continues as long as this umbilical cord is joined to it by the fetal membrane and the cotyledons through which the fetus is united to the mother. And this is the cause why a wish or exceedingly great desire or fright which the mother has, or any mental pain has more power on the child than on the mother, for there are many times when the child thereby loses its life etc.
>
> This discourse does not belong here, but is required in the composition of animal bodies. And the rest of the definition of the soul I leave to the minds of the friars, fathers of the people, who by inspiration know all secrets
> Let the crowned writings be, because they are supreme truth.[30]

Leonardo's reflections on the soul and on the spiritual powers instilled in nature are developed in his numerous texts on the fleeting identity of the "being of nothingness."[31] He wonders whether nothingness exists in nature. The "nothingness" to which he refers consists of the principles of Euclidean geometry: the point, line and surface, which Leonardo first tends to consider as purely intellectual entities, and as such to be excluded from the realm of nature:

Amid the vastness of the things among which we live, the existence of nothingness holds the first place; its function extends over all things that have no existence, and its essence, as regards time, lies precisely between the past and the future, and has nothing in the present. This nothingness has the part equal to the whole, and the whole to the part, the divisible to the indivisible; and the product of the sum is the same whether we divide or multiply, and in addition as in subtraction; as is proved by arithmeticians by their tenth figure which represents zero; and its power has no extension among the things of Nature.

What is called nothingness is to be found only in time and in speech. In time it stands between the past and the future and has no existence in the present; and thus in speech it is one of the things of which we say: They are not, or they are impossible.

With regard to time, nothingness lies between the past and the future, and has nothing to do with the present, and with regard to Nature it is to be classed among things impossible: hence, from what has been said, it has no existence; because where there is nothing there would necessarily be a vacuum.[32]

Leonardo's reasoning is based on the Aristotelian concept of the impossibility of a void existing in nature. Given that all of nature's bodies occupy space, existence cannot be attributed to that which, like the point, the line and the surface, has no extension. On a later folio we find, however, a different solution. Here Leonardo converses in his mind with an imaginary "adversary," who declares that "the nothingness and the vacuum are one and the same thing, but vacuum does not exist in nature [. . .]. The reply is that the nothingness exists apart from occupation of space [. . .], that the point exists in nature [. . .], that the point moves together with the place that it occupies [. . .], that the point with its motion describes the immaterial line."[33] Nothingness thus exists, and its power extends through all of nature:

The line has in itself neither matter nor substance and may rather be called an imaginary idea [*cosa spirituale*] than a real object [*sustanzia*]; and, this being its nature, it occupies no space.[34]

The principles of nature are thus a "spiritual thing," incorporeal reality instilled in matter, the informing principle of creatures. And it is not only the point, the line and the surface that comply with these characteristics, but all of the spiritual powers that animate nature: force, the visual species of objects, light, heat and, above all, time.

These powers exist and are active then, expressly because they occupy no place. For the mature Leonardo, the "being of nothingness" represents the spiritual component, which expresses the life-instilling energy of nature. The pictorial translation of this incorporeal energy may perhaps be seen in the *Leda*, a painting in which Leonardo intended to represent the extraordinary charge of sensuality of "nature-mother." Leonardo's conception of the spiritual power that moves the world is radically naturalistic. This power, although unfolded, is "infused" in matter. And this infusion is just as indispensable to matter, as is the latter to the spiritual principle, which in his texts never presents the characteristic of transcendency. The soul is formative energy, which however remains unexpressed unless it is joined to matter. For this reason the soul is fully satisfied to inhabit the body of man, so that separation from corporeal matter is for her a grievous event:

> And thou, man, who in this labor of mine consider the marvelous works of Nature, if you judgest it to be a wicked thing to destroy it, think how very wicked a thing it is to take the life of man; and if this his composition appears to thee marvelous construction, remember that it is nothing compared with the soul that dwells within that structure. For truly, whatever it may be, that is a thing divine. Leave it then to dwell in its work at its good pleasure, and let not thy rage or malice destroy such a life. For, in truth, he who does not value it, deserves it not.[35]

The concept is reiterated in another text: the soul is "reconciled to stay in its human prison on account of its vision of these works through the eyes, by means of which all the varieties of objects in nature are presented."[36]

It is probable that the ambiguity noticeable in Leonardo's description of the soul's properties and its mode of action derives from his awareness of touching on issues having the highest potential for conflict with religious orthodoxy. Nor is it difficult to imagine what reactions Leonardo's singular theories on the soul would have aroused, had they been divulged. In an extraordinary text, he goes so far as to state that the body of man is governed by different souls, at times in conflict with one another. This is the case of the spiritual principle that presides over the male organ. The text on the soul of the *verga* (the penis) represents one of the most paradoxical and provocatory results of Leonardo's investigation of the spiritual principles of nature:

> This dispute with the human intellect, and sometimes has intellect itself and although the will of a man may wish to stimulate it, it remains obstinate and goes its own way, sometimes moving on its own without the permission or thought of the man. Whether he is asleep or awake it does what it desires. And often a man is asleep and it is awake, and many times a man is awake and it is asleep. Many times a man wants to use it, and it does not want to. Many times it wants to and a man forbids it. Therefore it appears that this animal often has a soul and intellect separate from a man; and it seems that a man is wrong to be ashamed to name it and not to show it, always being anxious to cover it up and hide that which he ought to adorn and show off with solemnity like a minister of the human species.[37]

The soul inhabits and molds the body of man, conferring on him a specific character, and thus his individuality. The "souls" allow nature to infinitely vary its forms and to compose human beings who are always different; a diversity, which is the principle of individuality not only in outer appearance, but also in character traits and constitutions. The concept of the soul as molder of individuality is partially inspired by the 15th-century revival of medieval physiognomy, reinterpreted by Leonardo as an embryological doctrine. Given that both have been formed by the soul, there must necessarily be a mirror-like relationship between the body and the character: "it is true that the signs of faces display in part the nature of men, their vices and their temperaments."[38] Due to the necessary correspondence between the body and the soul, the motions of the soul can be read in the appearance of the body. This conclusion opened to the painter remarkable possibilities: portraying the motions of the mind in his personages through the expressions on their faces, the postures of their bodies, and the language of their hands. The most eloquent proof of this conviction may be found not only in many of Leonardo's texts, but also in his drawings and paintings. *The Last Supper*, above all, which he painted after profoundly studying the character of each of the twelve Apostles, in order to portray them in postures perfectly reflecting both their individual constitutions and the emotions they experienced at the fatal moment in which Christ announced the betrayal.[39]

The theory of the soul that molds the body and instills in it its character held still further implications for the artist. The painter's special affection for the form that the soul has conferred on his own body induces him to reproduce his own features in all of the persons he portrays. The painter's soul is thus revealed in the figures he paints. However, Leonardo judged the natural tendency to self-portraiture to be a most severe limitation, which threatened to cancel the infinite variety of forms created by nature.

> It is a fault in the extreme of painters to repeat the same movements, the same faces and the same style of drapery in one and the same narrative painting and to make most of the faces resemble their master, which is a thing I have often wondered at, for I have known some who, in all their figures seem to have portrayed themselves from the life, and in them one may recognize the attitudes and manners of their maker. If he is quick of speech and movement his figures are similar in their quickness, and if the master is devout his figures are the same with their necks bent, and if the master is a good-for-nothing his figures seem laziness itself portrayed from the life. If the master is badly proportioned, his figures are the same. And if he is mad, his narrative will show irrational figures, not attending to what they are doing, who rather look about themselves, some this way and some that, as if in a dream. And thus each peculiarity in a painting has its prototype in the painter's own peculiarity. I have often pondered the cause of this defect and it seems to me that we may conclude that the very soul which rules and governs each body directs our

judgment before it is our own. Therefore it has completed the whole figure of a man in a way that it has judged looks good, be it long, short or snub nosed. And in this way its height and shape are determined, and this judgment is powerful enough to move the arm of the painter and makes him repeat himself and it seems to this soul that this is the true way of representing a man and that those who do not as it does commit an error. If it finds someone who resembles the body it has composed, it delights in it and often falls in love with it. And for this reason many fall in love with and marry women who resemble them, and often the children that are born to such people look like their parents.[40]

The embryological theory of the soul and the naturalistic vision underlying it have led us to the heart of the meditations on nature and on the objectives of painting formulated by Leonardo in his *Book on Painting*.

Painting is "mental discourse" which is based on the same geometric principles employed by nature. Painting is the only human art able to imitate "all the manifestations of nature [. . .], and this truly is a science and the true-born daughter of nature, since painting is the offspring of nature. But in order to speak more correctly, we may call it the grandchild of nature; for all visible things derive their existence from nature, and from these same things is born painting. So therefore we may justly speak of it as the grandchild of nature and as related to God Himself."[41]

It is precisely because the principles of painting coincide with those of nature that it can depict "the beauty of the works of nature" and the "adornments of nature."[42]

But the painter goes even beyond perfect emulation: "the painter challenges nature and competes with it."[43] He depicts, in fact, not only all the forms of existing creatures, but produces at will an infinite number of others that will never exist in nature:

> The divinity, which is the science of painting, transmutes the painter's mind into a resemblance of the divine mind. With free power it reasons concerning the generation of the diverse natures of the various animals, plants, fruits, landscapes, fields [. . .].[44]

In a text where echoes of Ficino's concept of love appear evident, Leonardo stresses that imitation of the forms of nature in painting is the best tool man has at his disposal for understanding "the maker of so many wonderful things, and the way to love so great an inventor, for in truth great love is born of thorough knowledge of the beloved."[45]

Since love derives from knowledge, the painter's mind must make every effort to comprehend the processes through which nature operates: "necessity requires the mind of the painter to become the interpreter between nature and art, showing how the causes of her efforts are dependent on laws."[46]

In no other activity does man attain the heights of knowledge reached through "divine" painting, which is superior to poetry, sculpture and music. The poet possesses the same capacity for invention as the painter, but he is less effective, because he represents the phenomena of nature and the motions of the mind through words, the invention of man, while the painter re-creates them in images, the language of nature: "The same relation exists between the imagination of a thing and its actuality, as between a shadow and a body casting a shadow, and there is the same relation between poetry and painting. Poetry places things before the imagination in words, while painting really places the objects before the eyes, and the eye accepts the likenesses as though they were real [. . .]."[47]

The distinction between the indirect language of words and the direct one of images is emphasized in another text, which points out a further limitation of poetry. The poet represents human "stories" and the processes of nature in discursive form, that is, with the passing of time, where instead their beauty lies in the harmony of the proportions between their parts, which can be shown only through the synthetic representation inherent to the image. For this reason, "it is a sin against nature to send via the ear those things that should be sent via the eye."[48]

It is worthwhile to insist on the definition of the painter's mind as an "interpreter between nature and art." The painter's mind must transform itself into a mirror: "The painter ought to reflect upon what he sees [. . .] and choose the most excellent parts of the things he sees. He should be like a mirror, which is transformed into as many colors as are placed before him, and, doing this, he will seem to be a second nature."[49]

Apart from being a mirror, the painter's ability to transform himself into a "second nature" depends on the capacity of his imitations to arouse the same reactions produced by natural things in flesh and blood, so perfectly as to trick not only men but even animals: "I once saw a painting which deceived a dog by means of the likeness of the painting to its master. The dog made a great fuss of it."[50]

Having attained this level of perfection, the image takes on an almost magic character. It represents, in fact, the mimetic operation of the painter on nature, an operation no different from that of the magician.

The thesis of the semi-divine nature of painting also explains the complex relations between man and nature. Man's admiration for the life-instilling work of nature co-exists, in fact, with his inclination to hurl a challenge at her. For Leonardo, music is inferior to painting because the notes it is composed of die at the very moment they are born. Like musical notes, the things of nature are subject to the tyranny of time, which swiftly transfigures all beauty. Through pictorial imitation instead, man removes the beauty and harmony of nature from the ravages of time:

O marvelous science, you keep alive the transient beauty of mortals and you have greater permanence than the works of nature, which continuously change over a period of time, leading remorselessly to old age. And, this science has the same relation to divine nature as its works have to the works of nature, and on this account it is to be revered.[51]

And again:

[Painting . . .] is of such excellence that it keeps alive the harmony of those proportional parts which nature, for all her powers, cannot manage to preserve. How many paintings have preserved the image of a divine beauty, which in its natural manifestation has been rapidly overtaken by time or death. Thus the work of the painter is nobler than that of nature, its mistress.[52]

We have thus come back to the theme of time, which, along with necessity, imposes iron-clad rules on nature. And there re-emerges the image of a nature devoid of purpose, the theater of endless generation and corruption. A theater in which life presumes death, since only death can ensure the conservation of life. Among Leonardo's most beautiful and deeply felt texts should be numbered those that describe the continuous cycle of births and deaths and the passing of energy from one subject to another. It is a scenario observed with cold lucidity, almost with self-complacency, by one who is conscious of having fixed his gaze on reality, without indulging in pathetic illusions. For Leonardo, the ceaseless transition of life from one form to another represents the action of a cosmic force, of whose final outcome he is intuitively aware: the collapse of "nature's rules" and the return to chaos. Time extends its domain over all of the things of nature. Like the other spiritual principles, time too can exercise its demolishing mission only by making use of material instruments. The most diligent and effective of his servants is water, which Leonardo calls, not by chance, "nature's vector."

This [the water] wears away the lofty summits of the mountains, it lays bare and carries away the great rocks. It drives away the sea from its ancient shores [. . .], nor can any stability ever be discerned in it [. . .] And so it is sometimes sharp and sometimes strong, sometimes acid and sometimes bitter, sometimes sweet and sometimes thick or thin, sometimes it is seen bringing hurt or pestilence, sometimes health-giving, sometimes poisonous. So one would say that it suffers change into as many natures as are the different places through which it passes. And as the mirror changes with the color of its object, so it changes with the nature of the place through which it passes [. . .] sometimes it starts a conflagration, sometimes it extinguishes it [. . .], is the cause at times of life or death [. . .], nourishes at times and at times does the contrary [. . .], at times submerges the wide valleys with great floods. With time everything changes.[53]

This text, while written in the solemn tone of the prophecies, represents, in the author's intentions, simply a scientific description of the demolishing work of water, at the service of time, his overlord. Leonardo's explicit mention of floods evokes the specular visual narrations of his extraordinary drawings of the "Deluge series." These are, in fact, masterful representations of the final outcome of the process of corruption implemented by time and visualizations of the destiny of nature and of man.

I do not think that, through those drawings and the texts that mirror them, the artist intended to denounce the moral degeneration of mankind. The end of nature appears to Leonardo an inevitable outcome, an event self-evident to those who have investigated its mechanisms and laws. Nor does the end of human generation represent an event of special significance in this scenario. Nature is not the reign of man, nor does she heed his hopes and illusions. He who reads with clarity the history of nature, as shown us by geology, and reflects on the continuous oblivion of the works of man shown us by history, cannot fail to see the necessary succession of deaths and resurrections that marks the rhythm of worldly events. On a sheet of anatomical studies, admiration for the capacity of man's tongue to pronounce an infinite number of names is accompanied by the bitter awareness that even languages are subject to the harsh laws of time:

> And the languages themselves are subjected to oblivion; they are mortal like all created things; and if we concede that our world is eternal we shall say that these languages have been, and still must be, of infinite variety, through the infinite centuries, which are contained in infinite time.[54]

Along these lines of thought on nature and on man there clearly emerges Leonardo's distancing himself from the anthropocentric visions of the authors that have often been evoked as his sources: above all Cusano, Alberti and Ficino. I believe that his vision of nature, and above all of man, shows rather the influence of the Stoic doctrines, whose diffusion has often been underestimated. And, above all, that his conceptions—as Fabio Frosini has recently stressed[55]—appear extraordinarily close to those forms of crude, radical naturalism, immune from any otherworldly temptation, whose most lucid interpreter was in the same years Niccolò Machiavelli. That Machiavelli, whom Leonardo had occasion to frequent during the military campaigns of the Valentino in central Italy in 1502; in the war between Florence and Pisa of 1503; and lastly, at the time of the *Battle of Anghiari* between 1503 and 1505.[56] There still remains to be investigated Leonardo's relationship to Machiavelli and to those anti-conformist cultural tendencies to which the Secretary of the Florentine Republic gave masterly expression. To encourage research in this direction, I would like to cite, in conclusion, a passage from Machiavelli's *Discorsi sopra la prima deca di Tito Livio* that might have sprung from the pen of Leonardo himself.[57] Echoing through it are the same themes on which he insists: the eternity of the world, the oblivion of languages, and the cyclic extinction of civilizations caused by the evil nature of men and materially operated by floods. Machiavelli replies to his objectors that, if the world had always existed, we should have information of facts more remote than those of the 5,000 years of history of which we have testimony. Different causes—he insists—provoke the cancellation of memories. Mankind is responsible for the continuous destruction of religious sects and languages; but the cyclical dissolution of civilizations is most frequently caused by the inundation of water:

> Because, when a new sect springs up, that is, a new Religion, the first effort is, in order to give itself reputation, to extinguish the old; and if it happens that the establishers of the new sect are of different languages, they extinguish it (the old) easily. Which thing is known by observing the method, which the Christian religion employed against the Gentile heathen sect, which has cancelled all its institutions, all of its ceremonies, and extinguished every record of that ancient Theology [. . .] It is to be believed, therefore, that that which the Christian religion wanted to do against the Gentile sect, the Gentiles did against that which preceded them. And as these sects changed two or three times in five or six thousand years, all memory of things done before that time are lost [. . .] As to the causes that come from heaven, they are those that extinguish the human race and reduce the inhabitants of parts of the world to a very few. And this results either from pestilence, or famine, or from an inundation of water; and the last is most important, as much because it is the most universal, as because those who are saved are men of the mountains and rugged, who, not having any knowledge of antiquity, cannot leave it to posterity.[58]

Notes

*The text published here is the English translation of a reduced version of the paper (*La natura di Leonardo: "più tosto crudele matrigna che madre"*) read at the XII Colloquio Internazionale del Lessico Intellettuale Europeo (Rome, 4–6 January 2007). The Italian text is forthcoming in the Lessico Intellettuale Europeo Series.

1. Leonardo da Vinci, *I Codici Forster I–III del Victoria and Albert Museum di Londra*, Edizione facsimile, Trascrizione diplomatica e critica di Augusto Marinoni, 3 vols. (Florence: Giunti, 1992); *Forster III*, f. 20v. English translation by Jean Paul Richter, *The literary works of Leonardo da Vinci*, 2 vols. (London: Phaidon, 1970), §846 [from now onward quoted as Richter].
2. Leonardo da Vinci, *Il Codice Atlantico della Biblioteca Ambrosiana di Milano*, Trascrizione diplomatica e critica di Augusto Marinoni, transcriptions, 12 vols. (Florence: Giunti, 1975–1980), f. 393r: Richter §1293.
3. Ibid., f. 1032v: Richter §1296.
4. Ibid., f. 1033v: Richter §1296.
5. Ibid.
6. Leonardo da Vinci, *Il Codice Arundel 263 nella British Library*, A cura di Carlo Pedretti. Trascrizione e note critiche di Carlo Vecce, 2 vols. (Florence: Giunti, 1998) vol. II, P1r: Richter §1217.
7. Ibid., P 1v: Richter §1217.
8. Ibid., Richter §1218.
9. Ibid., Richter §1219.

10. Ibid., Richter § 1162.

11. Ibid., Richter §1339.

12. Leonardo da Vinci, *I Codici Forster I–III* cit., *Forster III*, f. 43v: Richter §1135.

13, Leonardo da Vinci, *Il Codice Atlantico* cit., f 398v: Richter §1148B.

14. Leonardo da Vinci, *Il Codice Arundel 263* cit., P 76v [author's translation].

15. Leonardo da Vinci, *Corpus of Anatomical Drawings in the Collection of Her Majesty the Queen at Windsor Castle*, edited by Carlo Pedretti and Kenneth Keele, 3 vols. (New York & London: Academic Press, 1979), K/P 156v [from now onward quoted as K/P].

16. Ibid., K/P 54r.

17. Ibid., K/P 80r.

18. Ibid., K/P 105r.

19. Ibid., K/P 107r.

20. Ibid., K/P 115r.

21. Ibid., K/P 136r.

22. Ibid., K/P 143r.

23. Ibid., K/P 50v.

24. Ibid., K/P 113r.

25. Ibid., K/P 72v.

26. Ibid., K/P 49v.

27. Leonardo da Vinci, *I Manoscritti dell'Institut de France*, edizione facsimile, Trascrizione diplomatica e critica di Augusto Marinoni, 12 vols. (Florence: Giunti, 1986–1990), *Ms. I*, f. 102r: Richter §1151.

28. Leonardo da Vinci, *Corpus of Anatomical Drawings* cit., K/P 153r.

29. Leonardo da Vinci, *Il Codice Arundel 263* cit., P 12r: Richter §1137.

30. Ibid., K/P 114v.

31. The most exhaustive analysis of the suggestive topic of the being of nothingness remains Augusto Marinoni's, *L'essere del nulla*, in *Leonardo da Vinci letto e commentato*. [Letture Vinciane I–XII (1960–1972)]. A cura di Paolo Galluzzi (Florence: Giunti, 1974), pp. 7–28.

32. Leonardo da Vinci, *Il Codice Arundel 263* cit., P 60r: Richter §1216.

33. Leonardo da Vinci, *Il Codice Atlantico* cit., f.784v [Author's translation].

34. Leonardo da Vinci, *Corpus of Anatomical Drawings* cit., K/P 118Br.

35. Ibid., K/P 136r.

36. Leonardo da Vinci, *Libro di Pittura. Codice Urbinate Latino 1270 nella Biblioteca Apostolica Vaticana*, A cura di Carlo Pedretti, trascrizione e note critiche di Carlo Vecce, 2 vols. (Florence: Giunti, 1995), §24. English translation by Martin Kemp and Margaret Walker, *Leonardo on Painting* (New Haven & London: Yale University Press, 1989), p. 21 [from now onward quoted as M/W].

37. Leonardo da Vinci, *Corpus of Anatomical Drawings*, K/P 72r.

38. Leonardo da Vinci, *Libro di Pittura* cit., §292: K/W p. 147.

39. Cfr. the catalogue of the Milanese exhibition *Il Genio e le Passioni. Leonardo e il Cenacolo*, a cura di Pietro Marani (Milan: Skira-Artificio, 2001). See also Domenico Laurenza, *De Figura Umana: Fisiognomica, anatomia e arte in Leonardo* (Florence: Olschki, 2001), and D. Laurenza, "Mental motions," in *The Mind of Leonardo: The Universal Genius at work*, catalogue of the exhibition (Galleria degli Uffizi, 28 March 2006–7 January 2007) a cura di Paolo Galluzzi (Florence: Giunti, 2006), pp. 292–310.

40. Leonardo da Vinci, *Libro di Pittura* cit., §108: K/W p. 204.

41. Ibid., §12: K/W p. 13.

42. Ibid., §21: K/W p. 18.

43. Leonardo da Vinci, *I Codici Forster I–III* cit., *Forster III*, f. 72r [Author's translation].

44. Leonardo da Vinci, *Libro di Pittura* cit., §68. English translation by A. Philip McMahon, *Treatise on Painting . . . by Leonardo da Vinci* (Princeton, NJ: Princeton University Press, 1956), §280 [from now onward quoted as McMahon].
45. Ibid., § 77: K/W p. 195.
46. Ibid., §40: K/W p. 40.
47. Ibid., §2: McMahon §21.
48. Ibid., §23: K/W p. 24.
49. Ibid., §58: K/W p. 202.
50. Ibid., §14: K/W p. 34.
51. Ibid., §29: K/W p. 35.
52. Ibid., §30: K/W p. 35.
53. Leonardo da Vinci, *Il Codice Arundel 263* cit., P 25r. English translation by Edward McCurdy, *The Notebooks of Leonardo da Vinci: Arranged, rendered into English and introduced by E. McC.*, 2 vols. (London [1st ed.], 1938 [*Of water*]).
54. Leonardo da Vinci, *Corpus of Anatomical Drawings* cit., K/P f. 50v.
55. Fabio Frosini, *Umanesimo e immagine dell'uomo: Note per un confronto tra Leonardo e Pico*, in *Leonardo e Pico: Analogie, contatti, confronti* (Atti del convegno di Mirandola, 10 maggio 2003), a cura di Fabio Frosini (Florence: Olschki, 2006), pp. 204–208.
56. On the relationship between Leonardo and Machiavelli, see Roger D. Masters, *Machiavelli, Leonardo and the science of power* (Notre Dame & London: University of Notre Dame Press, 1966).
57. Quoted in Frosini's essay (see footnote 55).
58. Niccolò Machiavelli, *Discorsi sopra la Prima Deca di Tito Livio*, a cura di Giorgio Inglese (Milano: Rizzoli, 1984), p. 307. English translation attributed to Henry Neville (*Tutte l'opere di Niccolò Machiavelli: Con una prefazione di Giuseppe Baretti*, Londra, Tommaso Davies 1772. Text, digitally edited by John Roland, available on the web: http://www.constitution.org/mac/disclivy_.htm).

How Torricelli Improved on Galileo's Laws of Free Fall and Projectile Motion

WILLIAM R. SHEA AND TIZIANA BASCELLI

During his comparatively brief life (he died at thirty-nine, like his contemporary Blaise Pascal), Evangelista Torricelli (1608–1647) devoted his unusual talents to the new Galilean science of motion. It is usually said that he expanded Galileo's laws, but in this essay we hope to show that he actually set two of them on a firm mathematical foundation. The first is the law of free fall that says that all bodies falling from rest, whatever their weight, move with the same acceleration and cover a distance that is proportional to the time squared (this is expressed in the familiar formula: $s = \frac{1}{2} at^2$, where s stands for space traveled, a for acceleration and t for time). The second law states that the path of any projectile, whatever the angle at which it is thrown, is a parabola.

We shall briefly introduce Galileo's *Discourses on Two New Sciences*, after which we will consider how Galileo presented the two laws and how Torricelli modified their proof.

The Publication of the Discourses on Two New Sciences

Galileo was condemned on 22 June 1633 for having violated in his *Dialogue on the Two Chief World Systems* the injunction that he had received in 1616 not to teach or write about Copernicanism. At the time of his condemnation, he was sixty-nine years old and in poor health, but as soon as he was told that he could leave Rome and accept the invitation of his old friend Ascanio Piccolomini, the Archbishop of Siena, he made a remarkable recovery. He set out within two weeks, and halfway along his journey he climbed out of his carriage and, as he later proudly informed a correspondent, "walked four miles on foot," no doubt to breathe the free air as well as to stretch his legs.[1]

Galileo found a congenial group of intellectuals in Siena, and they encouraged him to take up the great work on mathematics and mechanics that he had drafted in Padua during the eighteen years that he spent there between 1592 and 1610. As a professor, he had had time to investigate how bodies behave when allowed to fall straight down or made to roll along inclined planes, but after his return to Florence in 1610 he devoted himself mainly to astronomy. This was to be expected in the wake of his sensational telescopic discoveries that included such novelties as that the Moon has mountains and craters, that the stars are infinitely more numerous than what we see with the naked eye, and that Jupiter has not one but four satellites. It was only when he was barred from further astronomical speculation that he returned to his first love, and he did so with a zeal that is truly astonishing. A month after arriving in Siena, he was already able to send some of his work to Niccolò Aggiunti, a young mathematician in Florence,[2] and after he was permitted to go back to his house in Arcetri in December 1633, he immediately began writing a new book in the style that had served him so well in his *Dialogue on the Two Chief World Systems*. The same speakers, Salviati, Sagredo, and Simplicio congregate for four days in the same Venetian palace. Salviati remains Galileo's mouthpiece, Sagredo his sidekick, but Simplicio, the spokesman for the traditional worldview, is less of a buffoon than in the earlier *Dialogue*.

The new work, entitled *Discourses on Two New Sciences* (the first being on the cohesion and resistance of materials to fracture, and the second on uniform, accelerated and projectile motion) were completed by the summer of 1636, and the manuscript was sent to Fulgenzio Micanzio in Venice, who had agreed to interest the Dutch printer Louis Elsevier. Then arose the question of the required authorization. After his condemnation in 1633, the Holy Office had placed Galileo's name on the list of authors whose writings *de editis omnibus et edendas*, i.e. published or unpublished, were strictly forbidden, and so rigorously was this rule enforced that Micanzio, who had considered reprinting Galileo's *Discourse on Floating Bodies*, had not been permitted to do so.[3] In order to get around this obstacle, Galileo sent a copy of his work to Giovanni Pieroni in Vienna, only to be told that all books printed there had to be sanctioned by the Jesuits, among whom at the moment Galileo's old antagonist, Christopher Scheiner, was residing.[4] So Vienna would not do. Pieroni then appealed to Cardinal Franz Dietrichstein, the bishop of Olmütz, who had just set up a printing shop in the city.[5] The approbation of a Dominican father was secured to keep the business secret from Scheiner and the Jesuits. But soon thereafter, the Cardinal died on 19 September 1636 and, besides, Pieroni was not pleased with the Olmütz press, so the manuscript was brought back to Vienna. A new approbation was procured (Scheiner having gone meanwhile in Silesia) and the work was at the point of being sent to the printer when the dreaded Scheiner reappeared. Pieroni thought that in this case valor required flight and he took the manuscript to Prague where Cardinal Ernest Adalbert Harrach offered him the use of the University press, but here again difficulties cropped up.[6] Meanwhile Galileo, wearied with these delays, decided to rely on the Dutch printer Louis Elsevier and, finally, the work appeared in Leyden in July 1638. Distribution was slow and it was only in January of the following year that fifty copies arrived in Rome. They were immediately sold out at the high price of 2 *scudi*. Benedetto Castelli, who had been Galileo's best student and was now professor in Rome, only managed to get a copy in February. This is interesting for our story because Torricelli, who was working with Castelli at the time, probably borrowed the book shortly thereafter.

The Law of Free Fall

The law of free fall is discussed in the Third of the four Days of the *Discourses on Two New Sciences*. Unlike the first two days, which are introduced with elegant and witty exchanges between Salviati and Sagredo, the Third Day begins abruptly with a Latin treatise on local motion that is ascribed to "our Academician," whose name is never given but who is unquestionably Galileo himself. It is divided into three parts: the first is on uniform motion, the second on naturally accelerated motion, and the third on projectile motion. The first two parts are considered in the course of the Third Day, while the Fourth Day is reserved for projectile motion.

Uniform motion is dealt with briskly. It is defined as motion in which the distances traveled by a moving object during equal intervals of time are equal to each other. This will prove important to Galileo's discussion of accelerated motion, which is also in Latin, but is followed by objections in Italian. Galileo writes that after much effort, he is confident that he has discovered the way nature actually works because what he has found (by using mathematical reasoning) fits perfectly with what observation tells us. He was guided, he says, by the methodological assumption that nature always uses the simplest and easiest means. So once he noticed that a stone falling from rest accelerates steadily (or as he puts it, "acquires new increments of speed"), it was normal to ask what simple rule such additions follow. This all sounds easy but it is deceptive for, as a matter of fact, Galileo did not discover his law of free fall by applying a metaphysical principle about simplicity but by making experiments along an inclined plane.

The task was neither simple nor easy and it took him several years. Clearly he could not claim that the law obeyed the "simplest rule" by appealing immediately to the outcome of these difficult experiments. To show what he terms the "obvious" link between increments in speed and time, he cast his argument in the mathematical language that he considered rigorous. He begins by reminding his reader of the definition of uniform motion that he had supplied a few pages earlier: motion is uniform when equal distances are covered in equal times. Hence a falling body that receives equal increments of speed in equal times is an instance of uniformly accelerated motion. In practical terms, this means that the amount of speed acquired in the first two time-intervals is double that of the first alone; that acquired in the first three time-intervals is triple that of the first, and so on.

Galileo had discovered this correct law by 1604,[7] the year he wrote to his Venetian friend Paolo Sarpi, who is best remembered for his role as Venice's counselor in a bitter jurisdictional dispute with the Pope that came to a head in 1606 and was only resolved a year after. But Sarpi was also a gifted and stimulating scientist, and Galileo said "that no one in Europe was ahead of him in the mathematical sciences,"[8] the highest tribute he every paid anyone. By 1604, Sarpi had already been informed of the times-squared law, and the point of Galileo's letter was to announce a principle from which this law followed. "I have arrived at a proposition," he wrote,

> that is most natural and evident, and assuming it, I can demonstrate the rest: namely, that spaces traversed in natural motion are in the squared proportion of the times and, consequently, the spaces traversed in equal times are as the odd numbers beginning with unity . . . And the principle is this, that a naturally moving body increases its velocity in the proportion that it is distant from the origin of motion.[9]

This is a curious statement. For although the first part is right, it does not follow from the principle, which is wrong. Under uniform motion, velocity does not vary directly as distance, and any high school student learns the correct rule by rote as either or both of two equations:

$$s = \tfrac{1}{2} gt^2 \text{ and } s = \tfrac{1}{2} vt$$

Even when he finally did get it right, Galileo could not so express it, because algebra had yet to be adapted to a description of continuously developing quantities. But although he disposed only of the resources of ordinary language and of the geometry of Euclid and Archimedes, Galileo came to see that the increase of speed couldn't be proportional to the distance traversed because this would entail a contradiction. If speeds had the same ratio as distances traversed, the distances would be passed in equal intervals of time. For instance, if the speed with which a falling body traversed a space of eight feet were double that with which it covered the first four feet, then the time-intervals required for these passages would be equal. But for a body, initially at rest, to fall eight feet and four feet in the same interval of time would require motion to be instantaneous, which is contrary to fact. This is mentioned in the *Discourses on Two New Sciences*,[10] but the first objection that Sagredo raises is one that must have been persistent since it had already been discussed in the *Dialogue on Two Chief World Systems*.[11]

The objection runs as follows: since time is infinitely divisible, an instant of time may be as small as you wish. Now the smaller the time the shorter the distance traveled in free fall and the lesser the speed. Near the start of fall the body's speed could be so ridiculously slow that it might take a year or even a thousand years to cover a mile. But this is contrary to everyday experience, for we all see that heavy bodies fall rapidly as soon as they are released. Salviati counters this with an experiment of his own devising. Place a heavy stone, he says, on a flexible surface such as a soft lawn. You will note that the surface is depressed, and that the depression increases when the stone is dropped from successively greater heights. The impact (literally, the "blow" or *percossa* in Italian) is caused by the combined weight and the speed, and "the quality and quantity of the impact," he adds, "will enable us to estimate without error the speed of a falling body."[12] Salviati attempts to dispel the difficulty by inverting the direction of motion. Instead of dropping a ball, we are asked to throw one upwards and notice how its speed gradually decreases until it vanishes, and the ball begins to fall back down. In doing so, it passes through every degree of slowness before it reaches its maximum speed. Similarly, a body falling from rest, that is, from infinite slowness, must go through all degrees of slowness. Not that the body lingers, as it were, at any degree of speed, because the time can be divided to correspond to any number of degrees of speed or, as Salviati says, "to the infinite degrees of diminished velocity."[13]

Galileo's argument is not entirely rigorous, but it satisfies Sagredo, who endorses the definition of uniformly accelerated motion as that which, starting from rest, adds to itself equal moments of speed in equal times. Now comes the crucial assumption that will subsequently be studied by Torricelli: "I assume," writes Galileo, "that the degrees of speed of a moving body that rolls down differently inclined planes are equal when the heights of those planes are equal."[14] The speed considered here is the end speed acquired in fall through the vertical height of the plane. Sagredo finds this very plausible, provided the planes are perfectly smooth and the moving body is perfectly round, and Salviati takes things one step further by appealing to an experiment "that is little short of a necessary demonstration." This consists in observing that when a pendulum swings it takes the bob almost to the same height from which it was released, even when the swing beyond the perpendicular is shortened by a protruding nail (at E or F in Diagram 1) that catches

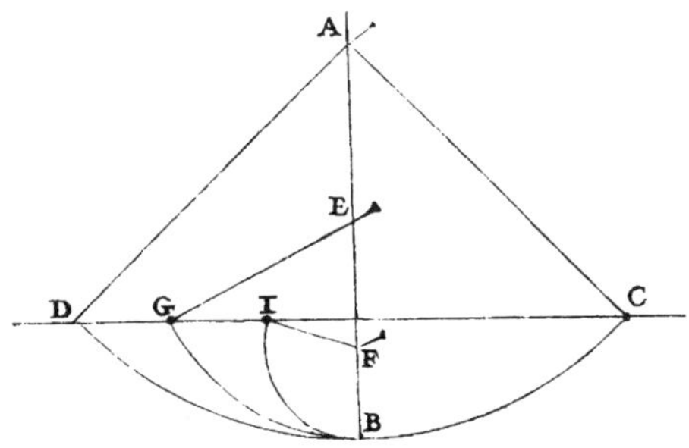

DIAGRAM 1.   The swing of a pendulum.

the string. This shows, says Salviati, that the moments of speed in all arcs such as BD, BG or BI are equal to the moments of speed in the fall to the perpendicular along CB. Now swinging a pendulum bob and seeing it rise to the height from which it was dropped is one thing.[15] Getting a ball to run down a slope and up another is another matter, for friction will dramatically slow its ascent, and Salviati accepts the assumption about inclined planes simply as a postulate.

The participants now move on to the exposition of Galileo's theorems. The first is a legacy of medieval discussions of latitude of forms and is known as "the Merton rule," since it was discovered at Merton College, Oxford, in the fourteenth century. This "mean speed rule" states that the distance covered by a uniformly accelerated motion in a given time is the same as would be covered in that time by a uniform motion at the mean speed of the accelerated motion. Galileo had already shown this in the *Dialogue on the Two Chief World Systems* with the aid of Diagram 2, where the relation between velocity, distance and time is expressed as a triangle. Falling from rest at A, the body picks up speed through "infinite degrees of velocity." The time is laid off on the vertical AC. Perpendiculars (DH, EI, etc.) represent the velocity after time AD, DE, etc. and the whole triangle is "the sum total of all the speeds with which such a distance was traversed in the time AC."[16] Or, to put it otherwise, the area of the triangle measures the distance traversed. To find the distances traveled by a body moving at uniform velocity (BC), the triangle may be doubled into a rectangle (ACBM). But although perfectly correct, this must still seem painful and clumsy to the modern reader. The velocity appears as one variable and time as the other, whereas it has become customary to think of velocity as a ratio of distance to time. Moreover, the linear distance is measured by an area, and the geometry does not yet allow deriving the law in the form that relates distance to acceleration.

In the *Discourses on Two New Sciences*, Galileo makes the "mean speed rule" his Theorem I, but he demonstrates it with an important variant. In Diagram 3, AB represents the time in which the space is traversed but the space itself is not mea-

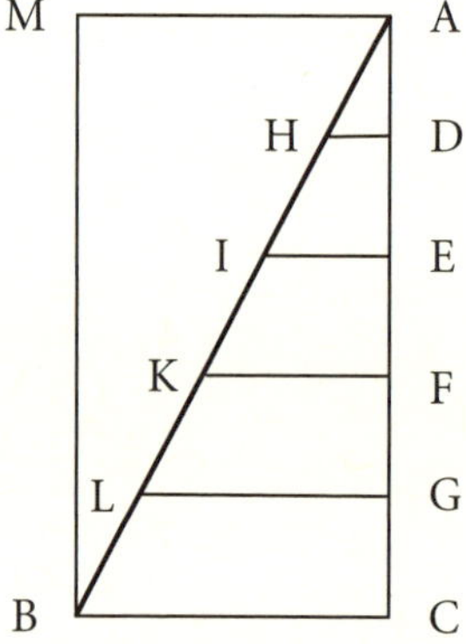

DIAGRAM 2.   First diagram of accelerated motion.

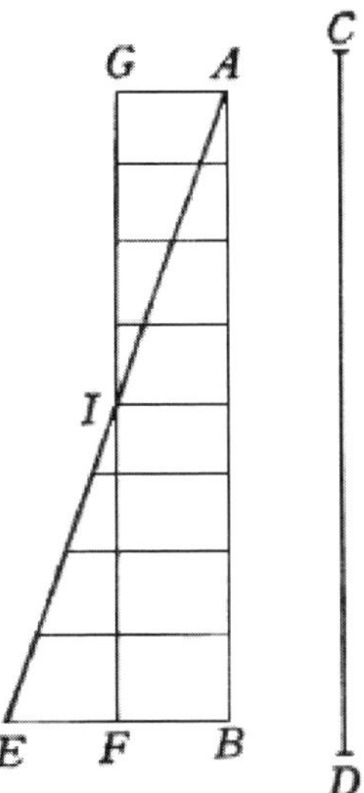

DIAGRAM 3.   Second diagram of accelerated motion.

sured by the area AEB but by the line CD, which lies outside the triangle, something that looks anomalous to the modern reader. The lines drawn at right angles to AB represent the increasing value of the speed up to the greatest, BE. From point F, which bisects BE, FG is drawn parallel to AB. It follows that rectangle AGFB is equal in area to triangle AEB because the side FG bisects the side AE at the point I. Since from any point on the time-axis AB, a parallel can be drawn, the sum of the parallel lines contained in the rectangle AGFB is equal to the sum of those contained in the triangle AEB. Now since the parallels drawn in the triangle AEB represent the increasing values of the uniformly accelerated body, while the parallels contained within the rectangle AGFB represent the values of a speed which is constant, it appears "in like manner," that the moments (*momenta*) acquired by the moving body may also be represented, in the case of accelerated motion, by the increasing parallels of the triangle AEB: "For what the moments lack in the first part of the accelerated motion AGI is made up by the *momenta* represented by the parallels of the triangle IEF."[17]

Theorem II follows immediately from theorem I and states that the spaces traveled by bodies falling from rest are to each other as the squares of the time-intervals employed in traversing these spaces. The first corollary (see Diagram 4) expresses the same law by saying that the distance traversed are to each other as the odd numbers beginning with unity: 1, 3, 5, 7, etc. Namely, the distances covered in three successive periods of time are like 1, 1 + 3 = 4, 1 + 3 + 5 = 9, 1 + 3 + 5 + 9 = 16, which are the squares of 2, 3, and 4. The second corollary makes explicit a relation that also follows from theorem I, namely that the times that two bodies fall

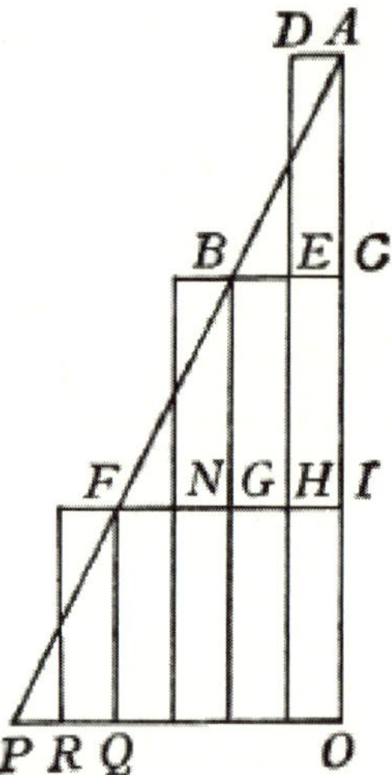

DIAGRAM 4.   Distance and time in free fall.

from rest are in the same ratio as one of the distances is to the mean proportional of the two distances.

It must be emphasized that these four propositions are purely geometrical, and that they are arrived at on the assumption that the speeds acquired along differently inclined planes are the same as that acquired along the vertical. Now if we move from pure geometry to applied mathematics and compare the motion of a ball that rolls down an inclined plane with one that falls straight down, the motions seem different: the two balls do not roll or fall through the same time-interval, they do not cover the same space, and they do not share the same speed. Vincenzo Viviani was intrigued by this when, in October 1638, he arrived in Arcetri, where he had come to act as an amanuensis for Galileo, who had become blind.

Although only seventeen years old, Viviani had a sharp mathematical mind and he made Galileo aware that his assumption that the speed along the inclined plane is the same as that along the vertical had to be demonstrated. Together they worked on a proof,[18] and the outcome was put in dialogue form by Viviani, but only published much later in 1655.[19] It is based on the *moment of speed*, a concept that was still fraught with conceptual and terminological ambiguity,[20] as we can see when Salviati refers to "the impetus, the power, the energy (*l'impeto, il talento, l'energia*) or, let us say, the momentum (*il momento*) of descent" as the property that varies with the steepness of the inclined plane (see Diagram 5). The moment is greatest along the vertical AB and decreases through AD, AE, and AF, until "it is completely extinguished" on the horizontal plane AC where the body "finds itself in a condition of indifference to motion or rest." Thus far we have an explanation that conforms with what we find in the treatise on *Mechanics* that Galileo wrote in Padua and to which he explicitly refers. Now comes a comment that is not in the original work and that may, as Stillman Drake conjectured, have been added in 1655 by Viviani:

> For just as a heavy body or a system of bodies cannot of itself move upwards or recede from the common centre toward which all heavy things tend, so it is impossible for these to move spontaneously other than with a motion that brings their own centre of gravity nearer to the aforesaid common centre. Hence, along the horizontal, by which we understand a surface, every point of which is equidistant from this same common centre, the body will have no momentum whatsoever.[21]

As we shall see, this conception is fundamental to the way Torricelli developed his own proof. We believe that he communicated it to Viviani, and that it was incorporated in the dialogue that he elaborated where it is followed by a discussion of the *impeto* acting on a body in descent. Salviati asserts that it is equal to the resistance or least force sufficient to hold it at rest. To measure the force and resistance he uses the weight of another body, H, which descends along the vertical while the body G moves up the plane (see Diagram 5 mentioned above). Because they are connected by a cord over a pulley, both must move equal distances. G does not move an equal vertical distance, however, and it exerts its resistance only along the vertical. It has no resistance along the horizontal because motion in that plane involves no change in its distance from the center of the earth. In a state of equilibrium, therefore, the *momenti*, the *velocità*, and the *propensioni al moto* must be in inverse ratio to the weights.

Salviati adds that this is demonstrable in every instance of mechanical motion. When the weights are in this ratio, they will have *momenti eguali*. Since it is agreed that *l'impeto, l'energia, il momento, o la propensione al moto* of a moving body is equal to the force or resistance sufficient to stop it, and since the *momento totale* of H is along the perpendicular, H is the exact measure of the *momento parziale* that

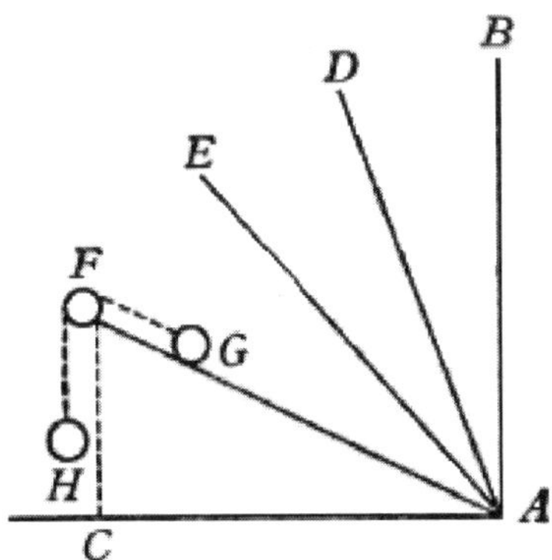

DIAGRAM 5.   Measuring the moment along an inclined plane.

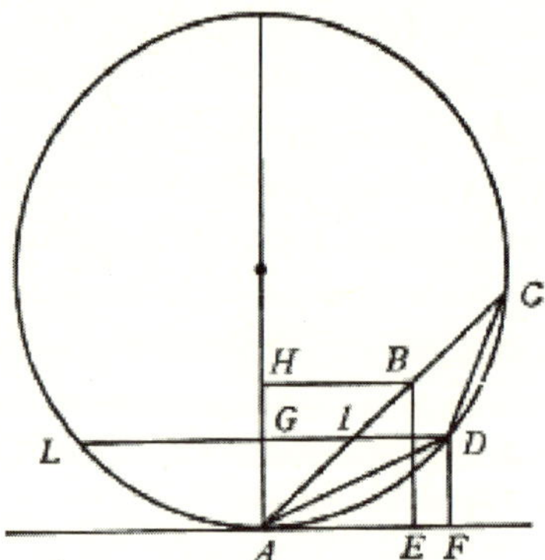

DIAGRAM 6.   The speeds along chords of a circle.

G exerts along the plane. But the measure of the total *momento* of G is its weight, and therefore the ratio of the *impeto o momento parziale* of G to its *impeto massimo e totale* is the ratio of H to G (which equals the sine of the angle of inclination).[22] Although this passage was written by Viviani, it was dictated by Galileo for its content is repeated in a later demonstration (Theorem VI) that appears in Galileo's original Latin. After demonstrating from kinematics that bodies will fall in equal times along all chords of a circle drawn to its lowest point, he adds that the same conclusion can be demonstrated another way "from mechanics." From the lowest point, A, of the vertical diameter of a circle, two chords are drawn such that AC is longer than AD (see Diagram 6), and AB, equal to AD, is laid off on AC. "From the principles of mechanics (*ex elementis mechanicis*)," writes Galileo, "it follows that the moment of weight (*momentum ponderis*) along ABC is to the total moment (*momentum suum totale*) as BE is to BA." Along the incline AD, the same ratio is DF/DA or DF/BA so the ratios on both chords represent the vertical rise of the chords to their total length. "Therefore," pursues Galileo, "the moment of the same weight (*eiusdem ponderis momentum*) along the inclined plane AD is to its moment along the inclined plane ABC and DA as DF is to BE. Hence the distances that the same weight traverses in equal times along the inclines ABC and AD will be to each other as the lines BE and DF, by Proposition II of Book I."[23] Next Galileo demonstrates that as BE is to DF, so AC is to AI, and so concludes that the times of descent along all chords are equal.

Galileo's argument "from mechanics" must be read against the background of early seventeenth-century mechanics when all investigation of motion started from the law of the lever.[24] By means of the principle of virtual velocities, Galileo extended the law to simple machines. In all instances, the governing principle is the equality of one end of the lever with that at the other. The *momento* (moment) of the lever thus easily transforms itself into the *momento* (momentum) of the moving body. A possibility of serious ambiguity is built into the lever, and Galileo slips into it unaware. Since both ends of the lever move in identical time without acceleration, it is immaterial whether one uses the virtual velocities of the two weights or their virtual displacements. Velocities must be in the same proportion as displacements. Although Galileo often uses displacement, it is clear that he means it as a substitute for velocity; and whenever he states the general principle of the lever, he states it in terms of velocity. The equivalence, however, holds only for the lever and analogous instances in which a mechanical connection insures that each body moves for the same time, and in which, because of equilibrium, the motion involved is virtual motion, not accelerated motion.

The cause of free fall is not identical with the conditions of equilibrium on the plane because the times involved are not identical and because two separate, accelerated motions take place. If there is an equality of the products of weight × distance, there cannot be an equality of momenta (mv) but rather of kinetic energies ($\frac{1}{2} mv^2$). From this ambiguity of the lever will spring the argument between quantity of motion and *vis viva* in which the second half of the century engages with such figures as Leibniz and his contemporaries. Galileo's failure to distinguish clearly between the *static concept of moment*, where what is considered is the weight and distance from the fulcrum, and the *dynamic concept of momentum* (which we write mv) is at the source of the dissatisfaction that Viviani experienced with Galileo's unproved assumption in the Third Day.

We now turn to Torricelli's attempt to do better. We have already mentioned that he probably saw a copy of the *Discourses on Two New Sciences* in February 1639. As a student of Benedetto Castelli, he must already have been familiar with the drift of Galileo's argument about uniformly accelerated motion, and by March 1641 he had written his own treatise, *De Motu Gravium Naturaliter Descendentium* (*On Naturally Falling Bodies*), and was busy drafting a second one on projectile motion.[25] When Castelli passed through Florence on his way from Rome to Venice in April 1641, he presented the manuscript to Galileo, and suggested that Torricelli be invited to assist him in drawing up the two "Days" that he had been thinking of adding to the *Discourses on Two New Sciences*. Galileo accepted with alacrity and sent him an invitation. Torricelli gratefully acknowledged receipt but explained that he was replacing Castelli at the University and could not leave before his return. The death of Torricelli's mother, who had moved to Rome with her other children, also interfered with his plans and it is only on 10 October 1641 that he arrived in Arcetri and took up residence in the house where Viviani was already living with Galileo. Unfortunately, from November onwards Galileo was confined to bed with fever and severe kidney pains. This was to be his terminal illness and he died on 9 January 1642, so what Torricelli was able to do was limited.

Torricelli's Procedure

Torricelli expresses admiration for the theorems in the Third Day of the *Discourses on Two New Sciences* that follow from Galileo's initial postulate that the speed along an inclined plane is the same as along the vertical, but he stresses that unless the postulate itself is demonstrated, the whole edifice is open to challenge. Torricelli restructures the argument on accelerated motion on the basis of a different postulate that puts the center of gravity in the foreground[26]: "Two heavy bodies," he writes, "that are conjoined cannot move unless their common center of gravity descends."[27] Applying this postulate, Torricelli derives five propositions that culminate in the proof that the speed along the inclined is the same as along the vertical.[28]

Proposition I. The postulate about the center of gravity is applied to a system of two bodies, A and B, that lie on the line AB and are connected by the string ACB (see Diagram 7). Torricelli shows that when the ratio of their weights (A/B) is equal to the ratio of their inclines (AC/CB), the bodies have the same momentum and are in equilibrium.

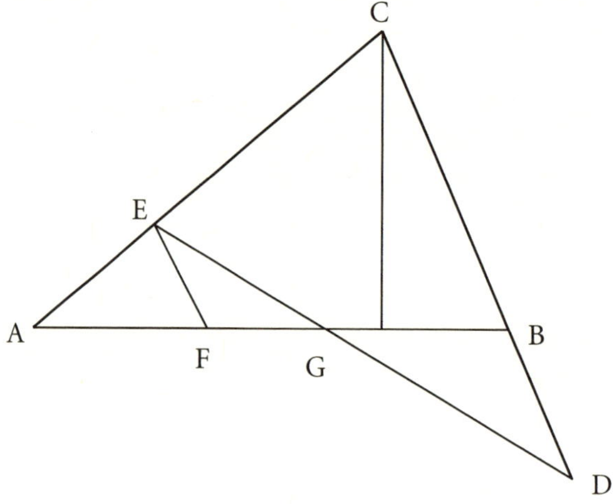

DIAGRAM 7.   Diagram illustrating the way the center of gravity descends.

Proposition II. There follows the demonstration that on differently inclined planes equal weights are in the same ratio as the length of the planes. Torricelli introduces an alternative proof "*ex ipsis mechanicis principiis,*" by which he means the law of the lever, which says that the weight at one end of the fulcrum is balanced by the weight at the other end when their ratio is inversely proportional to their respective distance from the fulcrum. But this law is given a new twist by Torricelli, who expresses it as stating that equilibrium obtains when the ratio of the *moments* of the weight are inversely proportional to the distances. He also shows how the law applies to rolling spheres or bodies of any shape whatsoever.

Proposition III. Since Torricelli is working with ratios and not absolute values, he can "measure" the moment. With the aid of Diagram 8, he shows that the moment

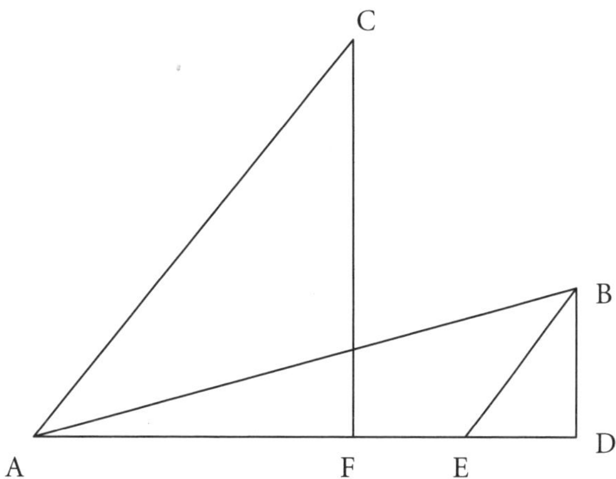

DIAGRAM 8.   Inclined planes of different elevations.

along BA is to the moment along CA as BD/CF. This result leads to the realization that the momentum depends on the sine of the angle of inclination, namely the ratio of the vertical height and the length of the incline. For instance, in Diagram 8, sine of angle BAD = BD/BA. At the end of the book, Torricelli supplies tables that embody this insight.

Proposition IV. The previous propositions are extended to show that the times of fall from rest along inclined planes are proportional to the lengths of these planes. In diagram 8, this means that the time along AB is to the time along AC as AC/AB.

Proposition V. This is the crucial step, for Torricelli now provides the proof of the assumption initially made but not demonstrated by Galileo, namely that "the degrees of speed that the same moving body acquires on differently inclined planes are equal when the heights of the planes are equal."[29]

   The root of the problem is that what Torricelli, and before him Galileo, refer to as *momento* is *force* in Newtonian physics. We now say that what is required to keep a given weight from sliding down an inclined plane is a force that is a function of the sine of the angle between the inclined plane and the horizontal, and we express this as $F = \sin \theta \, mg$ (where mg stands for the weight, the product of mass × gravitational acceleration). Since we also know, from Newton's law that $F = ma$, we can equate the two expressions:

$$ma = (\sin \theta) \, mg$$

from which we obtain:

$$a = (\sin \theta) \, g$$

and, hence, the realization that the acceleration of bodies is independent of their mass and only depends on the sine of the angle of inclination and the value of the gravitational force at a given location. But neither Galileo nor Torricelli had traveled that far.

Projectile Motion

The Fourth Day is the culmination of Galileo's study of motion: what was said about uniform motion and uniformly accelerated motion are combined, and the resulting path is shown to be a semi-parabola. That there could be such a composite motion, with neither motion interfering with the other, seemed impossible to Galileo's contemporaries, who supposed that a cannonball, or any other projectile, moves forward in a straight line until the force that impels it is exhausted and it begins to drop.

How horizontal and vertical motions combine requires some knowledge of geometry, and both Sagredo and Simplicio feel out of their depth. Salviati, Galileo's *alter ego*, gives them a crash course on what they need to know from Apollonius of Perga's *Conics* in order to understand a parabola. Simplicio has a hard time following but he can still make shrewd comments that enable Salviati to describe more clearly just what his new science of motion can achieve. Simplicio objects that if we represent the horizontal plane by a straight line, then all its parts are not equidistant from the center, and a body moving from the middle of the line will really be going uphill. Hence, its motion cannot be perpetual nor uniform, but will always diminish until it stops. Furthermore, the resistance of the medium will destroy the uniformity of the horizontal motion and alter the rule of acceleration in fall.

This objection was not uncommon, and Galileo answers it here as he had in his juvenile *De Motu* from which we quote,

> Now I am not unaware that someone at this point may object that for the purpose of these proofs, I am assuming as true the proposition that weights suspended from a balance make right angles with the balance—a proposition that is false, since the weights, directed as they are to the centre [of the universe], are convergent. To such objectors I would answer that I cover myself with the protecting wings of the superhuman Archimedes, whose name I never mention without a feeling of awe. For he made the same assumption in his Quadrature of the Parabola.[30]

So the way out is to appeal to the authority of Archimedes, who assumed that weights suspended from the ends of a horizontal lever are perpendicular to the lever. In the *Discourses on Two New Sciences*, Galileo repeats that "the authority of Archimedes alone should satisfy everyone," but he also makes an attempt to justify the procedure on pragmatic grounds:

Taking this liberty is excused by some because, in practice, our instruments and the distances involved are so small in comparison with the enormous distance from the centre of the terrestrial globe . . . If in actual practice one had to consider such small quantities, it would be necessary first of all to criticize the architects who presume, by use of a plumb line, to erect high towers with parallel sides. I may add that, in all their discussions, Archimedes and the others considered themselves as located at an infinite distance from the centre of the earth, in which case their assumptions were not false, and therefore their conclusions were absolutely correct.[31]

Coming from a mathematician, this is a curt dismissal especially since Guidobaldo del Monte had argued at great length in his *Mechanics* that the assumption lacked rigor and was therefore unsatisfactory.[32] Galileo may have realized early on that Guidobaldo's quest for absolute mathematical precision in mechanics had kept him from accepting what Galileo made a principle, namely that the products of weight (or force) and virtual displacements at the ends of any system in equilibrium are equal. He never shared Guidobaldo's insistence that, strictly speaking, the lines of descent of suspended weights are never mathematically parallel. Galileo was not interested in geometrical minutiae but with the fruitful application of mathematics to mechanics, and he struggled to bridge the gap between static and dynamics in the case of simple machines.

In projectile motion, one specific horizontal motion compounds with the uniformly accelerated motion of a falling body. If we adopt the vertex of the parabola as our origin, the latter motion, starting from rest, is wholly defined by the law of freely falling bodies. But the nature of the individual parabola will depend on the horizontal velocity, which is unchanging, and accelerated vertical motion, which is increasing constantly. From our point of view, the horizontal velocity can easily be specified in feet or meters per second, but such an apparently straightforward solution simply does not present itself to Galileo, who works within a geometrical rather than an algebraic framework and deals with ratios rather than numbers. He arrives instead at the concept of "sublimity," which we briefly present.

Galileo wants to define some common standard by which to estimate the velocity or momentum of both motions, and he proposes the terminal speed of a body falling from a given height (see Diagram 9):

For the sake of clearness, draw the vertical line *ac* to meet the horizontal line *bc*. The height is *ac*, and *bc* the amplitude of the semi-parabola *ab*, described by the compounding of the two motions, one that of a body falling with naturally accelerated motion through the distance *ac* from rest at *a*, the other a transverse uniform motion along the horizontal *ad*. The impetus acquired at *c* by a fall through the distance *ac* is determined by the height *ac*; for the impetus of a body falling from the same elevation is always one and the same. But innumerable degrees of speed may be assigned to equable motion on the horizontal. However, in order that I may select one out of this multitude and separate it from the rest in a perfectly definite manner, I will extend the height *ca* upwards to *e* just as far as is necessary and will call this distance *ae* the sublimity.[33]

Because the acceleration of falling bodies is constant, the horizontal velocity can be stated in terms of the distance through which the body must fall to acquire that velocity. In this way, the "sublimity" furnishes a standard of horizontal velocity! But projectiles are set in motion by some force such as a throwing arm, a bowstring, or gunpowder. Galileo knows this, of course, but he never suggests that such sources of motion be subjected to the same analysis as free fall. He merely accepts the resultant motion and measures it by a sublimity.

From our Newtonian point of view, motion in a straight line is not problematic since it is inertial and requires no force to continue at the velocity that it has acquired after an initial impulse. For Galileo, matters are very different. Uniform motion, he frequently repeats, is on a horizontal plane, where there is neither a tendency to motion nor a repugnance to it, in other words, where the force of weight does not operate to produce any change. So far so good, but Galileo also maintains that the gain or loss of motion depends only on the vertical displacement to and from the center of the Earth. His horizontal plane is not a perfectly straight line, but a small part of the circumference of the sphere that is the Earth. Inertial motion is not rectilinear but circular.

In the *Dialogue on the Two Chief World Systems*, Galileo had even attempted to do away with rectilinear motion altogether. In Diagram 10,[34] BI is the surface of the Earth, and BC the height of a tower that is carried along by the Earth and marks out

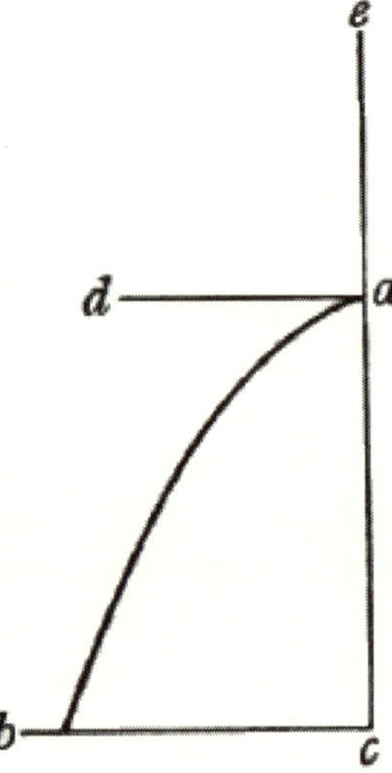

DIAGRAM 9.   Galileo's diagram of a parabola.

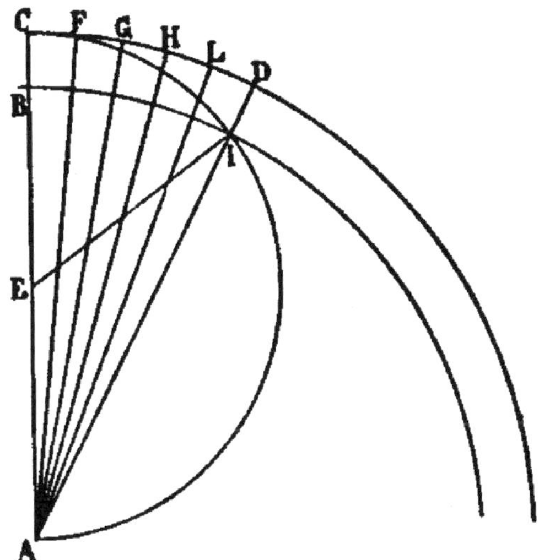

DIAGRAM 10.   Circular motion wins every time.

with its top the arc CD. Assuming that the acceleration is constant and that the line of compound motion must terminate at the center of the Earth, Salviati uses E, the mid-point of CA as centre and the length of EC as radius to draw the semicircle CIA, "along which I think it most probable that a stone dropped from the top of the tower would move." The very slow speed of fall at the beginning, the constant acceleration and the tendency towards the center of the Earth are clear from the diagram. Further analysis reveals three more surprising features. First, the stone always follows a circular path, whether it is at rest or in motion. Secondly, the distance traveled by the stone in falling is equal to the one it would have traversed if it had remained on top of the tower, because the arcs CF, FG, GH, HL, and LD are equal. Thirdly, the true motion of the stone is not accelerated but uniform since all the arcs marked on the circumference are passed over in equal times. Sagredo is delighted and exclaims: "If only the demonstration of the philosophers had half the probability of this one!"[35] but Salviati remains cautious: "I do not wish to declare at present that the descent of heavy bodies takes place in exactly this way, but if the line described by a falling body is not exactly this one, it is very near to it."[36]

Galileo had overreached himself and, in a letter to Pierre Carcavy in 1637,[37] he tried to pass off his description as jest, but it is one that he is careful not to repeat in the *Discourses on Two New Sciences*. And how could he, since he was arguing that the correct path of a projectile is a parabola not a circle! In the absence of a realization that uniform rectilinear motion is inertial, this leaves us with the embarrassing question of how exactly the parabola fits actual projectile motion. The more since Galileo repeats in the Fourth Day of the *Discourses on Two New Sciences* what he had written in the First Day of the *Dialogue on the Two Chief World Systems*, where he had banished motion in a straight line from the privileged category of "natural" motion. In the *Dialogue*, Salviati stresses that rectilinear motion is strictly transitional and merely serves to return bodies to their natural place where they spontaneously resume their motion in the circle. Salviati even suggests, on the authority of Plato, that after their creation the planets

> were for a certain time moved in a straight line by their Maker and, on reaching certain predetermined places, were made to revolve one by one, passing from straight to circular motion in which they maintained themselves and continue to maintain themselves. A sublime concept and worthy indeed of Plato, which I remember having heard discussed by our friend, the Lyncean Academician.[38]

When Salviati refers to the Academician, namely Galileo himself, it is always to stress that he was the first to make a major discovery. Now introducing a rectilinear motion prior to circular motion seemed perfectly natural to Galileo. It is important to stress this, for from the viewpoint of Newtonian physics, which we naturally adopt, the reverse is true and the Platonic cosmology produces no gain. Rather, it implies two major miracles. First, it involves changing instantaneously the direction of the movement of the falling planets, which is as difficult as conferring instantaneously a determined velocity to a body. Indeed, in the natural order of things it is impossible. Secondly, it implies that the force of attraction of the Sun is doubled at the very moment when a circular motion is substituted for a downward one. For Galileo, these objections do not hold. He considered the operation of conferring motion on a body at rest and that of changing its direction as altogether different. In the first case, something new has to be produced, but in the other the change is merely accidental. As to the doubling of a force of attraction, Galileo has no need for it whatsoever because circular motion is inertial and does not engender centrifugal forces, so that no force of attraction from the Sun is necessary to make the planets describe their particular orbits and stay in them. Furthermore, Galileo does not work on the assumption that the Sun attracts the planets; they move towards the Sun by virtue of a natural inclination that has its origins in their bodies.

Torricelli's Strategy

Galileo's speculation is fascinating, but as we have seen, it leaves us with a radical dichotomy between naturally accelerated motion that tends towards the center of the Earth and other kinds of rectilinear motion headed elsewhere. In the parabola that Galileo describes, the accelerated motion is vertical (along what we would call the y-axis) and the other motion along the horizontal (our x-axis). But what if this second motion points in some other direction, say upwards or downwards? Galileo knows, of course, that cannon balls are not usually fired point-blank but at an angle and that the maximum range is obtained when the angle of elevation of the gun is 45°. He even provides laboriously calculated tables of the height and distance that a projectile will reach when shot at different elevations.[39] What he fails to do is to show that the path of the projectile is a parabola regardless of the direction in which it is set in motion. By effectively treating the "forced" motion (along the x-axis) as if it were truly inertial, Torricelli demonstrates that the path is indeed parabolic for all projectiles. This is important in the history of mechanics because it will create the impression, shared by Newton, that Galileo and his School (the most prominent member being Torricelli) had arrived at the correct law of inertia.

Notes

1. This is reported by the Florentine Ambassador to Rome, Francesco Niccolini, in a letter to Andrea Cioli, the First Secretary of the Granduke of Tuscany, dated 10 July 1633 in Galileo Galilei, *Opere di Galileo Galilei*, ed. A. Favaro, 20 vols. (Florence: G. Barbèra, 1890–1909), XV, p. 174. This edition will be quoted as *Opere di Galileo* followed by the volume in Roman and the page in Arabic numbers. An Italian mile is roughly the equivalent of an English mile.
2. See Niccolò Aggiunti's letter of acknowledgment of 10 September 1633, *Opere di Galileo*, XV, p. 257.
3. Letter of Fulgenzio Micanzio to Galileo, 10 February 1635, *Opere di Galileo*, XVI, p. 209.
4. Letter of Giovanni Pieroni to Galileo, 11 August 1635, *Opere di Galileo Galilei*, XVI, p. 301.
5. Letter of Antonio Minati to Giovanni Pieroni, 28 January 1636, *Opere di Galileo*, XVI, p. 386.
6. Letter of Giovanni Pieroni to Galileo, 9 July 1637, *Opere di Galileo*, XVII, pp. 130–131.
7. For a thorough investigation of the stages of Galileo's discovery of the law of free fall see Jürgen Renn et al., "Hunting the White Elephant: When and How Did Galileo Discover the Law of Fall?" *Science in Context* 13 (2000), pp. 299–419.
8. *Opere di Galileo*, II, p. 549.
9. Letter of Galileo to Paolo Sarpi, 16 October 1604, *Opere di Galileo*, X, p. 115.
10. Third Day, *Discourses on Two New Sciences* in *Opere di Galileo*, VIII, pp. 203–204. English translation by Stillman Drake, *Galileo Two New Sciences* (Madison: University of Wisconsin Press, 1974), where the pagination of the *Opere di Galileo* is given.
11. First Day, *Dialogue on the Two Chief World Systems* in *Opere di Galileo*, VII, pp. 46–52. In the English translation by Stillman Drake, *Dialogue Concerning the Two Chief World Systems* (Berkeley & Los Angeles: University of California Press, 1962), pp. 21–28.
12. Third Day, *Discourses on Two New Sciences* in *Opere di Galileo*, VIII, p. 199.
13. Third Day, *Discourses on Two New Sciences* in *Opere di Galileo*, VIII, p. 201.
14. Third Day, *Discourses on Two New Sciences* in *Opere di Galileo*, VIII, p. 205.
15. Paolo Palmieri is carrying out a number of interesting experiments on Galileo's procedure. See Paolo Palmieri, "Galileo's construction of idealized fall in the void," in *History of Science*, vol. 43 part 4, no. 142 (2005), pp. 343–389.
16. *Dialogue on the Two Chief World Systems*, in *Opere di Galileo*, VII, p. 256; *Drake*, p. 229.
17. Third Day, *Discourses on Two New Sciences* in *Opere di Galileo*, VIII, p. 209.
18. See Galileo's letter to Benedetto Castelli, 3 December 1639, *Opere di Galileo*, XVIII, pp. 125–126.
19. The text is reprinted as an extended footnote in the *Opere di Galileo*, VIII, pp. 214–219.
20. On this concept the essential work is still Paolo Galluzzi, *Momento: Studi Galileiani* (Rome: Edizione dell'Ateneo & Bizzari, 1979). A briefer but remarkably lucid account can be found in

Richard S. Westfall, "The Problem of Force in Galileo's Physics" in *Galileo Reappraised*, ed. Carlo L. Golino (Los Angeles: University of California Press, 1996), pp. 67–95.

21. Third Day, "Discorsi sopra due nuove scienze," in *Opere di Galileo*, VIII, p. 215. For Drake's conjecture, see Drake's translation on the *Dialogue on the Two Chief World Systems*, p. 172, n. 27.

22. *Opere di Galileo*, VIII, p. 216.

23. *Opere di Galileo*, VIII, p. 222. Proposition II of Book I refers to the section on local motion (*Opere di Galileo*, VIII, p. 193). It reads, "If a moving body traverses two distances in equal times, these distances will bear to each other the same ratio as the speeds. And if the distances are as the speeds then the times will be equal." The statement contains no restriction to equable or uniform motion and can be applied, as here, to accelerated motion.

24. Galileo uses material from the *Mechanics* that he had written in Padua around 1607 ("Le mecaniche," in *Opere di Galileo*, II, pp. 173–175).

25. *De Motu Gravium Naturaliter Descendentium et Prorectorum Libri Duo* was published in Florence in 1644 as part of Torricelli's *Opera Geometrica*, his only work to have been published during his lifetime.

26. This postulate was generally recognized by contemporary mathematicians. For instance, Guidobaldo del Monte, who had been Galileo's mentor and to whom he owed his appointment as university professor, gives it pride of place in his *Mechanicorum Liber* (Pesaro, 1577), which was translated in Italian with commentaries by Filippo Pigafetta as *Le Mecaniche* (Venice, 1581). An abridged translation from the Italian was published in Stillman Drake and I.E. Drabkin, *Mechanics in Sixteenth-Century Italy* (Madison: University of Wisconsin Press, 1969) where the postulate that the measure of descent is given by the descent of the center of gravity appears on p. 259.

27. We quote the standard edition of Torricelli's *Opere* edited by Gino Loria and Giuseppe Vassura (Faenza: G. Montanari, 1919), Vol. II, p. 105.

28. The five propositions are ibid., pp. 105–114.

29. Ibid., p. 112. Torricelli's Latin wording is practically identical with that of Galileo in the *Discourses on Two New Sciences* (*Opere di Galileo*, VIII, p. 205).

30. "De Motu," in *Opere di Galileo*, I, p. 300, translation by I.E. Drabkin in Galileo Galileo, *On Motion and On Mechanics* (Madison: University of Wisconsin Press, 1960), p. 67.

31. Ibid., pp. 274–275.

32. In the translation from the Italian published in Stillman Drake and I.E. Drabkin, *Mechanics in Sixteenth-Century Italy* (Madison: University of Wisconsin Press, 1969), Guidobaldo's lengthy discussion of the balance is given on pp. 260–298 (from 2r–34v of *Le Mecaniche*).

33. *Opere di Galileo*, VIII, pp. 282–283.

34. *Opere di Galileo*, VII, p. 191; *Drake* p. 165.

35. *Opere di Galileo*, VII, p. 192; *Drake* p. 166.

36. *Opere di Galileo*, VII, p. 193; *Drake* p. 167.

37. Letter of Galileo to Carcavy, 5 June 1637, in *Opere di Galileo*, XVII, p. 89.

38. *Opere di Galileo*, VII, p. 44; *Drake* p. 20.

39. *Opere di Galileo*, VIII, p. 304, p. 307.

Solomon's Houses

Francesco Bianchini's Academic Places

JOHN L. HEILBRON

Nemo solus satis sapit.

Tore Frängsmyr has attended to the organization of the sciences not only as a historian but also as an activist. In this second capacity, he helped bring forth the International Summer School in the History of Science. Its session held in Uppsala in 2006 considered "The two cultures in the Republic of Letters." The present paper is adapted from the lecture I gave there. It makes use of the career of an individual, Francesco Bianchini, to illuminate the place of learned academies in the Republic of Letters in Italy around 1700. Bianchini belonged to many, and helped found several, academies of diverse size, scope, organization, and social function. I take up these types in their order of geographical inclusiveness—local, metropolitan, national, and international—which also corresponds roughly to the chronological order of Bianchini's engagements with them. I use the terms *eruditi* and *letterati* interchangeably to signify the productive scholars and men of science in his diverse "academies."[1]

Local

At the University of Padua, which he attended from 1680 to 1684, Bianchini, though intending an ecclesiastical career, spent much of his time with its professor of mathematics, Geminiano Montanari. Together they detected several comets and followed up Montanari's discovery of variable stars. Astronomy was a leading subject of an "academy" at which Montanari also showed up-to-date experiments in physics to his students and a few interested people from the community, perhaps a dozen "members" in all.[2] It might best be conceived as a course of semi-public lectures in natural philosophy of the sort that interpreters of Descartes' physics were then popularizing in Paris.[3] Montanari had set up a similar group of similar size when teaching at the University of Bologna, the *Accademia della Traccia*, the Academy of Vestiges, presumably of natural knowledge. Its staple of experiments resembled those of the Florentine *Accademia del Cimento*, whose meetings Montanari had attended before beginning his university career in Bologna.[4] Its program for 1675 has come down to us. The *Traccia* posed questions about vacuum, fluid pressure, and viscosity; and, "for use in an academician's particular studies," did experiments on light, vision, and sound, and heard wide-ranging "physico-mathematical discourses" by Montanari.[5]

Montanari's *Traccia* propagated the pioneering practices of the *Cimento*—systematic exploratory experiments designed, performed, and evaluated by a group—in two ways. One was multiplication of itself in "academies" established by professors or by gentlemen close to a university to teach natural philosophy by experiment and to demonstrate the latest inventions. Sometimes, as in Montanari's lectures on capillary action and *larmes de verre*, the organizer of the academy or someone appointed by him would bring in original work and the academy would function as a research seminar.[6] Usually these small pedagogical academies did not outlive their founders.

The second way in which the legacy of the Cimento moved via the *Traccia* produced an important academy of sciences that still exists. Under Montanari's successor Eustachio Manfredi the *Traccia* flourished as the *Accademia degli Inquieti* but remained within the established parameters. It became something more after Count Luigi Fernando Marsili, a senator of Bologna and a general in the papal army, decided to give his collection of books, artefacts, and specimens to his native city on condition that it establish an "Institute" for their study and preservation. With the support of Pope Clement XI (1700–21), acting with the advice of his friend and chamberlain Bianchini, the city fathers of Bologna agreed to implement the arrangement. The new institute had two classes, one for art, the other for science. With continuing papal patronage, the scientific class, whose original nucleus was Manfredi's *Inquieti*, provided for the Papal States the same program in "physico-mathematics" and experimental philosophy that the royal societies of London and Paris provided in their national contexts.[7]

Most of Clement's immediate predecessors would not have been willing to support science and learning in an institution largely controlled and staffed by laymen. The reigning pope when Bianchini transferred to Rome in the summer of 1684, the Blessed Innocent XI (1676–89), was not a man of letters or a patron of arts like Clement, but a close student of the determinations of the Council of Trent, which he strove to enforce as effectively as his austere, reclusive, vacillating, hypochondriac character permitted. A strict constructionist, he cracked down on Jesuit laxity, Gallican and other challenges to papal authority, quietism, nepotism, and natural philosophy not anchored in the system of Thomas Aquinas, whom the Tridentine fathers had appointed, retrospectively, doctor of the church.[8] Innocent's retrograde reformism affected the programs of Rome's learned academies in two main ways.

The more obvious consequence was to discourage development and even discussion of the dangerous modern philosophies of Gassendi and Descartes. Despite the condemnation in 1663 of Descartes' mechanical universe as destructive of the Tridentine-Thomistic account of the eucharist, it had prospered in Rome and Naples under the rule of Innocent's predecessor Alexander VII (1655–67).[9] Alexander was a man of letters. Perhaps his greatest contribution to them, and to Roman politics, was enabling the conversion to Catholicism of the exigent intellectual former Queen of Sweden, Christina. Her *Accademia reale*, which included the future Clement XI, discussed every respectable subject, including atomism.[10] A similar regime of relative tolerance in respect to letters continued under popes Clement IX and Clement X, who together filled the space between Alexander and Innocent.[11]

The second and subtler consequence for letters of Innocent's reformist program was guarded support for the study of early church history. The field had become dangerous. French savants, particularly the Benedictines of Saint Maur in Paris, had developed a critical approach to medieval documents and applied their learning to argue the Gallican cause against Rome. Their leader, Jean Mabillon, and their bible, Mabillon's *De re diplomtica* (1681), set the standard of critical evaluation of sources. Innocent wanted to domesticate Maurist technique to counter learned Gallicans with their own weapons, but feared, quite rightly, that critical history would escape control and eventually be turned against Roman interests.[12] The *eruditi* of Italy quickly became enamored of Maurist methods and received Mabillon enthusiastically when he travelled through the peninsula looking for documents in 1685. Although he met a few scholars he admired, notably the librarian of the Grand Duke of Tuscany, Antonio Magliabecchi, the historian (and fellow Benedictine) Benedetto Bacchini, and the up-and-coming Bianchini, he returned to France with a higher opinion of Italian manuscripts than of Italian scholarship. Innocent's reluctant endorsement of Maurist methods would help right the balance.[13]

The impresario of Roman academies in 1684 was Giovanni Giusti Ciampini, a rich prelate high in the Vatican's legal bureaucracy, who had founded a permanent *Conferenza dei concili* (an academy for the study of church history) in 1671 and an *Accademia fisico-matematica romana* (*AFR*) in 1677. Bianchini attended meetings of both the *conciliari* and the *fisicomatematici*. That did not exhaust him or Roman academic opportunity. In furtherance of his interest in anatomy and medicine, which he had pursued in Padua, he also attended the *Congresso medico romano*, a creation of Christina's doctor, Girolamo Brasavola, and the later papal archiater, Giovanni Maria Lancisi. Bianchini became a prominent member of Ciampini's academies after an interval in Verona, where he spent the years 1686–88 on family and academic business.

There was no academy of experimental science in Verona in 1686. Bianchini therefore set about creating one, in collaboration with several young physicians discontented with the medical establishment and its outmoded Galenic teaching. Following the example of Descartes' earliest Dutch disciples, they pushed the mechanical philosophy, the Cartesian concept of the body, and experiments designed to undermine medical orthodoxy to advance their professional ambitions. Bianchini was a great help to these disaffected doctors. Fresh from the metropolis, he could bring a Roman gloss to the *fisicomatematica* he had learned from Montanari. He seems to have been the life of the experimental program of the *Aletofili* ("truth-seekers") of Verona, for after his return to Rome they devoted themselves primarily to their fight with the local medical college, which came to its moment of truth with their victory in 1700.[14]

Bianchini had left the *Aletofili* with instructions for carrying on the missionary work of experimental academies and an emblem (a magnetic compass) and slogan (*Aut docet aut discit*) to keep them steady. Both the slogan, "[the academy] either learns or teaches," and the compass signified that much remained to be added to what others had discovered.[15] Building on the work of their great predecessors, the Galileans, and the virtual Italian virtuoso, Robert Boyle, who was present in Florence at the time of Galileo's death, the *Aletofili* should follow the sound method of philosophizing discovered by the moderns. This consisted of explaining all physical appearances in terms of the motions of invisible particles, without, however, insisting upon the details of the motions, the particles, or the explanations.[16] By renouncing this impossible task, Bianchini advised the truth seekers, they would escape the errors and braggadocio of school philosophy while gathering knowledge sufficient for the improvement of medicine.[17]

Metropolitan

Ciampini served his academic apprenticeship as a member of Christina's *Accademia reale*. Soon he saw how to better her arrangements in the interests of science and erudition. He would have no rhetorical declamations, no musical interludes, no high-flying philosophy; but he would follow as best he could Christina's practice of co-opting princes of the church as patrons and participants.

Ciampini's first academic creation, the *Conferenza dei concili*, gained the favor of Innocent XI, who wanted to make it public and papal in furtherance of the domestication of critical study of the early church. Consquently Ciampini lost control of his creation. Although it continued to boast real scholars like Bianchini, Emmanuel Schelstrate (prefect of the Vatican Library), and Ciampini himself, it ended advancing curial careers more effectively than science. Some 28 of its 186 known members were or became cardinals; another 51, bishops or other prelates; two became pope. In 1704, the *Conferenza* came under the supervision of a papal appointee and new regulations limited membership to 48 souls, all of whom must have studied theology for at least two years.[18]

When Ciampini established the *AFR* in the time of Alexander VII, he named it the *Accademia dell'esperienze filosofiche e naturale* in reference to its model, the *Accademia del Cimento*, from which it took its initial activities and protocols. At each meeting, after prior consultation with Ciampini, one member would lecture and show experiments "with apt mechanical instruments." After further "curious physical demonstrations," the company, which averaged two dozen, would proceed to philosophize. The two-hour sessions took place weekly and, after the first exuberance, monthly. The original scope comprised philosophical, medical, mathematical, and mechanical subjects, which Ciampini understood in some surprising ways. Under "philosophical" he grouped speculations about the elements and the natural history of humans, plants, and animals; under "medical," anatomy of all creatures, transmutation of metals, and new chemical inventions; under "mathematical," cosmological speculation (geography, hydrography, meteorology, astronomy) and recent discoveries in arithmetic, geometry, and music; and under "mechanical," optics, horology, painting, sculpture, civil and military architecture, etc.[19] The natural science in this tall order soon dwindled to *fisicomatematica*, that is, Galilean subjects. The *AFR* did not abandon the mechanical arts, however, or the study of antiquities (under sculpture and architecture). Thus Ciampini created a bond between his physico-mathematical and historical academies. Since several members of both, like Bianchini, also had an interest in medicine, and since the *Congresso medico romano* concerned itself also with mathematics and natural philosophy, the three institutions reinforced one another.[20]

When Bianchini began to frequent the *AFR* in the 1680s, its experimental program balanced between Galilean subjects as developed by the *Cimento*—mechanics (the *Two new sciences*), pneumatics (experiments in spaces void of air), and optics (improvements of telescopes)—and practical projects such as carriage design (a fixation of Ciampini's) and embellishment of Rome (a fixation of the *AFR*'s engineer Cornelius Meijer). In the early 1680s, in the time of Innocent XI, the *AFR*'s leading experimenter was Franesco Eschinardi, S.J., Professor of Mathematics at the Roman College. He suffered from the usual schizophrenia of the able Jesuit fisicomatematico: he helped himself to Galileo's principles in his treatises on mechanics while engaging freely in the pious work of discrediting them. His technique, still used and useful in academic battle, was to carp at unimportant slips or over-simplifications in the master's arguments. Pursuing this strategy, Eschinardi supported his confrère Francesco Vanni, librarian of the Collegio Romano, in a protracted argument intended to show that the "moment" of a weight on an inclined plane could not be proportional to the sine of the angle of inclination, as had Galileo taught.[21]

Ciampini's academy divided over the merits of Vanni's argument.[22] This made slippery footing for Bianchini, who knew from Montanari how to explode it, and from prudence and common sense that it ill behoved a beginner without patronage to take sides in debates among people who could be helpful to him. Furthermore, he knew better than to expound his modest and mechanical philosophy, even with Montanari's instrumentalist twist, in Innocent Rome. It seemed advisable to retreat to the neutral plane of astronomical observation, in which his sharp eyes and Montanari's instruction gave him a competitive edge. That is the character in which Leibniz met him in 1690 and esteemed him a likely successor to Tycho or Hevelius.[23]

One of Ciampini's academicians was Giuseppe Campani, Europe's master maker of lenses of long focal length. Campani happened to have several of these on hand in 1684, returned by Louis XIV, whose minister Colbert, then lately deceased, had bought them as suitable jewels for his monarch's scientific crown but had neglected to pay for them. Ciampini commissioned a telescope incorporating one or more of them. It is a measure of his wealth as well as of his commitment that he could shop in the same market as the King of France. Unfortunately, he could not buy academicians as Louis could and, as no prince came forward to shoulder the burden, the *AFR* did not survive Ciampini's death in 1698. However, in 1684, with Campani's lenses and Bianchini's technique, the *AFR* could hope to contribute something to astronomy that Louis' academicians might notice.[24] In any case, the gigantic Ciampini-Campani telescopes, with which, as one academician wrote, "no one doubts that new and wonderful things are to be seen," made a fit spectacle for baroque Rome. They turned out to be more remarkable than anything observed through them.[25]

The *AFR* quickly found a small place for itself in the fast-paced international cultivation of astronomy. Shortly after arriving in Rome in 1684, Bianchini had the good fortune to spy a new comet before anyone else in Italy had reported it. Ciampini informed the Royal Society of London of the discovery made by the "lynx-eyed Abb. Bianchini, disciple of Geminiano Montanari."[26] The international reputations of Bianchini and the *AFR*, "composed of minds so sublime that . . . it flourishes to the amazement of the entire universe," rose together.[27]

Leibniz tried to direct these sublime minds, especially Ciampini's and Bianchini's, toward cancelling the Catholic church's condemnation of Copernican astronomy. Leibniz presented the irenic argument that, since motion is relative, all and none of the contending cosmologies represented the true system of the world.[28] Unfortunately, Leibniz developed his argument and launched his appeal during the reign of Alexander VIII (1689–91), when Rome suffered from one of the Vatican's periodic crackdowns on modern philosophizing. Bianchini responded equivocally that he doubted he could convince the relevant authorities; "but if things fall out as I wish, I will take great care that the old importance is extended to the science I cultivate so eagerly." Ciampini backed away altogether. Worried about charges of atomism as well as of Copernicanism, he avoided scientific pitfalls and guided his *fisicomatematici* through the troubles of the time.[29]

The complete learned society of the later seventeenth century informed the world of its work through an invention of the time, the scientific journal. The first steps were timid. The Royal Society of London began to publish its *Transactions* in 1665, with the caveat that it took no responsibility for them. The Académie des sciences of Paris could not go even that far; it made use of the *Journal des sçavans* until its reorganization in 1699, when it started to publish its *Mémoires*. Ciampini planned a bolder course: he intended to issue the proceedings of the *AFR* under its name and responsibility.[30] That did not happen. There already existed an Italian version of the *Journal des sçavans*, the *Giornale dei letterati di Roma* (*GL*), founded in 1668 to acquaint Italians with the latest foreign work and to defend Italian work from foreign plagiarists.[31] The *GL* had a strong Galilean emphasis in natural science and a bias in favor of atomism purged of paganism.[32] Two years before the founding of the *AFR*, the *Giornale* split into two journals, both with the original name, one edited by Ciampini and several of his academicians including Eschinardi.[33] Ciampini's *GL* included book reviews, reports taken from the *Journal des sçavans*, and items from members of the *AFR*. He had to be content with the partial reporting of the academy's meetings because Eschinardi opposed issuing proceedings in which pure Jesuit physics would sleep under the same covers with meretricious Galilean doctrine.[34] Even this partial reporting did not last long, however, since Ciampini stopped publishing in 1681 (the rival *GL* had already ceased) probably because he did not want to flog his wares in Rome during the bear market of Innocent XI.[35]

National

Despite the political fragmentation of the peninsula in the late seventeenth century, Italian culture, expressed in a common religion and conveyed by a common literary language, provided a meaning other than "geographical hodgepodge" to the word "Italy." Regard for religion and language marked the activities of many academies throughout the peninsula. It took a foreigner, however, new to the language and the religion, to plant the seed that grew into the first Italian national academy. The foreigner was Queen Christina.

Christina's *Accademia reale* dissolved at her death in 1689. One of its members, Giovanni Crescimbeni, designed a partial replacement that would continue to function as an arbiter of literary taste and a resort of powerful prelates. He went about his work with a reformer's ardor. He persuaded fourteen others who shared his dislike of the formulaic pomposity of baroque Italian poetry to join in establishing an academy to (as one of its founders later put it) "restore correct and good taste to literature." They called themselves "Arcadian pastors" in identification with the simple and supposedly wholesome ways of the shepherds of ancient Arcadia; and they took bizarre names, such as (in Bianchini's case) Selvaggio Afrodisio, "the wild Aphrodisian," to emphasize their occasional departures from ordinary life. The community of Arcadians grew rapidly by addition of the great churchmen who had frequented Christina's academy, and, following their example and allure, young nobles, lesser aristocrats and clerics, lawyers looking for briefs, poets looking for audiences, abbés looking for places, and, at least at first, men of true erudition and science.[36]

During the 38 years of Crescembeni's stewardship, 2,619 people joined the *Arcadi*. Although Rome remained the main center, "colonies" spread all over the peninsula; in1710 there were 21 of them, two in monasteries. By 1728, anyone interested could read of Arcadian accomplishments in nine volumes of *Rime*, three of *Prose*, and three of *Vite* of defunct pastors, all chosen and edited by Crescimbeni. These exemplary *Vite* taught the ingredients and rewards of a correct blend of literature, erudition, and religion.[37] The obituary of the highest flier of all the Arcadians, Clement XI, traces his advance from his studious youth, which ended with doctorates in both laws and in philosophy at the age of eighteen; through his frequenting of Christina's and Ciampini's academies, in which he distinguished himself equally for his "perfect erudition in profane matters [...] and the profundity [of his knowledge] of sacred things;" to the recognition of his talents by the Vatican. Despite the increasing demands of the offices assigned him, he kept up his academic life. Queen Christina recommended him to Alexander VIII, a former member of her *Accademia reale*, who promoted him to cardinal though he was not a priest.[38] We need not follow Crescimbeni's account of Albani's triumphal course further. The lesson is clear: the right academies gave men of talent access to the highest ecclesiastical preferment.

The great influx of members during the first years of the *Accademia degli Arcadi* included Cardinal Albani and his friend Bianchini among other *eruditi* and men of science, Ciampini, Lancisi, Malpighi, Redi. Bianchini's contributions included a learned paper on Olympiads to anchor the Arcadian calendar and some clever riddles in rhyme, which earned him the reputation of a humorist.[39] After becoming pope, Albani continued to smile upon the *Arcadi* as a low-risk way to bring together the intellectuals of the peninsula.[40] Crescimbeni organized Arcadia hierarchically, centralized its decision-making in Rome, and promoted Clement's cultural and religious policies. The Pope rewarded this service with offices and benefices.[41] According to one stern student of Crescimbeni's stewardship, he allowed high officers of the church (all pastors no doubt) to set the goals and methods of an increasingly passive organization. Their goal was to secure "hegemony" over Italian intellectuals, and to "marginalize and control theoretical initiatives."[42]

One unsubmissive pastor, "the wise and learned" Leucoto Galeate (in real life the scholar-librarian Ludovico Antonio Muratori), tried to set up a stronger and broader Arcadia.[43] He complained that Crescimbeni's pastors did not move fast enough and that their ridiculous names and often silly verse merely reinforced the prejudices of the rest of Europe. He had in hand two remedies: lengthy instructions for reforming poetry, which he published after some delay as *Della perfetta poesia italiana* (1706), and a plan for a *Repubblica letterata d'Italia* (*RLI*), drawn up in 1703, which would cultivate the natural sciences and Maurist historiography. *Perfetta poesia* would instruct the *Arcadi*; the *RLI* would make it difficult for the rest of Europe to ignore or appropriate solid science from the peninsula.[44]

The turning points of the short life of the *RLI* are soon told. In December 1703 Bernardo Trevisan, a Venetian patrician, received from an unknown correspondent Lamindo Pritanio (that is, from Muratori) the first plan ("Primi disegni") for the *RLI*, which Trevisan had agreed to print and distribute. Owing in part to the secrecy

Pritanio imposed on the undertaking, the first mailings did not go out until the following October. It met with little response. Pritanio (who had not yet revealed his identity) then asked Trevisan to repeat and extend the distribution, this time with the notice that all correspondence should be addressed to Bianchini as "depositario," or acting secretary, of the nascent republic. Trevisan refused, on the scruple that he would not embroil someone in the project who knew nothing about it. Pritanio consequently mailed out the second round himself, in January 1705. On January 31, Muratori and his former teacher Bacchini duly sent their opinions to Bianchini. He replied on February 7 that he would have nothing to do with the Pritanian republic. That prompted (in March) the publication of a "Lettera apologetica" from Pritanio that explained his odd conduct and effectively killed the project, although Trevisan and others kept it alive briefly as they sought a substitute "depositario."[45] Muratori himself nursed the hope that Pritanio's ineptness had not destroyed the prospects of a defensive league of Italian *eruditi*.[46]

Pritanio's ostensible objective, to correct Italian scholarship and foreign evaluation of it, became colored by the French occupation of Modena beginning in 1703 and the consequent temporary exile of its duke and the duke's librarian Pritanio-Muratori. The coloring may be seen in references to the *RLI* as a league or confederation, as if it were an alliance in the War of the Spanish Succession, which had brought the French to Italy. Did he have in mind a confederation of *eruditi* as a forerunner of a neutral protective league of Rome and the northern Italian states?[47] The confederation would have little chance of success even as the Italian province of the Republic of Letters if the participating states did not support it financially and morally. It needed money and freedom from the Roman censorship. At the time, the censorship was the bigger problem for Muratori. His teacher Bacchini had a book awaiting an imprimatur, a scholarly edition of Andrea Agnello's eighth-century *Liber pontificalis* (lives of the early bishops of Ravenna). Agnello was a wild and irresponsible writer and his main theme, the independence of the bishops of Ravenna from the See of Rome, was disagreeable to the Vatican. Had Bacchini glossed Agnello severely enough to destroy Agnello's credibility altogether? The censors asked the opinion of Monsignore Bianchini. He advised against publication. The censors agreed and prohibited the book in March 1705 just as Pritanio-Muratori was composing his response to Bianchini's rejection of the *RLI*.[48]

For architects of Solomon's houses, Pritanio's plan is more important than his politics. In the "Primi disegni" of 1703, he specified "a union, a republic, a league" of the most distinguished Italian *letterati*, irrespective of condition or degree, provided they had published at least one solid book and would work for the "reform and increase of the arts and sciences in furtherance of the Catholic religion, the glory of Italy, [and] public and private welfare." To inspire emulation, the republic should offer prizes; and to exclude the merely ambitious, it should have gatekeepers and censors, experts who would diagnose the defects of scientific work in all fields and suggest ways to correct them.[49] The main fields of study would be languages ancient and modern, with an emphasis on good taste; and natural philosophy, with its endless prospects. In pursuing these objectives, the *RLI* should try for the middle ground between obstinate scholasticism and immoderate modernism. "Acute minds can then discover a thousand ways to benefit physics and the truth." Italians have an advantage here: inheritors of the science of Galileo and the critical methods of the *eruditi*, they are positioned to correct the errors of the new ways and the old, and to lead the advance. "We must place our best hope for national glory in the philosophy we call experimental."[50]

Like all well-run republics, Pritanio's would have police powers. It would instruct booksellers and publishers not to print the twaddle of literary academies. It would appoint scholars to review all worthwhile books, domestic and foreign. It would build up libraries (behold the librarian behind Pritanio!). But above all, it should see to the restoration of good taste in the schools of all religious orders dedicated to study, whence should issue a constant supply of manpower for the advancement of science. To assure quality in the supply, the Republic's disciplinary powers should extend to expelling dilettantish students and incompetent professors. With more benign coercion, the Republic should demand civility, criticism, competition, and harmony; "open warfare in opinions, but not in hearts." In this impossible way, "the empire of science and arts will increase with the reputation of the learned and to the advantage of everyone."[51]

All citizens of Pritanio's republic were not equal. At the top were *arconiti*, acknowledged experts, authors of books, whom the recipients of Pritanio's "Primi disegni" would nominate. The arconites would co-opt the rest of the membership, *studiosi* (upcoming scholars) and *candidati* (beginners). Bacchini offered important and politic suggestions about this organization. He suggested a reduction to two classes and a shift of emphasis. The first class would include important patrons of science as well as *arconiti*; the second class, mechanics and instrument makers as well as scholars. The *RLI*'s best chance lay not in natural science, because (in Bacchini's opinion) it was tolerated in Italy only as a helpmate to medicine. Ancient and modern literature, the subjects of historians and Arcadians, were closest to the developed Italian genius. Above all, Bacchini presciently warned Muratori, do not privilege the increase of Italian prestige over the quest for truth among the Republic's objectives.[52]

As for "the immensity of the expense with which anyone who wants to advance in scholarship in our age is inevitably faced," Bacchini recommended lowering the qualification for aconites so as to include rich ecclesiastics, philanthropists like Marsili, and communicators like Magliabecchi. The class of *studiosi* should be eliminated; the class of *candidati* should have space for representatives of all the trades necessary to the republic's publishers, also for instrument makers and cartographers, and men able to help with expenses. To proceed, Bianchini should choose thirty arconites from the names proposed by Pritanio's correspondents, and the thirty should appoint the rest of the Republic.[53]

The charade unravelled when, toward the end of 1704, Muratori wrote Bianchini asking for help to clear Bacchini's *Agnello*. He mystified the situation by writing about Pritanio as a third party ("it would take all the astrologers to identify the author"), by hinting that the initiative came from Rome ("and not without the knowledge of the Holy Father"), and congratulating Bianchini on his election to "depositorio" ("perhaps inspired by the magnanimous zeal of the reigning pope").[54] This colossal effrontery concludes with the hope that *Agnello* had been approved. "If not, I must say that I could not be very sanguine about the proposed republic."[55]

Bianchini replied that personal and national ambition should not be confounded with erudition, that a literary competition among nations was neither Christian nor philosophic, and that (sliding off his high horse) the proposed organization, with its fancy names and classes, was as infantile as Pritanio's approach was dishonest. As Bianchini knew very well, Clement had not sanctioned the plan, and he doubted whether any other prince, including the Duke of Modena, had done so either.[56] Muratori blamed the failure of his republic on Bianchini the censor. "Since Bacchini's *Agnello* has been prohibited [. . .] I don't see how the project to make any literary republic can go on." Still, still, it was a good idea, too good to drop. Muratori reprinted the "Primi disegni" in the first part of his treatise on the correct method in everything, *Reflessioni sopra il buon gusto nelle scienze e nelle arti* (1708). He believed that ultimately Pritanio's plan would be realized, with the right sponsors and statutes. The current crop of prospective arconites just did not have what it took. They should have studied the substance, not the manner, of Pritanio's proposal. The *oltremontani* were right. "O, you lazy Italians!"[57]

International

Bianchini's "punctilio" (as Muratori put it) in appealing to cosmopolitan values against Pritanio's Italian republic was not feigned. He took pride in his status as corresponding member of the Paris Academy of sciences, which he had enjoyed since 1699. In that year the Academy adopted new statutes that provided for non-resident correspondents and also for no more that eight foreign associates. Holding one of these eight positions was a very high distinction in the international Republic of Letters. No one becomes a member of so exclusive a club as the Paris Academy on merit alone. Bianchini's connection was Gian Domenico Cassini, head of he Royal Observatory. The two had met in 1695, during a visit by Cassini to Italy, when they worked together comparing the accuracy of meridian lines.[58] No doubt Bianchini owed his corresponding membership to Cassini. In this formal capacity, he reported astronomical observations made in Italy and discussed reforms of the Gregorian calendar proposed in Rome and elsewhere at the beginning of the eighteenth century. Cassini published some of the information Bianchini sent in the Academy's *Mémoires*.[59]

Bianchini's promotion in 1706 to foreign associate was a laissez-passer throughout the learned world. For this reason, as well as his reliability and diplomacy, Clement chose him for the delicate task of delivering the badge of office—the crimson hat—to a newly appointed French Cardinal.[60] The mission was delicate owing to the continued strain in relations between Louis and Clement and because Bianchini had the additional charge of reporting on negotiations among the great powers to end the war. Clement was concerned particularly with the chances of placing the man he and Louis recognized as "James III"—the Catholic son of the deposed James II Stuart—upon his father's throne. Bianchini met James and other important actors and gleaned information sensitive enough to send to the Vatican in cipher.[61]

During the peace negotiations, Louis abandoned the cause of James III. After his ill-fated attempt in 1715 to oust the new occupant of his throne, George I of Hanover, James retired to Italy. Bianchini acted as his cicerone. They got on well together. Bianchini helped blunt the boredom of James' sojourn in the pope's family palace in Urbino by teaching him astronomy. Together they viewed an eclipse of the moon there that furnished observations for Bianchini's last contribution to the *Mémoires* of the Paris Academy.[62]

Bianchini's northern tour of the Republic of Letters included England. He arrived early in 1713, apprehensive about the welcome he would receive as a Catholic and a Jacobite. He need not have worried. The learned heretics spoke French or Latin and were eager to show the Pope's man their collections, libraries, and instruments. He could be of considerable use to them for his blend of knowledge and power: as an *érudit* he could appreciate, identify, and date problematic antiquities; and, as superintendent of all ancient inscriptions found in Rome (a charge given him by Clement in 1703), he knew the latest finds and had the authority to allow their removal. He visited Oxford as the guest of the Jacobite Chancellor, the Duke of Ormonde. The Bodleian opened its treasures to him; the colleges entertained him; the most famous professors waited in line to see him. Among them was John Keill, professor of astronomy, disciple of Newton, and Fellow of the Royal Society.[63]

Bianchini had met the Jacobite Keill in London soon after arriving in England. Keill took him to see Newton. Though a fierce Whig and anti-papist (indeed, anti-Trinitarian), Newton acted the perfect host. No doubt he warmed under the news that his visitor had succeeded in reproducing the principal demonstrations in his book on light and colors, as set forth in its Latin edition (*Optice*, 1705). This achievement, perhaps the first in Europe, was the work of an "academy" that gathered in the cell of Celestino Galiani in Rome. Galiani's group met to study contemporary work in natural science. It decided to try to master Newton's *Principia*, and applied through the British minister in Florence for a copy from a London bookseller. The seller sent the then-new *Optice* instead. Within a year or two Galiani's group had confirmed the contested experimental basis of Newton's optical theories and Galiani had become expert enough to perform the chief experiments before other academies, notably the "Alessandrina," which had the Pope's nephew Alessandro Albani as patron and Bianchini as program director. Newton rewarded the cleverness of the Italians by giving their herald his portrait, a glass of canary wine, and a promise of copies of his books when Bianchini returned from his excursion to Oxford.[64]

On his return to London, Bianchini attended a meeting of the Royal Society, greeted its president Newton again, and witnessed experiments on electricity performed by Francis Hauksbee. A week later, on Newton's recommendation, Bianchini became a Fellow of the Society. Hauksbee again provided the evening's entertainment, this time with experiments on refraction and capillarity. Bianchini judged correctly that Hauksbee's experiments were of great importance. Among the many books he brought back to Italy were Hauksbee's *Physico-mechanical experiments* (1709), of which he commissioned an Italian translation, and several copies of the Royal Society's attack on Leibniz, *Commercium epistolicum* (1712), which he distributed among Italian mathematicians according to Newton's directions.[65]

Bianchini had made small contributions to the work of the Royal Society long before his election to its fellowship. We know Ciampini's report of the observations by his lynx-eyed protégé of the comet of 1684. These observations came to the attention of the Astronomer Royal, John Flamsteed, whom Bianchini would visit during his London trip, and Flamsteed's archenemy, Edmond Halley, who used them in his book on comets.[66] The following year, 1685, Bianchini sent his observations of Jupiter's satellites, which Flamsteed published in the *Transactions*, and news of Montanari's recent work on hydraulics.[67] After this flurry, Bianchini directed his astronomical reports to the *Acta eruditorum*, with which his then patron, Schelstrate, had a connection. There followed long years of work on history and chronology, and observations at the meridian line in Santa Maria degli Angeli, which Bianchini did not publish.[68] But he occasionally sent the Society products of his lynx-eyed viewing; his collaborator in Portugal, the Jesuit Giovanni Battista Caprone, supplied their joint determinations of longitude; and Halley possessed a list of observations Bianchini made during the last seven year's of his life.[69]

Universal

Bianchini could participate actively in so many academies because he had few obligations other than to work at and talk about science. A bachelor, he had neither the care nor the social life a family brings. A cleric but not a priest, he had no parochial obligations or monkish duties. A beneficiary of several lucrative sinecures, including room and perhaps board, he did not need to bother himself about creature comforts, to which in any case he was indifferent. In short, he had his evenings free. Attendance at academies provided much of his social and intellectual life. And as one academy catered to some subjects and another to others, so did each provide a distinctive mix of social utilities.

Montanari's *Traccia*, Manfredi's *Inquieti*, and the small groups convened by Bianchini and Galiani amounted to advanced seminars for teaching and learning up-to-date science and sometimes to research seminars. Bianchini belonged to many of them, usually to several at once, for most of his adult life, some specialized in natural science, others in ecclesiastical history and archaeology. Verona's *Aletofili* and its stronger southern counterpart, the *Congresso medico romano*, added an explicitly professional, even careerist, objective to academic life. Their members worked to improve medicine by injections of natural philosophy and to enlarge opportunities for physicians by opening up the medical establishment. These several sorts of academy cultivated modern philosophy, that is, an eclectic undogmatic corpuscularism and experimentalism, as sound science and good strategy. Bianchini's directions to the *Aletofili* well represent the higher aspirations of the academies of experimental science.

Ciampini's academies avoided strong commitment to modern philosophy because several members found it uncongenial and the changing level of tolerance in Rome made it dangerous. Nonetheless, the *Conferenza dei concili* and the *AFR*, though eclectic in interests and viewpoints, pushed the programs of Maurist history and *fisicomatematica*, and, in benign times, collaborated with the on-again off-again *Giornale dei letterati* to spread both of them. As we know, the swollen membership of the *Conferenza* came to have a greater interest in promoting careers than in advancing science. As it took this unexpected course, Ciampini organized small shadow academies that met in his house on most evenings for serious conversation and sober refreshment. Bianchini frequently attended these soirées, which provided a social life agreeably combined with wide-ranging talk.[70] They were the functional equivalents of the senior common rooms in Oxbridge colleges before the admission of married fellows.

The *AFR* contributed another element to the academic mix besides science, sociability, and career advancement. This was "felicità pubblica," the body politic equivalent of domestic bliss. Ciampini was a philanthropist as well as a maecenas. He gave to the poor, succoured the sick, and willed half his estate to support the education of the underprivileged.[71] Some of the practical applications pursued by the *AFR* aimed to ameliorate urban life. In a word, a breeze of Christian charity stirred the atmosphere of Ciampini's noble academy. A similar breeze, or rather a wind, had emanated from the original *Solomon's House*, wherein the Christian brethren of the New Atlantis went about acquiring knowledge and power, and wherefrom they issued the fruit of their labors for the benefit of all. Their paternalism, which extended to withholding inventions they deemed dangerous or unhealthy, went beyond the programs of practical improvements undertaken by the academies that claimed their mantle, notably the Royal Society of London. In Ciampini's *AFR*, as in Bacon's utopia, means for increasing public bliss came from the philanthropist, not the entrepreneur.[72] Bianchini would have found this circumstance both congenial and familiar—congenial because charity figures high among the cardinal virtues and familiar because Montanari, who advised the Venetian state on hydraulics and coinage and attacked astrology as a public menace, had emphasized the obligation of the Christian philosopher to work for the common good.[73]

Although its main initial purpose was reform of secular Italian literature, the *Arcadia* did not long escape the embrace of the church. Charity here was not in question on either side. The founders of the academy of pastors wanted to assert the poetical merits of Italy against attacks or dismissals by *oltremontani*. The mighty churchmen who belonged to it wanted to assure that its program remained poetical and pastoral, that in defending against foreign critics it did not learn how to oppose authority at home. The academy of pastors became a diffuse society of poetasters, distributed around the peninsula and directed, in so far as their Arcadian life was concerned, by Crescimbeni. Bianchini at first enjoyed the pastoral play and then seems to have appreciated the *Arcadia* as a means of social control. In addition to assisting identification of and with a national literary culture, it gave intellectuals and fellow travellers a supervised forum for venting discontent without attacking the true pan-peninsular cultural power, the *Accademia delle accademie*, the Roman Catholic Church.

The Catholic Church in Italy around 1700 discharged many of the functions of Pritanio's *Repubblica letteraria d'Italia* through the libraries and academies of its learned cardinals and prelates, its patronage systems, its teaching orders, and its censorship. Considerable revenues derived from personal wealth as well as from church offices went into these enterprises. Pritanio dreamed of combining papal resources with those of the lay princes to better the material circumstances, and establish the scientific authority, of the proposed Republic. He also wanted to blunt the censorship of books by assigning the job to competent *aconiti* who would consider only the scientific value, never the contemporary political relevance, of the works they examined. And he wanted to oversee the pedagogy of the teaching orders. The Republic's police would inspire or remove incompetent professors and lazy students and see to the inculcation of good taste in the arts and sciences.

Nothing of the kind was going to happen. The head of the Academy of Academies, the Pope, could not have descended to the level of the Duke of Modena as patron of the *RLI* even if he had approved of Pritanio's objectives. The congregations responsible for the spiritual wellbeing of literate Christians would scarcely have relinquished the Lord's work to peer review by an institution championing freedom of thought. No more would the teaching orders, chief among them the Jesuits, submit to the supervision of even the wisest aconites. No wonder Muratori wrote under an assumed name!

Bianchini understood that despite the progressive elements in the *RLI* plan, the propelling cause, the insistence on Italy's return to the van of European culture, was retrograde and unrealizable. Leadership had passed beyond the Alps. To achieve international distinction, Italians had to adhere to the norms and expectations of the larger Republic of Letters.[74] And they had to do so while avoiding the handicaps of the Academy of academies. Bianchini was a master at this game. With mild self-censorship he could do astronomy as a Copernican, discourse on natural philosophy as an atomist, write history as a Maurist, admire the scholarship of heretics and Frenchmen, and occupy a position of honor in the Pope's household.

Bianchini's elevation to foreign associate of the Paris Academy and his election to the fellowship of the Royal Society give a final example of the utilities of academies. By rendering favorable judgments of the work of foreigners, particularly by the award of foreign memberships, they could and still can strengthen the position of savants in places where the freedoms and norms of science are difficult to maintain. In the hope of this sort of endorsement, Vico addressed the first edition of his *Scienza nuova* (1725) to "the academies of Europe."[75]

The mix of institutions that have fulfilled the many functions of the old academies has changed over time. Today, universities, national laboratories, and industrial research organizations do most of the research in the natural sciences and their applications. National and international agencies have replaced the church, princes, and maecenases of yesteryear. Commercial publishers and university presses publish most of the research that the agencies support. Associations for the advancement of science and societies for the advancement of individual sciences have taken on most of the work of protecting, promoting, and policing scientists. Although big national academies with some of the powers Muratori granted his *RLI* exist, even the strongest of them, the Royal Society and the U.S. National Academy of Science, face competition from other domestic academies and professional societies. On the international level, the certification of grandee status in science has become more international than Bianchini could have imagined: the Nobel institution has reduced foreign academic memberships to consolation prizes in the fields it honors. No doubt the existence of a Nobel Prize in Arcadian productions and the absence of one for Maurist disciplines would also surprise him. However, encouraging the study of history is now an important activity in many of the old academies, whose libraries preserve much of the histories of the sciences, natural and historical.[76]

However the social and intellectual functions of the academies of Bianchini's day are split among the institutions of our time, Tore Frängsmyr can be proud of his part in supporting many of them. He did much better than Muratori in mobilizing a dispersed cultural community—in his case, Scandinavian historians of science and learning—to assume their rightful place in the world. He used instruments unavailable to Muratori, like the jet plane, to bring foreigners to Sweden and to distribute his Scandinavian newsletter widely and gratis. Pritanio would have approved. Frängsmyr did as well as Queen Christina in the exercise of sociability, as everyone who attended international schools and conferences he arranged knows; and he did better than Ciampini in assisting the careers of scholars, notably the many students whom he brought to the Uppsala Summer Schools. He built Solomon's Houses in Uppsala and Stockholm—a department and seminar in the one that would make Bacchini's mouth water, and an air-conditioned archive in the other that would arouse the envy of those rare librarians Bianchini and Muratori. He has always had a generous concept of the scholarly life and the community that nurtures it. Although he is a very wise man, he knows well that, as Bacchini put it, *nemo solus satis sapit*, no one is wise enough alone.

Notes

1. The usage corresponds to that in Jean Boutier et al., "Les milieux intellectuels italiens comme problème historique," in *Naples, Rome, Florence. Une histoire comparée des milieux intellectuels italiens (XVIIe–XVIIIe siècles)*, ed. Jean Boutier et al. (Rome: Ecole française de Rome, 2005), pp. 1–31, on p. 20.

2. J. L. Heilbron, "Bianchini as an astronomer," in *Francesco Bianchini (1662–1729) und die europäische gelehrte Welt um 1700*, ed. Valentin Kockel and Brigitte Sölch (Berlin: Akademie Verlag, 2005), pp. 57–82, on pp. 58–64. Recent studies of Montanari include Salvatore Rotta, "Scienza e 'pubblica felicità' in Geminiano Montanari," *Miscellanea seicento,* 2 (1971), pp. 64–210; Susana Goméz Lopéz, *Le passioni degli atomi. Montanari e Rossetti: Una polemica tra galileiani* (Florence: Olschki, 1997); and Mario Umberto Lugli, *Geminiano Montanari* (Modena: Il Fiorini, 2004).

3. Paul Mouy, *Le développement de la physique cartésienne* (Paris: Vrin, 1934), pp. 98–101, 108–113, 145–147.

4. Marta Cavazza, *Settecento inquieto. Alle origini dell'Istituto delle scienze di Bologna* (Bologna: Il Mulino, 1990), pp. 44–46, 124, 135; Rotta (note 2), pp. 98–99, 131–133; Marialaura Soppelsa, *Genesi del metodo Galileiano e tramonto dell'Aristotelismo nella scuola di Padova* (Padua: Atenore, 1974), pp. 129–130.

5. *Scienziati del seicento,* ed. Maria-Luisa Altieri-Biagi and Bruno Basile (Milan/Naples: Ricciardi, 1980), p. 513n.

6. Outlined in J. L. Heilbron, "Bianchini and natural philosophy," in *Unità del sapere, molteplicità dei saperi: l'opera di Francesco Bianchini* (to appear).

7. Cavazza (note 4), pp. 173–175, 216–219; John Stoye, *Marsigli's Europe, 1680–1730* (New Haven: Yale University Press, 1994), chapt. 10; J. L. Heilbron, "The contributions of Bologna to Galvanism," *Historical studies in the physical sciences,* 22:1 (1991), pp. 57–85, on pp. 61–65. Maria Pia Donato, *Accademie romane. Una storia sociale (1671–1824)* (Naples: Edizioni scientifiche italiane, 2000), pp. 44–45, counts the Bologna academy as Roman owing to the support of Clement XI; Cavazza (note 4), pp. 160, 171, and *Museo dalla specola,* ed. Enrica Baiada et al. (Bologna: Bologna University Press, 1995), p. 101, indicate Bianchini's involvement in it.

8. Bruno Neveu, "Culture religieuse et aspirations réformistes à la cour d'Innocent XI," in *Accademie e cultura: Aspetti storici tra sei e settecento* (Florence: Olschki, 1979), pp. 1–38, on pp. 7–8, 12–13, 34.

9. J. L. Heilbron, "Censorship of astronomy in Italy after Galileo," in *The church and Galileo,* ed. Ernan McMullin (Notre Dame: University of Notre Dame Press, 2005), pp. 279–322, on pp. 286–288; Maurizio Torrini, "L'Accademia degli Investiganti, Napoli 1663–1670," *Quaderni storici,* 48 (1981), pp. 845–883, on pp. 869, 875.

10. Overview by Antonio Clericuzio and Maria Conforti, "Christina's patronage of Italian science," in *Siderius nuncius and stella polaris: The scientific relations between Italy and Sweden in early modern history,* eds. Marco Beretta and Tore Frängsmyr (Canton, MA: Science History, 1997), pp. 25–36.

11. Neveu (note 8), p. 4; Antonella Romano, "A l'ombre de Galilée? Activité scientifique et pratique académique à Rome au XVIIe siècle," in Boutier et al. (note 1), pp. 209–242, on pp. 231–233, 237. Cf. W. E. K. Middleton, "Science in Rome, 1675–1700, and the Accademia fisicomatematica of Giovanni Giusti Ciampini," *British journal for the history of science,* 8 (1975), pp. 138–154, on pp. 139–140, 153–154.

12. Neveu (note 8), pp. 3, 10–11, 20.

13. Ibid., pp. 21–23, 25, 29–30, 38; Pier-Alessandro Paravia, "Vita di M. Francesco Bianchini," in Francesco Bianchini, *Storia universale,* 5 vols. (Venice: G. Batagia, 1825), vol. 1, pp. xv–xliv (p. xviii).

14. Silvano Benedetti, "L'Accademia degli Aletofili di Verona tra sei e settecento," in *Accademie e cultura* (note 8), pp. 223–226; Ivano Dal Prete, "Scienza e società nella terraferma veneta: Il caso Veronese 1680–1796" (Ph.D. thesis, University of Verona, 2004), pp. 147–150, 159–161, 167; Salvatore Rotta, "L'Accademia fisicomatematica Ciampiniana: un'iniziativa di Cristina?" in *Cristina de Svezia: Scienza ed alchemia nella Roma barocca*, eds. W. di Palma et al. (Bari: Dedolo, 1990), pp. 96–186, on pp. 151–154.

15. Francesco Bianchini, *De emblemate* (1686), as excerpted in Gaetano Gasperoni, *Scipione Maffei e Verona settecentesca* (Verona: Valdonega, 1955), pp. 28–29.

16. Francesco Bianchini, "Dissertazione . . . da lui recitata nella radunanza dell'Accademia degli Aletofili," *Nuova raccolta di opuscoli scientifici e filologici*, 41 (1785), pp. 3–37, on pp. 12–18, 21 (text of 1687).

17. Ibid., pp. 28, 35–37 (quote).

18. Donato (note 7), pp. 16–25, and "Le due accademie dei concili a Roma," in Boutier et al. (note 1), pp. 243–255, on pp. 245–249.

19. G. Toschi to A. Rocca, 23 July 1677, in Middleton (note 11), pp. 144–145; and Jean-Michel Gardair, *Le 'Giornale de' letterati' de Rome (1668–1681)* (Florence: Olschki, 1984), pp. 145–146; Susanna Åkerman, *Queen Christina of Sweden and her circle* (Leiden: Brill, 1991), p. 254; Romano (note 11), p. 321, from Piazzi, *Eusevologio romano* (1698).

20. Donato (note 7), pp. 27–28 (quotes from Toschi), 29, 34; Rotta (note 14), pp. 124–125, 132–133, 152.

21. Heilbron (note 6). Eschinardi could be useful and helpful; Middleton (note 11), pp. 150–151.

22. Maurizio Torrini, *Dopo Galileo: Una polemica scientifica (1684–1711)* (Florence: Olschki, 1979), pp. 45n, 80–84, 90–95. Vanni argued that a sphere resting between touching planes of inclinations α and β would require by Galileo's principle of momento (M) a counter force $g(\sin\alpha + \sin\beta)$. In the case Vanni considered, α and β were complementary so that $M_\alpha + M_\beta = g(\sin\alpha + \cos\alpha) > g$.

23. Leibniz, in André Robinet, *G.W. Leibniz: Iter italicum* (Florence: Olschki, 1988), p. 55.

24. Giuseppe Monaco, "Un parare di Francesco Bianchini sui telescopi di Giuseppe Campani," *Physis*, 25 (1983), pp. 413–431, on pp. 415–418.

25. J. L. Heilbron, *The sun in the church: Cathedrals as solar observatories* (Cambridge, MA: Harvard University Press, 1999), pp. 248–250; Carlo di Napoli [G.G. Ciampini?], *Nuova invenzioni di tubi ottici dimostrate nell'Accademia fisicomatematica romana l'anno 1686* (Rome: G. Komarek, 1686), pp. 10–17; Schelstrate to Otto Mencke (editor of the *Acta eruditorum*), 1664, in Rotta (note 14), p. 133n (quote).

26. Rotta (note 14), pp. 163–165; [G.G. Ciampini], "An extract of a letter . . . concerning a late comet seen at Rome," Royal Society, *Philosophical transactions*, 15 (1685), pp. 920–921.

27. Quote from a work of 1683 cited by Ferdinando Fabiani, *Il merito applaudito . . . nelle ponderazione dell'Illustriss. e reverendiss. Monsig. Ciampini* (Fermo: G.F. Bolis, 1694), p. 15.

28. Robinet (note 23), pp. 96–118.

29. Bianchini to Leibniz, [1690], in Robinet (note 23), p. 97; Rotta (note 14), pp. 173–174; Donato (note 7), p. 43.

30. Gardair (note 19), p. 147.

31. Ibid., pp. 34, 49, 53, 86–87, 253–255.

32. Ibid., pp. 145–146, 182–192, 206–220, 250–252.

33. Ibid., pp. 119–120, 143.

34. Ibid., pp. 127–128, 154.

35. Ibid., pp. 51, 271–274.

36. Donato (note 7), pp. 68–69; Michel Giuseppe Morei, *Memorie istoriche dell'adunanze degli Arcadi* (Rome: Rossi, 1761), pp. 1 (quote), 16–21. Anna Maria Giorgetti Vichi, *Gli arcadi del 1690 al 1800: Onomasticon* (Rome: Arcadia, 1977), gives the real and Arcadian names of all members from 1690 to 1800.

37. Amadeo Quondam, "L'istituzione Arcadia. Sociologia e ideologia di un'accademia," *Quaderni storici*, 8:2 (1973), pp. 389–438, on pp. 392–393, 403, 408–413; Morei (note 36), pp. 189–197.

38. Giovan Maria Crescimbeni, "Clemente XI," in Crescimbeni, *Notizie istoriche degli Arcadi morti*, vol. 3 (Rome: A. de Rossi, 1721), pp. 361–377, on pp. 361–365; Donato (note 7), pp. 72–73: the *Vite* stress the "ideal of a scholar in perfect equilibrium among erudition, good taste, sociability, and moderate religiosity."

39. Morei (note 36), p. 75; Vichi (note 36), p. 229 ("academico umorista"); Francesco Bianchini, *Opuscola varia*, 2 vols. (Rome: Barbiellini, 1753), Vol. 2, pp. 83–87, 108–111, 121–122 (texts of 1703/4).

40. Bianchini (note 39), Vol. 2, p. 176; Donato (note 7), p. 60.

41. Donato (note 7), pp. 71–72.

42. Quondam (note 37), pp. 395, 402, 405, 419 (quote). Cf. Donato (note 7), pp. 73–74.

43. "Sage et savant," Voltaire's appraisal as quoted by Alphonse Dupont, *L. A. Muratori et la société Européenne des pré-lumières* (Florence: Olschki, 1976), p. 8.

44. Lodovico Antonio Muratori, *Opere*, ed. Giorgio Falco and Fiorino Forti, 2 vols. (Milan: Ricciardi, 1964), vol. 1, pp. 179–180 (text of 1703). Françoise Waquet, *Le modèle français et l'Italie savante (1660–1750)* (Rome: Ecole française de Rome, 1989), gives an excellent account of French views of Italian letters (and vice versa) around the turn of the eighteenth century.

45. Anna Burlini Calapaj, "I rapporti tra Lamindo Pritanio e Bernardo Trevisan," in *Accademie e cultura* (ref. 8), pp. 73–94, on pp. 75–76, 80–83, 93; Gian-Francesco Soli Muratori, *Vita del proposto Lodovico Antonoio Muratori*, in L. A. Muratori, *Opere tutte*, 13 vols. (Arezzo: Bellotti, 1767), vol. 1, pp. 23–27.

46. Alberto Vecchi, "La nuova academia letteraria d'Italia," in *Accademie e cultura* (ref. 8), pp. 39–72, on pp. 65–67.

47. Argument adopted from Vecchi (note 46), pp. 39, 41, 44, 46, 48–51; Calapaj (note 45), p. 73.

48. Aldo Andreoli, "Il Bacchini e il Muratori," *Benedictina*, 6 (1952), pp. 59–84, on pp. 60–63; Arnaldo Momigliano, *Dizionario biografico degi Italiani*, 5 (1963), s.v. "Bacchini" (pp. 127–129); Vecchi (note 46), p. 65; Calapaj (ref. 45), p. 91.

49. Muratori (note 44), vol. 1, pp. 182–184.

50. Ibid., pp. 186–188.

51. Ibid., pp. 195–197.

52. Tommaso Sorbelli, "Benedetto Bacchini e la Repubblica lettereria del Muratori," *Benedictina*, 6 (1952), pp. 85–98, on pp. 93–95.

53. Sorbelli (note 52), pp. 95–98; Muratori-Pritanio to Scipione Maffei, 1704, in L. A. Muratori, *Epistolario*, ed. Matteo Campori, 14 vols. (Modena: Società tip. Modenese, 1901–22), vol. 2, p. 739.

54. Muratori to Bianchini, 20 December 1704, ibid., p. 744.

55. Ibid., pp. 744–745. Cf. Waquet (note 44), pp. 380–381.

56. Bianchini to Muratori, 7 February 1705, in Soli Muratori (note 45), pp. 232–236. Cf. Sergio Bertelli, *Erudizione e storia in Lodovico Antonio Muratori* (Naples: Nella sede dell'Istituto, 1960), pp. 81–82; Alberto Carraciolo, *Domenico Passionei tra Roma e la Repubblica delle lettere* (Rome: Storia e letteratura, 1968), pp. 199–201.

57. Muratori to Gatti, 16 April 1705, in Muratori (note 53), vol. 2, p. 757, first and third quotes; Muratori, "Lettera apologetica," in Soli Muratori (note 45), p. 247, second quote.

58. Heilbron (note 25), pp. 152–153.

59. Francesco Bianchini, "Observation d'une comète du mois d'avril de cette anneé 1702," Académie royale des science, Paris, *Mémoires*, 4 (1702), pp. 118–120, and "Refléxions sur des mémoires touchant la correction grégorienne," ibid., 6 (1704), pp. 142–145.

60. Francesco Uglietti, *Un erudito veronese del settecento. Mons: Francesco Bianchini 1662–1729* (Verona: Biblioteca capitolare, 1986), pp. 74–83.

61. Ibid., pp. 83–86, 129–132; J. L. Heilbron, "La merdiana di Bianchini e la politica estera di Clemente XI," *Giornale di astronomia*, 32 (2006), pp. 46–53.

62. Francesco Bianchini, "Observation de l'éclipse de Lune, faite à Urban le 9 septembre 1718," Académie royale des sciences, *Mémoires*, 20 (1718), pp. 327–329, and "Notizie, e pruove della corografia del Ducato d'Urbino," in *Memorie concernenti la Città di Urbino dedicate all sagra Real Maestà di Giacomo III*, ed. Annibale Albani (Rome: Salvioni, 1724), pp. 133–147, on pp. 140–144.

63. Details from Bianchini's travel diary transcribed in Salvatore Rotta, *Francesco Bianchini in Inghilterra* (n.p.: Paideia, 1966), pp. 39–49.

64. Ibid., p. 39; Hanns Gross, *Rome in the age of the Enlightenment* (Cambridge: Cambridge University Press, 1990), pp. 252–253; Vincenzo Ferrone, *Intellectual roots of the Italian Enlightenment: Newtonian science, religion, and politics in the early eighteenth century* (Atlantic Highlands, NJ: Humanities Press, 1995), chapts. 2–4.

65. Rotta (note 63), pp. 575–578; Claudia Pighetti, *L'influsso scientifico di Robert Boyle nel tardo seicento italiano* (Milan: Francoangeli, 1988), pp. 225–255; J. L. Heilbon, *Electricity in the seventeenth and eighteenth centuries* (Berkeley: University of California Press, 1979), pp. 231–238. The Italian translation by Thomas Dereham, *Esperienze fisico-matematiche* (1716), differs from Bianchini's unpublished version.

66. Ciampini to Croone, 20 August 1684, in John Flamsteed, *Corresponence*, ed. Eric Forbes et al., 3 vols. (Bristol: IOP, 1995–2002), vol. 2, p. 191; Edmond Halley, *Astronomicae cometicae synopsis* (Oxford: Oxford University Press, 1705); William Whiston, *Praelectiones physico-mathematicae cantabridgiae* (London: B. Motte, 1726), p. 357.

67. Flamsteed, Royal Society of London, *Philosophical transactions*, 15:177 (December 1685), pp. 1215–1217; Flamsteed (note 66), Vol. 2, p. 261; Rotta (note 14), pp. 177–182.

68. Heilbron (note 2), pp. 61–63, and (note 25), pp. 148–166.

69. RBC.12.479 (comets) and 14.273 (observations); Cl.P. IV(i)81 (auroras), all at the Royal Society Library; RGO 2/10 (Halley Papers, Cambridge University Library); G. B. Carbone, "Observationes astronomicae," Royal Society, *Philosophical transactions*, 35 (1727/8), pp. 471–476, on pp. 473–476, solar eclipse; Francesco Bianchini, "Observations of the eclipses of Jupiter's satellites," ibid., 36 (1728/9), pp. 35–36; Rómulo de Carvalho, *A Astronomia em Portugal no sécolo XVIII* (Lisbon: Instituto de cultura e lingua portuguesa, 1985), pp. 47–50.

70. Lina Montalto, "Un ateneo internazionale vagheggiato in Roma sulla fine del secolo XVII," *Studi romani*, 10 (1962), pp. 660–673, on pp. 663–664.

71. Ibid., pp. 665–671.

72. Francis Bacon, *The new Atlantis* (1627), in Bacon, *The major works*, ed. Brian Vickers (Oxford: Oxford University Press, 1996), pp. 457–458, on pp. 487–488.

73. Rotta's seminal paper (note 2) has the title, "Scienza e 'pubblica felicità' in Geminiano Montanari."

74. Waquet (note 44), pp. 382–386.

75. Giambattista Vico, *The first 'New science,'* ed. Leon Pompa (Cambridge: Cambridge University Press, 2002), p. 3.

76. J. L. Heilbron, "Historical pointers to the future of academies," in *L'esperienze delle accademie e la vita morale e civile dell'Europa*, ed. Edouardo Vesenti and Leopoldo Mazzarolli (Venice: Istituto Veneto di scienze, lettere ed arti, 2006), pp. 123–36.

The Origin of Species from Linnaeus to Darwin

NICOLAAS RUPKE

Frängsmyr's Linnaeus: A Man of His Time

During the heyday of the "forerunners" approach, historians of science cast Carl Linneaus (1707–1778) in the role of one of the great precursors of Charles Darwin (1809–1882). The Swedish taxonomist and author of the *Systema naturae*, having early on given support to creationist fixism, had come to accept an origin of new species by means of hybridization. Moreover, towards the end of his life, Linnaeus had conjectured that God's work of creation had remained restricted to the ancestral form of each genus or even just of each order, and that natural diversification within the limits of these original types had produced today's many species, which could thus be described as "the work of time." Draw a straight line from early to late Linnaeus, extrapolate this line and you arrive at Darwin—or so the story of the Darwin commemorative literature of 1959 went.[1]

Was Linnaeus indeed one of the great eighteenth-century pioneers of evolutionary biology who paved the way for Darwin? It is to the credit of Tore Frängsmyr that he has resisted pinning this forerunner badge on the lapel of his famous Uppsala colleague and Swedish national hero. In *Linnaeus: the Man and his Work* (1994), Frängsmyr in no uncertain terms urged us to think of Linnaeus as a man of his time, a botanist firmly rooted in eighteenth-century physico-theology, devoutly respectful of the biblical narrative of creation and with a special interest in the location and nature of Paradise. Frängsmyr additionally drew our attention to the fact, however, that Linnaeus had shown an appreciation of process, of change over time, especially with respect to geological change. For example, Linnaeus proposed in his conferment address of 1743, *Oratio de Telluris Habitabilis Incremento* (On the Growth of the Habitable Earth), that the surface area of dry land had gradually increased, as the primordial ocean subsided; initially, only a small island had stuck out above the surface of the waters, and this island had been the Garden of Eden of Genesis. In some detail, Frängsmyr showed that these Linnaean conjectures were part of contemporaneous preoccupations.[2]

Still, by emphasizing "process" in the writings of Linnaeus, Frängsmyr provided a link, if not with the next-but-one generation of Darwin, then with the immediately connecting one of the Göttingen professor of medicine, Johann Friedrich Blumenbach (1752–1840). The latter, in addressing the question of the origin of species, altogether abandoned creationist thinking and replaced miraculous creation by a naturalistic process. This process was not evolution, however. A third theory dominated European biology before Darwin's *The Origin of Species* brought about the ascendancy of organic evolution. During the period covering the late-eighteenth to middle-nineteenth century, in mainstream biology, a non-miraculous, naturalistic mode of species origins became widely adopted that supplanted the metaphysical notion of creation yet equally rejected evolution. This third theory stated that species, or the root forms from which present-day species developed, had originated naturally by the spontaneous aggregation of their germ masses.[3]

Today's history of evolutionary biology is to a large extent a Darwinian narrative. Right from its beginning—from the appearance of Darwin's own "An historical sketch of the progress of opinion on the origin of species," which was added to the third and, in an expanded form, to the fourth and later editions of the *Origin of Species*—the third theory of spontaneous, naturalistic species generation was by and large ignored. Darwin put forth the story that when his *Origin of Species* was first published, in 1859, the major issue in biology was creation (supernatural) versus evolution (natural) and that he—Darwin—had settled the controversy in favor of the latter.[4] Ever since, this story has been retold and elaborated. Nevertheless, it is inaccurate.

The Third Theory

Let us recall that from the time of the classics, from Aristotle to Lucretius, there had existed some four views of the origin of species: species are eternal; species are created; species originate spontaneously; and species evolve by the transmutation of one into another—evolution.[5]

By the time the *Origin of Species* appeared, and within mainstream scientific thought, only the first of these four was no longer seriously considered. Species eternalism had all but vanished, one of its few last representatives having been the military surgeon and philosopher Heinrich Czolbe (1819–1873) who, however, in the mid-1850s was uncompromisingly taken to task for his views by, among others, Berlin's founder of medical pathology, Rudolf Virchow (1821–1902).[6] The second view—creationism—by contrast, continued to flourish, especially among Anglo-American biologists and paleontologist, famous among whom were Oxford's

William Buckland (1784–1856), Cambridge's Adam Sedgwick (1785–1873) and Harvard's Louis Agassiz (1807–1873). The fourth view of the origin of species—evolution or rather the transmutation of species—found few adherents and remained marginal until Darwin was forced out of his evolutionary closet and published his *magnum opus*.

Wiped from the slate of our collective memory since then is the fact that—let me reiterate for emphasis—during the immediately preceding seven or so decades the majority of the great minds of the earth, life, and biomedical sciences, especially in the German-speaking world, believed neither in creation nor evolution but in the third theory of the origin of species—spontaneous origin or *Urzeugung*.

The reader might be excused for objecting that the theory of spontaneous generation has not been quite so forgotten as here maintained. Don't we have—one may well ask—a substantial body of secondary literature on the seventeeth- to nineteenth-century controversy over the spontaneous origin of "infusoria" and "entozoa," which controversy was settled to the satisfaction of most biologists by Louis Pasteur's classic experiments that appeared to disprove the phenomenon?[7] Is this story not well and widely known? However, two theories of spontaneous generation existed that, although related and in some publications interwoven, were distinct. One of these was the familiar theory about the contested observation that today, under our very eyes, primitive forms of life spontaneously originate from lifeless matter.

Yet spontaneous generation also had a second meaning. It could refer to the past origin of species, when the very first specimens of any fixed form of life, low and high, and including humans, must have come into existence—it was postulated—by non-parental generation. This theory stated that plants, animals and humans had originated, not by miraculous, special creation but by the natural generation of one, two or many of the very first individual representatives of each species. Various explanations were put forward why in taxonomically higher organisms the original spontaneous generation of its members had not continued but been superseded by a process of sexual reproduction in perpetuating the species.

Obviously, the first of these two theories could have a bearing on the second; but it was perfectly possible for biologists not to believe in the spontaneous generation of primitive life at present and still fully endorse the theory of the *generatio spontanea* of species in the past, and quite a few of them did.[8] If simple organisms originate spontaneously today, this enhances the likelihood that also more advanced species similarly came into being long ago; but it is also imaginable that the origin of each permanent form of life was unique to certain moments in geological history, because only at those moments, temporarily, the right conditions for the original aggregation of their germs or seeds had occurred.

About a dozen different terms were used to describe spontaneous generation (e.g., *generatio spontanea, primigenia, cosmica, primitiva, originaria, automatica, aequivoca, heterogenea,* "Urzeugung," "Urerzeugung," "autogene Zeugung") and, in order to avoid confusing the two theories of spontaneous generation, I shall refer to the second one by using a synonym, namely autogenous generation/autogeny. It is this theory that has not been entered into the annals of the history of biology. Yet autogeny was integral to four major research programs: (1) the physiology of generation; (2) the palaeontology of extinction and origins/renewal; (3) the geography of plant and animal distribution; and (4) anthropogeography. Let us briefly look at the place of autogenous generation in each of these fields of scientific theory and practice.

PHYSIOLOGY OF GENERATION

First and foremost, the notion of autogeny was part of the biomedical study of physiology, especially the study of generation, regeneration and nutrition. More specifically, it was intimately connected with the theory of epigenesis, which stated that the germ material of an organism gains its form gradually, as opposed to the "nested boxes" view of "preformation." Preformation with its near-infinite complexity of pre-existing individuals favored creation, whereas epigenesis made it possible to think of organic origins in terms of aggregation, similar—some conjectured—to the process of crystallization. If the origin today of an individual by sexual reproduction could be thought of as the effect of a *nisus formativus* working on undeveloped substances to produce and develop an egg, then it was imaginable that the very first production of an individual of any species had taken place in like manner, by the aggregation of the required chemicals to form a primordial germ. Conversely, sexual reproduction could be thought of as a repeat process of the primordial moment of aggregation or coagulation of the very first "motherless," non-parental germ. Once sexual generation had taken over, spontaneous generation had become obsolete and today this phenomenon, if indeed it still occurs, is restricted to organisms that typically do not procreate by means of sexual reproduction—mainly primitive micro-organisms, it was thought.

Thus autogenous generation could, and did, become an integral part of biomedical research and, from the 1810s on, was taught in various physiology textbooks, for example by the Leipzig professor of obstetrics Johann Christian Gottfried Jörg (1779–1856). Another highly regarded obstetrician, who held the chair of medicine at Giessen, Ferdinand August Ritgen (1787–1867), attributed the origin of the human species to the motherless development of ova-like beginnings in "Uferschlamm," mud along the banks of bodies of water. Also, the Bern professor of physiology and comparative anatomy Gabriel Gustav Valentin (1810–1883) used the concepts and language of autogeny to discusss human generation.[9]

A central figure in the development of these speculations was Blumenbach, who, as author of the pro-epigenesis booklet *Über den Bildungstrieb* (1781; 2nd edn 1791), was also one of the early naturalists cautiously to express the likelihood of the non-miraculous, autogenous generation of species. This he did in 1791 in a footnote to the fourth edition of his *Handbuch der Naturgeschichte* (1791, 2–3)— also later, in his *Beyträge zur Naturgeschichte* (second edition of the first part, 1806,

19–20). The *Handbuch* went through many editions and several translations, and it exerted considerable influence throughout learned Europe. "Although Blumenbach tended to follow the Linnaean system, this work ushered in a new era in natural history. It contains an abundance of new or hitherto insufficiently evaluated morphological and ecological findings, from which Blumenbach drew conclusions that led to a more modern (biological and evolutionary) concept of the plant and animal kingdom."[10] In the first three editions of the *Handbuch*, Blumenbach referred to a first "creation;" but in the fourth and later editions, he unambiguously naturalized the word creation by adding the following footnote to a mention that a species consists of individuals that form an uninterrupted series going back "bis zur ersten Schöpfung:"

> Oder wenigstens bis zu ihren ersten Stammältern hinauf.—Denn ich habe im ersten Theile meiner *Beyträge zur Naturgeschichte* Facta angeführt, die es mehr als bloß wahrscheinlich machen, daß auch selbst in der jetzigen Schöpfung neue Gattungen von organisirten Körpern entstehen, und gleichsam nacherschaffen werden; wohin namentlich auch die erste Entstehungsweise mancher sehr einfachen und mikroskopischkleinen organisirten Körper, wie z.B. der mehrsten sogenannten Infusionsthierchen gehören scheint.[11]

One of Blumenbach's many students, Gottfried Reinhold Treviranus (1776–1837), the person to introduce the term and the subject of biology as a separate discipline in Germany, in his main work *Biologie, oder Philosophie der lebenden Natur* (1802–1822), broke a lance for the spontaneous generation today of primitive organisms, and approvingly discussed the theory of the autogenous origin of species. More explicitly than had his Göttingen teacher, he attributed the appearance of higher forms of life to the same abiogenetic, spontaneous generation that—he believed—still today continues to produce such lower organisms as "zoophytes":

> Wir dürfen daher, gestützt auf die Analogie der Zoophyten, Pflanzen und niedern Thierclassen, annehmen, dass auch die Urformen der Säugthiere und Vögel einst auf dieselbe Weise erzeugt wurden, worauf in jeztigen Zeiten meist nur noch Zoophyten gebildet werden, [... und] dass der Weg, worauf die ganze lebende Natur gebildet wurde, derselbe war, auf welchem jetzt noch die plastischen Kräfte bey der Erzeugung aus formloser Materie wirken.[12]

Others in biomedical physiology followed suit and treated the theory of the autogenous generation of species as integral to the physiology of generation and related phenomena. "Urzeugung" or "Urerzeugung" became a major topic, ranking substantial entries in the new encyclopedias and encyclopedic series of the period.[13] No lesser figure than the Königsberg physiologist Karl Friedrich Burdach (1776–1847), who counted some of the greatest names of biomedical physiology and anthropology among his pupils, such as Karl Ernst von Baer (1792–1876) and

Martin Heinrich Rathke (1793–1860), in his physiological and anthropological works explicitly advocated species autogeny and, as had Blumenbach, allowed for a certain degree of species variability, yet modifications had not gone so far as a transformation of one major form into another.[14] Also the Berlin physiologist Johannes Müller (1801–1855) in his *Handbuch der Physiologie des Menschen* (1844), and Virchow in a major discussion of "Alter und neuer Vitalismus" (1856), used concepts and language of autogeny in discussing the origin of species, although Müller indicated that the origin of species lay outside human experience and would better be dealt with by philosophy than physiology.[15]

PALEONTOLOGY OF EXTINCTION AND REGENERATION

An autogenous origin of species was readily integrated into the new geology of the time that recognized the extinction of many species in the course of a long earth history as well as the repeated origin *de novo* of new organic communities or "worlds." Species had come and gone. Also here Blumenbach led the way with early speculations about extinctions, past and present. In his *Beyträge zur Naturgeschichte*, he stressed "die Veränderlichkeit" (changeableness) of nature as shown, for instance, by the fossil record. Neither a belief in Providence nor any supposed chain of being disallowed the fact of extinction. Blumenbach recognized some three periods of pre-Adamic geological history. The rock formations corresponding to the oldest contained unfamiliar organic remains that had become extinct as the result of a major global revolution.

But how had the earth become repopulated? Blumenbach tried to answer this question without invoking a supernatural power. The life force we know from the phenomena of generation and regeneration, the *Bildungstrieb*, had restarted life on earth, producing new species that differered somewhat from the old ones. As an example, Blumenbach cited the whelk *Murex contrarius*, considered extinct, which looks identical to its living relative *Murex despectus*, except that in the former the shell is left-turned and in the latter right-turned. Such changes were not caused by a mere degeneration of early forms into later ones, but were typical instances of changes in direction of the *Bildungstrieb*. So, following geological revolutions, nature had produced new species, different from the ones that had been wiped out because altered physico-chemical conditions made the *Bildungstrieb* deviate from its previous course. While retaining biblical vocabulary, Blumenbach changed the meaning of "creation" from a miraculous event to a natural process, stating,

> dass der *Bildungstrieb* nach dem durch eine solche Totalrevolution freylich wohl anders modificierten Stoffe auch bey Erzeugung der neuen Gattungen eine von der vormaligen mehr oder weniger abweichende Richtung hat nehmen müssen. [. . .] So dass die bildende Natur bey diesen Umschaffungen zwar auch zum Theil wieder Geschöpfe von ähnlichen Typus, wie die in der Vorwelt, von

neuem reproducirt, die bey weitem allermehrestem aber mit andern der neuen
Ordnung der Dinge zweckmässigern, Formen hat vertauschen müssen."[16]

At the occasion of the fiftieth anniversary of the start of Blumenbach's leg-
endary career as a Göttingen University teacher, his pupil K.E.A. von Hoff
(1771–1837), by then himself a famous Gotha geologist, saluted his mentor for
having solved the problem of the origin of species: "die Lösung des Räthsels" had
been provided by Blumenbach and the new understanding of the history of life on
earth had been generated, "erst von Seinen Zeitgenossen, die Seine Ideen benutzt
haben, und zu einem nicht geringen Theile von Seinen Schülern."[17]

A majority of those who further developed the paleontology of extinction and
regeneration adhered to the theory of autogeny. Many assumed that, following a
worldwide catastrophe that had wiped out life on earth, special environmental conditions may have temporarily prevailed that favored the spontaneous generation of
germs and the autogenous origin of life and species in repopulating the earth. The
Halle professor of zoology and later director of Argentina's National Museum of
Natural History, Hermann Burmeister (1807–1892), in his semi-popular *Geschichte
der Schöpfung* (1848) summed it up:

> Wollen wir also nicht zu Wundern und Unbegreiflichkeiten unsere Zuflucht
> nehmen, so müssen wir die Entstehung der ersten organischen Geschöpfe auf
> der Erde durch die freie Zeugungskraft der Materie selbst einräumen und die
> Gründe, warum diese Zeugungskraft jetzt nicht mehr für höhere Organismen
> fortdauert, aus allgemeinen Naturgesetzen, denen zufolge nur das Noth-
> wendige, nicht das Ueberflüssige statuirt worden ist, deduciren.[18]

Among the influential names to work with the notion of autogeny were two of
Germany's leading earth scientists, the geologist Leopold von Buch (1774–1853),[19]
and the paleontologist Heinrich Georg Bronn (1800–1862). Bronn's magisterial
*Untersuchungen über die Entwickelungs-Gesetze der organischen Welt während der
Bildungs-Zeit unserer Erd-Oberfläche* (1858), which won the prize of the Academy
of Sciences of Paris in answer to a prize question, set in 1850 and repeated in 1854,
what the nature of the fossil record was, may be regarded as the defining monograph of the paleontology of autogeny. Like many of his contemporaries, Bronn
recognized that a progressive development manifests itself throughout geological
history. Yet he did not visualize the succession of fossil groups from low to high in
terms of genetic descent but holistically: species had been produced autogenously
in conformity with and as an expression of the level of development of nature (the
surface of the earth, the atmosphere, flora and fauna) as a whole.[20]

A spread of opinion existed, however. Some, such as Lorenz Oken (1779–
1851), a fervent nature philosopher, dreamt up speculations about the earth being
a living entity that had exuded a primordial organic slime that had functioned as
the substratum from which plants, animals and humans had emerged.[21] Others,

such as the materialist Carl Vogt (1817–1895), ridiculed Okenian nature philosophy and reduced life—and mind—to nothing more than a constellation of matter. New forms of life had originated from the earth following geological revolutions, just as progressive changes in contemporaneous society were the results of political revolutions.[22]

Also, the non-catastrophist, gradualist picture of earth history as put forward by Charles Lyell (1797–1875) was believed to provide evidence of abiogenetic species origins. Lyell calculated that on the basis of a gradual uniformitarian rate of extinction and origin of new species, in a region the size of Europe, only once every 8,000 years or more a mammal would disappear and emerge.[23] Nearly all these leading figures, Lyell prominently among them, rejected or additionally ridiculed the notion of species transformation—evolution.

GEOGRAPHY OF PLANT AND ANIMAL DISTRIBUTION

Like geology, Humboldtian physical geography was one of the cutting-edge fields of early-nineteenth-century science. One of its major accomplishments was the discovery and delineation of global provinces of distribution of plants and animals and the investigation of the environmental parameters that condition these distributions. The discrete geological worlds in time seemed mirrored by equally discrete provinces of distribution in space, each definable by a characteristic assemblage of species and by dominant forms of life.

Probably more than any other field, biogeography adopted autogenous generation as a central organizing concept. These provinces had come into existence—it was widely believed—because the species that made up their separate communities had originated *in situ*. Provinces of distribution were "specific centers," where species had originated. Some species may have migrated, but the existence of discrete areas over which plants and animals are spread by and large reflected nativeness. Thus autogeny became autochthony, part of the environmental determinism of the times, prominently advocated by Carl Ritter (1779–1859).[24] A few scientists such as Agassiz, who endorsed the work on provinces of distribution, resorted to "the intervention of a Creator" at several geological stages and in many geographical places.[25] Others, however, argued that the Bible allows for only a single center of creation, the Garden of Eden, and that dispersal from there or dispersal after the deluge from Mount Ararat could not possibly account for the present-day distribution of species that perforce reflected autochthonous origins.

Plants in particular came to be thought of as species with a well-defined area of distribution, vertically from low to high altitude in mountainous regions, and horizontally/latitudinally from equator to pole across the globe.[26] Moreover, in the context of Humboldtian geography, the interest that species have as entities of Linnaean taxonomy was significantly increased by species being additionally considered interdependent members of communities of vegetation. Single species or single individuals of species could not have survived on their own—a community had been necessary for survival. Consequently, the process of autogeny had involved the origin at more or less the same time of many species and many members of each species. The multiple origins of members of a single species seemed corroborated by the discovery of identical species in wholly separate, unrelated provinces of distribution between which it seemed impossible that migration had taken place.

These and related implications added theoretical significance to where species are found, and for several decades an enormous amount of work went into elaborate statistics of species distribution. Janet Browne, in her pioneering study of the history of biogeography, drew attention to "Pflanzenstatistik," and to the problem of why people considered such numerical tables of interest.[27] I believe that the answer is provided by the fact that these statistics had a direct bearing on the theory of the autogenous/autochthonous origin of species.

An entire generation of famous Humboldtian biogeographers advocated autochthony, early among them the Copenhagen botanist Joakim Frederik Schouw (1789–1852) who famously defined over twenty provinces ("Reiche").[28] Among the others were the Berlin botanist Franz Meyen (1804–1840), the Genevan phytogeographer Alphonse de Candolle (1806–1893), the Austrian zoogeographer Ludwig Schmarda (1819–1908) and, until well after Darwin's *Origin of Species*, Göttingen's professor of natural history and director of the botanical garden, August Grisebach (1814–1879), who as late as 1872 argued for the autogenous origin from "scattered germs" of each characteristic flora.[29] As Meyen stated:

> So wie nun die Natur in der gegenwärtigen Zeit nur niedere Gebilde ohne Keime ihres Gleichen zu erzeugen vermag, so hat sie einst, als sich die jetzige Erde mit Pflanzen belebte, auf eine ähnliche Art die höheren Pflanzen und Thiere erschaffen, deren Fortpflanzung wir gegenwärtig nur durch Keime oder Eier vor sich gehen sehen.[30]

ANTHROPOGEOGRAPHY

Most dramatic in its consequences was the adoption of autogeny as an organizing concept in Romantic anthropology. Nearly to a man, its star representatives in the German-speaking world believed in the *Urzeugung* of humans. As in biogeography generally, in anthropogeography particularly, the phenomenon of provinces of distribution was attributed to autochthonous origins and autogenous generation. Phyto- and zoogeography appeared to indicate that a species need not go back to a single individual or a single pair but that it could be traced to multiple first individuals, each originated from non-living matter in one and the same area or even in widely separated parts of the world. Applied to the human species, this conclusion had a potentially explosive effect, because it made possible to think of human races as autochthons—as varieties of humankind that each had a distinct geographic origin, even though they formed a single species *Homo sapiens*. Human races were autochthons—many thought—and polygeny was allied to autogeny well before it became part of evolution.

At this time, when physical anthropology was giving scientific precision to the concept of race and, in the political arena, bringing this to bear on the practice of slavery, the model of autogenous polygeny was grist to the mill of the pro-slavery camp. An abolitionist such as Blumenbach, however, who in his doctoral dissertation *De generis humani varietate nativa liber* (2nd edn of 1781 and later ones) had defined five varieties, each belonging to a particular part of the world, famously insisted that humans share a common, single point of origin.[31] Over time they had migrated and due to processes of "degeneration" (a term that meant departures from the original form) became diversified into several "races."

Several fellow anthropologists agreed with Blumenbach, among whom James Cowles Prichard (1786–1848), who in his *Researches into the Physical History of Man*, identified a single "centre of creation" for humans. Other autogenists sided with Blumenbach and Prichard when it came to *Homo sapiens*. One of them was Burdach, who saw the distribution of human varieties across the globe as a product of migration and put the issue as follows:

> Da nun jede Gegend unsres Planeten die ihrer Eigenthümlichkeit entsprechenden Pflanzen und Thiere erzeugt hat, und da die verschiednen Menschenstämme, so weit unser Wissen in das Alterthum heraufreicht, immer dieselben Eigenthümlichkeiten gehabt haben, durch welche sie sich jetzt noch von einander unterscheiden, so ist die Behauptung aufgestellt worden, dass sie ursprünglich verschieden gewesen und in verschiednen Gegenden aus den Händen der schaffenden Natur hervorgegangen seien. . . . Der Mensch ist nicht wie Pflanze und Thier an die Scholle gebunden, und seine verschiedne Stämme können fern von der Heimath auf jedem Puncte der Erde sich behaupten, sind also nicht die Erzeugnisse eigner Klimate.[32]

Equally great names stood on the other side of the issue. Karl Asmund Rudolphi (1771–1832) argued that human races are true "aborigines" whose historical distribution is not a result of migrations but due to separate origins *in situ*.[33] A similar view was championed by Carl Gustav Carus (1789–1869), who pleaded ignorance about the origin of species and of the human species in particular—primordial origins lie outside the range of our observational possibilities—but who then specified nevertheless how humans must have come into existence by summing up the various ways in which this could not have happened. Humans had neither originated as told in the Bible nor had they evolved from apes. Specifically, *Homo sapiens* had not made its first appearance in adult form, on land, as a single individual or a single pair, like Adam and Eve in the Garden of Eden; nor had it originated as some evolutionists speculated by the transformation of monkeys into men. The human species—he asserted—had its origins in primordial vesicles that had developed in enormous numbers in water under mild and stable climatic conditions:

> Nicht mag aber gedacht werden die Entstehung des Menschen als durch plötz-
> liches Hervortreten eines oder mehrerer vollendeter Organismen bedingt; ...
> wir [dürfen] einen Zustande der Erde denken, wo bei gewaltigerem allge-
> meinen Bildungsleben auch höhere, ja die höchsten epitellurischen Organis-
> men aus ... Urbläschen ... hervorgingen. Nicht darf man sich aber dann die
> Entwicklung derselben im trocken, sondern nur im feuchten Elemente
> denken; und hat man einmal jene Ansicht deutlich erfaßt, so wird es drittens
> nicht zu denken sein, daß nur einfach oder zweifach jene Entwiklung ur-
> sprünglich erfolgt sei, sondern in gewaltiger Menge; ... Viertens mag es nicht
> gedacht werden, daß anders als unter Einwirkung einer höhern gleichmäßigen
> Wärme—eines mildesten Clima's, jene Entwicklung erreicht worden sei ...
> Endlich aber und hauptsächlich darf es nicht gedacht werden, daß der Mensch
> entstanden sei, indem sich ein Thier (ein Affe etwa ...) in seiner Entwicklung
> gesteigert und so Mensch geworden sei.[34]

This process could and would have taken place in different regions of the globe, and Carus was an advocate of polygeny, combining it with the explicit contention that the different races are unequal with respect to their intellectual abilities. In a famous "Denkschrift" on the occasion of the centenary of Goethe's birth in 1849, Carus produced a map of the world on which he plotted the climatic zone that had produced the highest forms of life—unsurprisingly, this included his own part of the world—and zones that had given rise to lower levels of humanity.[35]

The third theory had its largest following in Germany; but also in France it had outspoken representatives, ranging from Jean-Claude de Lamétherie (1743–1814)[36] to Félix Archimède Pouchet (1800–1872); the latter's major work, *Hétérogénie ou traité de la génération spontanée* (1859), summed up the debate about both the spontaneous generation of primitive forms of life today and the autogenous generation of species in the past. In the German literature, the British authors Lyell and Edward Forbes (1815–1854) were cited in the context of autogeny, but although they are likely to have thought in terms of the third theory, they stopped short of explicitly expressing support.[37] Nevertheless, the theory was familiar to British scientists either directly through Continental literature or via Ray Society translations into English, such as Meyen's outspokenly autogenist and autochthonist *Outlines of the Geography of Plants* (1846) and Oken's more bizarrely holistic-vitalistic *Elements of Physiophilosophy* (1847).

Taking off our Darwinian Reading-glasses

In recent decades, historians of biology have been studying both Linnaeus and Darwin as men of their time whose scientific work was integral to contemporaneous culture. Above, in the introduction, Frängsmyr's persuasive contribution to this revisionist approach for Linnaeus has been cited. Here I should like to add that ever since Robert Young, Darwin scholars have admirably succeeded in connecting the *Origin of Species* to the public and private conditions of its Victorian making.[38] Yet we have not always been dealing in the same way with the biologists from the intervening period but continue to apply a Darwinian hermeneutic, interpreting their scientific contributions as foreshadowing the coming of the Messiah of organic evolution—an approach that is still atuned to Bentley Glass' hymnic preface to the classic *Forerunners of Darwin: 1745–1859*:

> By the middle of the eighteenth century, theories of evolution were being heatedly debated. As the evidence in favor of organic evolution grew, the struggle between two seeming alternatives, the one, Divine Plan and Providence in an Original Creation and the other, a godless mechanism of chance and of blind cause and effect, became not so much a debate that divided scientists and philosophers into two camps as a cleavage in the heart of each individual man. Thus, as the individual often resolves an implacable conflict by repressing it into the subconscious, so human thought in the first half of the nineteenth century stubbornly and blindly repressed the implications of the growing evidence in regard to the origin of species, including his own. Darwin was the outburst of those repressed conclusions, the victory of that submerged scientific conviction. His was the magnificent synthesis of evidence, all known before, and of theory, adumbrated in every postulate by his forerunners—a synthesis so compelling in honesty and comprehensiveness that it forced such men as Thomas Henry Huxley to say: How stupid not to have realized that before![39]

We thus have fallen in line with Darwin's self-assessment that everything great and good in nineteenth-century biology came together in his theory of organic evolution and that there existed no scientific alternative, the only opposition to evolution being the non-scientific belief in divine creation.[40] In this chapter, I have shown that a scientific alternative did exist in the form of the theory of the spontaneous origin of life-and-species. My recovery of this "third theory" is not based on new source materials but on a mere rereading of the familiar primary, published literature. Its authors are well known to historians of biology. Few of us, however, in examining their works have previously noticed the theory of autogeny, even less its prevalence and significance. Owsei Temkin in 1959 came closest to recognizing the third theory, yet in the end he yielded to the forerunners of evolution approach.[41] A number of other historians, recently among them Thomas Junker and Uwe Hossfeld, have briefly indicated a clear knowledge of the third theory.[42] All the same, just as Temkin, they treat autogenists as quasi-evolutionists.

Rereading the historical sources from the period Linnaeus to Darwin with a clear notion of the third theory in mind produces fundamental changes in the history of the biology of origins. Instead of seeing proto-Darwinians battling antiquated creationists, we notice communities of scientists that were contributing to a significant innovation in scientific epistemology, characteristic of the first half of the nineteenth century, namely methodological naturalism or rather its application to the issues of the origin of life and species—even to the origin of *Homo sapiens*. The result was not evolution but the theory of autogenous generation which, in contrast to Darwinism, treated as a single problem the origin of life and the origin of species, allowing for only limited species variability. Here I have tried to restore to Blumenbach, his colleagues and disciples their own convictions about species origins, which may well have owed much to Linnaeus and may well have contributed to Darwin yet which were distinct and constituted a separate, significant chapter in the history of the life sciences.[43]

This new reading of "the origin of species from Linnaeus to Darwin" presents us with a number of interesting questions: "Why was the theory more strongly represented in Germany than in France or Great Britain?" "How come Darwin ignored the theory?" "Which were the reasons for the autogenists' mass conversion to Darwinism"? "What if autogeny had prevailed, or at least had survived in a modernized, quasi-evolutionary form?" Answers to these questions will require the same contextualizing of Blumenbach and the other autogenists as has been done for Linnaeus and Darwin.

Notes

1. See for example Bentley Glass, "Heredity and variation in the eighteenth century concept of the species," in *Forerunners of Darwin: 1745–1859*, ed. Bentley Glass, Owsei Temkin and William L. Straus, Jr. (Baltimore: The Johns Hopkins University Press, 1959), pp. 144–151.

2. Tore Frängsmyr, "Linnaeus as a Geologist," in *Linnaeus. The Man and his Work*, ed. Tore Frängsmyr, revised edition (Canton, MA: Science History Publications, 1994), pp. 110–155. Among other biographical studies see also Lisbet Koerner, *Linnaeus: Nature and Nation* (Cambridge, MA: Harvard University Press, 1999).

3. A discourse of species *Urzeugung* had started well before Blumenbach. Among the contributors were Benoît de Maillet (1656–1738), Pierre Louis Moreau de Maupertuis (1698–1759), George Louis Leclerc, Comte de Buffon (1707–1788), Denis Diderot (1713–1784), as well as Immanuel Kant (1724–1804) and Johnn Gottfried Herder (1744–1803). See Thomas Junker and Uwe Hossfeld, *Die Entdeckung der Evolution* (Darmstadt: Wissenschaftliche Buchgesellschaft, 2001), pp. 35–48. See also Peter Hanns Reill, *Vitalizing Nature in the Enlightenment* (Berkeley and Los Angeles: University of California Press, 2005), *passim*. A major source of relevant information continues to be Jacques Roget, *Les sciences de la vie dans la pensée française du XVIIIe siècle* (Paris: Colin, 1963), *passim*. With Blumenbach, however, the discourse shifted from a predominantly philosophical to a biological arena.

4. Charles Darwin, *The Origin of Species by Means of Natural Selection*, in *The Works of Charles Darwin*, vol. 16 (London: Pickering, 1988), p. xiii, p. 441.

5. For an overview see Junker and Hossfeld, *Die Entdeckung der Evolution*, pp. 24–48.

6. Heinrich Czolbe, *Neue Darstellung des Sensualismus* (Leipzig: Costenoble, 1855), pp. 161–185. Rudolf Virchow, "Alter und neuer Vitalismus," *Archiv für pathologische Anatomie und Physiologie und für klinische Medizin*, vol. 9 (1856), pp. 3–55, on pp. 23–25.

7. From among the several monographs on the subject see Otto Taschenberg, *Die Lehre von der Urzeugung sonst und jetzt* (Halle: Niemeyer, 1882); Edmund O. von Lippmann, *Urzeugung und*

Lebenskraft (Berlin: Springer, 1933); John Farley, *The Spontaneous Generation Controversy from Descartes to Oparin* (Baltimore and London: The Johns Hopkins University Press, 1974); Iris Fry, *The Emergence of Life on Earth: Historical and Scientific Overview* (New Brunswick, NJ, & London: Rutgers University Press, 2000); James Strick, *Sparks of Life: Darwinism and the Victorian Debates over Spontaneous Generation* (Cambridge, MA: Harvard University Press, 2000).

8. See for example Heinrich Georg Bronn, *Handbuch einer Geschichte der Natur*, vol. 2 (Stuttgart: Schweizerbart, 1843), p. 30: "Lässt sich jene [the spontaneous origin of simple organisms today] aber, wie es scheint, nicht erweisen, so müssen wir gleichwohl die Urerzeugung, wenn auch als eine jetzt völlig erloschene Zeugungs-Kraft der Erde zu Hülfe rufen, um die erste Entstehung der Arten zu erklären."

9. Johann Christian Gottfried Jörg, *Grundlinien zur Physiologie des Menschen* (Leipzig: Kummer, 1815), pp. 1–32; Ferdinand August Ritgen, *Probefragment einer Physiologie des Menschen* (Kassel: Krieger, 1832), pp. 44–50; Gabriel Gustav Valentin, *Grundriß der Physiologie des Menschen*, 3rd ed. (Braunschweig: Vieweg, 1851), p. 664.

10. Walter Baron, "Blumenbach, Johann Friedrich," *Dictionary of Scientific Biography*, ed. Charles C. Gillispie, vol. 2 (New York: Charles Scribner's Sons, 1981), pp. 203–205, on p. 204.

11. Johann Friedrich Blumenbach, *Handbuch der Naturgeschichte* (Göttingen: Dieterich, 1830), pp. 2–3.

12. Gottfried Reinhold Treviranus, *Biologie, oder Philosophie der lebenden Natur für Naturforscher und Ärzte*, vol. 2 (Göttingen: Röwer, 1803), pp. 377–378.

13. F. S. Leuckart, *Allgemeine Einleitung in die Naturgeschichte* (Stuttgart: Schweizerbart, 1832), pp. 46–55.

14. Karl Friedrich Burdach, *Anthropologie für das gebildete Publicum* (Stuttgart: Baltz, 1837), pp. 726–727, 742–744.

15. Johannes Müller, *Handbuch der Physiologie des Menschen*, vol. 1, 4th ed. (Coblenz: Hölscher, 1844), pp. 8–17, 23–24; vol. 2 (Coblenz: Hölscher, 1840), p. 769. Virchow, note 6 above, pp. 25–26.

16. Blumenbach, *Beyträge zur Naturgeschichte*, vol. 1, 2nd ed. (Göttingen: Dieterich, 1806), pp. 19–20.

17. K. E. A. von Hoff, *Erinnerung an Blumenbach's Verdienste um die Geologie* (Gotha: Engelhard-Rehyer, 1826), pp. 16–17.

18. Hermann Burmeister, *Geschichte der Schöpfung: Eine Darstellung des Entwickelungsganges der Erde und ihrer Bewohner: Für die Gebildeten aller Stände*, 3rd ed. (Leipzig: Wigand, 1848), p. 314.

19. Leopold von Buch, *Physicalische Beschreibung der Canarischen Inseln* (Berlin: Koenigliche Akademie der Wissenschaften, 1825), 128–136.

20. Heinrich Georg Bronn, *Untersuchungen über die Entwickelungs-Gesetze der organischen Welt während der Bildungs-Zeit unserer Erd-Oberfläche* (Stuttgart: Schweizerbart, 1858).

21. Lorenz Oken, *Die Zeugung* (Bamberg: Goebhardt, 1805). See also [Oken, Lorenz], "Entstehung des ersten Menschen," *Isis oder Encyclopädische Zeitung von Oken*, 2 (1819), pp. 1118–1123.

22. Carl Vogt, *Lehrbuch der Geologie und Petrefactenkunde: Zum Gebrauche bei Vorlesungen und zum Selbstunterrichte*, 2 vols., 2nd ed. (Braunschweig: Vieweg, 1854), pp. 338, 386–387. See also footnote commentary to his translation of *Vestiges*, [Chambers, Robert], *Natürliche Geschichte der Schöpfung des Weltalls, der Erde und der auf ihr befindlichen Organismen, begründet auf die durch die Wissenschaft errungenen Thatsachen* (Braunschweig: Vieweg [translated into German from the 6th English ed. by Carl Vogt; 2nd ed. 1858]).

23. Charles Lyell, *Principles of Geology, being an Attempt to Explain the Former Changes of the Earth's Surface, by Reference to Causes now in Operation*, 3 vols. (London: Murray, 1830–33), vol. 2, pp. 182–183.

24. Carl Ritter, "Ueber geographische Stellung und horizontale Ausbreitung der Erdtheile," *Abhandlungen der historisch-philologischen Klasse der Koeniglichen Akademie der Wissenschaften zu Berlin*, 1826 (Berlin, 1829), pp. 103–27.

25. Louis Agassiz, *Ueber die Aufeinanderfolge und Entwickelung der organisirten Wesen auf der Ober-fläche der Erde in den verschiedenen Zeitaltern* (Halle: Gräger, 1843); "Sketch of the natural provinces of the animal world and their relation to the different types of man," in J. C. Nott and George R. Gliddon, *Types of Mankind* (Philadelphia: Lippincott, Grambo & Co., 1854), pp. lviii–lxxvi; *Essay on Classification* (1857), ed. Edward Lurie (Cambridge, MA: Belknap Press, 1962), pp. 34–40, 64, 104.

26. Alexander von Humboldt, *De distributione geographica plantarum secundum coeli temperiem et altitudinem montium prolegomena* (Lutetiae Parisiorum: in Libraria Graeco-Latino-Germanica, 1817).

27. Janet Browne, *The Secular Ark. Studies in the History of Biogeography* (New Haven & London: Yale University Press, 1983), pp. 58–85.

28. Joakim Frederik Schouw, *Grundzüge einer allgemeinen Pflanzengeographie* (Berlin: Reimer, 1823).

29. Alphonse de Candolle, *Géographie botanique raisonnée: ou exposition des faits principaux et des lois concernant la distribution géographique des plantes de l'époque actuelle*, 2 vols. (Paris: Masson; Geneva: Kessmann, 1855); August Heinrich Rudolf Grisebach, *Die Vegetation der Erde nach ihrer klimatischen Anordnung: Ein Abriss der vergleichenden Geographie der Pflanzen*, 2 vols. (Leipzig: Engelmann, 1872); Franz Julius Ferdinand Meyen, *Grundriss der Pflanzengeographie* (Berlin: Haude & Spener, 1836); Ludwig K. Schmarda, *Die geographische Verbreitung der Thiere* (Vienna: Gerold und Sohn, 1853).

30. Meyen, *Grundriss der Pflanzengeographie*, p. 312.

31. The point was central enough for the French translation of Blumenbach's *De generis humani varietate nativa liber* to be entitled *Sur l'unité du genre humain et de ses varietés* (Paris: Allut, 1806).

32. Burdach, *Anthropologie*, p. 742.

33. Karl Asmund Rudolphi, *Grundriss der Physiologie*, vol. 1 (Berlin: Ferdinand Dümmler, 1821), pp. 50–57.

34. Carl Gustav Carus, *System der Physiologie für Naturforscher und Ärzte* (Dresden & Leipzig: Fleischer, 1838), part I, pp. 112–113.

35. Carus, *Denkschrift zum hundertjährigen Geburtsfeste Goethe's: Ueber ungleiche Befähigung der verschiedenen Menschheitstämme für höhere geistige Entwickelung* (Leipzig: Brockhaus, 1849).

36. Jean-Claude de Lamétherie, *Théorie de la terre*, vol. 3 (Paris: Maradan, 1795), pp. 160–166.

37. For Lyell see note 23 above; also Peter Bowler, *Evolution: The History of an Idea* (Berkeley, Los Angeles, London: University of California Press, 1984), p. 133. Edward Forbes, "On the connexion between the distribution of the existing fauna and flora of the British Isles, and the geological changes which have affected their area, especially during the epoch of the Northern Drift," *Memoirs of the Geological Society of Great Britain*, vol. 1 (1846), pp. 336–432. See also Forbes, "Abstract of the theory of specific centres," in Baden Powell, *Essays on the Spirit of the Inductive Philosophy* (London: Longman, Brown, Green, Longmans, & Roberts, 1855), pp. 498–500.

38. Robert M. Young, *Darwin's Metaphor: Nature's Place in Victorian Culture* (Cambridge: Cambridge University Press, 1985). Adrain Desmond and James Moore, *Darwin* (London: Joseph, 1991); Janet Browne, *Charles Darwin*, 2 vols. (New York: Knopf, 1995, 2002).

39. Glass, *Forerunners of Darwin*, pp. v–vi.

40. Note 4 above.

41. Owsei Temkin, "The idea of descent in post-Romantic German biology: 1848–1858" in Bentley Glass, Owsei Temkin and William L. Straus, Jr., eds., *Forerunners of Darwin: 1745–1859* (Baltimore: The Johns Hopkins University Press, 1959), pp. 323–355.

42. Junker and Hossfeld, *Die Entdeckung der Evolution*, p. 18.

43. See also my "Neither creation nor evolution: The third way in mid-nineteenth century thinking about the origin of species," *Annals of the History and Theory of Biology*, vol. 10 (2005), pp. 143–172; "The origin of life before and after *The Origin of Species*," in *The Uses and Abuses of Biology*, eds. Denis R. Alexander and Ronald L. Numbers (in preparation).

The Natural Economy of Households

Charles Darwin's Account Books

JANET BROWNE

or many years now, Tore Frängsmyr has perceptively explored the structure of scientific knowledge as an integral part of European culture, working across time, from the Renaissance to the present day, and across institutional structures, from the early organization of science to the award of its highest prizes. Such a cultural approach is undoubtedly the best way to understand the full range and nature of Charles Darwin's success. This article takes as its theme the links between Darwin's theory of natural selection, his personal life, and the prevailing economic structure of his period. It follows a path that has been codified elsewhere by Theodore M. Porter, and yet surprisingly neglected by Darwin scholars, by focusing on a series of financial ledgers and account books that Darwin kept for the greater part of his life.[1] While financial archives for scientists hardly exist in sufficient volume to make this line of investigation a regular opportunity for historians, in a few lucky cases such as Darwin's, interesting correlations can emerge. It is argued here that the practice of keeping accounts has many points of similarity to those other forms of record-keeping that dominated Darwin's scientific activity. Indeed, his search for operative regularities in his financial life may well have served as a significant organizing motif for his ideas about natural selection. Taking my cue from Frängsmyr's longstanding interest in locating the history of scientific concepts in a humanistic framework, I suggest that the natural economy of households helped Darwin envisage the natural economy of species.

The significance of financial matters in Darwin's life can be vividly characterized by a comment he made in 1873 in response to a questionnaire sent to him by Francis Galton.[2] In the questionnaire, Darwin was asked to list the mental qualities that he thought contributed to his success. The same questionnaire had been sent by Galton to selected Fellows of the Royal Society and in itself was an early use of printed survey forms for gathering information.[3] Galton received 104 responses that supplied the basis of his hereditarian study *English Men of Science, Their Nature and Nurture*, published the following year. The questionnaire was seven pages long, covering family background, health, mental traits, education, politics, and religious beliefs. An additional column requested the same information for the respondent's father. At the end, Galton asked each man of science if he thought he possessed any special talent. We might nowadays expect Darwin to have said that he was good at observing, or very patient in collecting evidence, or that he experienced a strong love for natural history from his childhood.[4] Instead, he answered that he felt he had no special talents:

> None, except for business as evinced by keeping accounts, replies to correspondence, and investing money very well. Very methodical in all my habits.[5]

What should we infer from this curious self-evaluation? What sort of man was this who rated his skill in "investing money" higher than any other quality he might possess?

Galton's questionnaire brings a bourgeois, economic Darwin out from the historical shadows and invites us to explain his existence. Indeed, this was the driving imagery behind the important work conducted by Robert Young in the 1960s, in which Young conclusively demonstrated that Darwin's system of biological change was founded on economic metaphors of competition, struggle and survival.[6] My account will, with gratitude, build on Young's synthesis. Darwin's account books provide an unsuspected opportunity to look again for those metaphors being worked out in actual pounds, shillings, and pence.

Charles Darwin's financial affairs have of course been addressed in passing by a multitude of scholars, especially now that historians of science routinely look to contextual factors in assessing any scientist's background and position in society.[7] There we learn that Darwin was born in 1809 into a prosperous family in Shrewsbury, England, at a time of great industrial, political, social, economic, and agricultural change. Darwin's father had married Susanna Wedgwood, of the Staffordshire pottery family, in 1796, and become esteemed as a local physician. The Darwin-Wedgwood circle lived in comfort, their combined prosperity deriving primarily from commercial acumen rather than the inheritance of patrimony.[8] This emerging category of an affluent, commercially-minded middle class has been well described by Harold Perkin, who shows that wealth originating in eighteenth-century manufacture was, over the course of two or three generations, frequently converted into land, estates, education, political activity, and property.[9] New forms of investment opportunity emerged during the English industrial revolution too, such as part-shares in transport schemes, mainly the roads, canals, bridges, and harbors that provided a communication network for progress, followed in the nineteenth

QUESTION.	YOURSELF.			YOUR FATHER.	
Specify any interests that have been very actively pursued	Science, and field sports to a passionate degree during youth.				
Religion ?	Nominally to Church of England.			Nominally to Church of England.	
Politics?	Liberal or Radical.			Liberal.	
Health ?	Good when young—bad for last 33 years.			Good throughout life, except from gout.	
Height, &c. ?	Height ?	Figure, &c. ?	Measurement round inside of hat.	Height ?	Figure, &c. ?
	6 ft.	Spare, whilst young rather stout.	22¼ in.	6 ft. 2 in.	Very broad and corpulent.
	Colour of Hair ?		Complexion ?	Colour of Hair ?	Complexion ?
	Brown.		Rather sallow.*	Brown.	Ruddy.
Temperament ?	Somewhat nervous.			Sanguine.	
Energy of body, &c. ?	Energy shown by much activity, and whilst I had health, power of resisting fatigue. I and one other man were alone able to fetch water for a large party of officers and sailors utterly prostrated. Some of my expeditions in S. America were adventurous. An early riser in the morning.			Great power of endurance although feeling much fatigue, as after consultations, after long journeys; very active—not restless—very early riser, no travels. My father said his father suffered much from sense of fatigue, that he worked very hard.	
Energy of mind, &c. ?	Shown by rigorous and long-continued work on same subject, as 20 years on the ' Origin of Species,' and 9 years on ' Cirripedia.'			Habitually very active mind—shown in conversation with a succession of people during the whole day.	
Memory ?	Memory very bad for dates, and for learning by rote ; but good in retaining a general or vague recollection of many facts.			Wonderful memory for dates. In old age he told a person, reading aloud to him a book only read in youth, the passages which were coming—knew the birthdays and death, &c., of all friends and acquaintances.	
Studiousness ?	Very studious, but not large acquirements.			Not very studious or mentally receptive, except for facts in conversation—great collector of anecdotes.	
Independence of Judgment ?	I think fairly independent ; but I can give no instances. I gave up common religious belief almost independently from my own reflections.			Free thinker in religious matters. Liberal, with rather a tendency to Toryism.	
Originality, or Eccentricity ?	—— thinks this applies to me ; I do not think so—*i.e.*, as far as eccentricity. I suppose that I have shown originality in science, as I have made discoveries with regard to common objects.			Original character, had great personal influence, and power of producing fear of himself in others. He kept his accounts with great care in a peculiar way, in a number of separate little books, without any general ledger.	
Special talents ?	None, except for business as evinced by keeping accounts, replies to correspondence, and investing money very well. Very methodical in all my habits.			Practical business—made a large fortune and incurred no losses.	
Strongly marked mental peculiarities, bearing on scientific success, and not specified above ?	Steadiness—great curiosity about facts and their meaning. Some love of the new and marvellous.			Strong social affection and great sympathy in the pleasures of others. Sceptical as to new things. Curious as to facts. Great foresight. Not much public spirit—great generosity in giving money and assistance.	

FIGURE 1.   Galton's questionnaire, distributed in 1873. Taken from Francis Darwin, ed., *The Life and letters of Charles Darwin* (London, 1887), vol. 3, pp. 178–179. This is the only known version of Darwin's response.

century by railway companies, government stock issues, and entrepreneurial banking ventures. Darwin's father, Dr. Robert Waring Darwin, understood the local significance of these social and financial shifts. From his private capital, Dr. Darwin offered loans in the form of mortgages (secured by property) to a number of nearby gentry, shopkeepers, and landowners who were asset-rich but cash-poor.[10]

Few scholars, however, have made a special investigation into Charles Darwin's personal finances, with the notable exception of Hedley Atkins, who in 1974 described Darwin's married life at Down House in comprehensive detail.[11] Yet Atkins did not probe particularly far, despite the existence of marvelous archival resources. Since then, the Darwin Correspondence Project at Cambridge University Library has made extensive use of Darwin's account books to confirm dates for letters and extend on numerous points that crop up in the correspondence. These documents have added immeasurably to the richness of the editorial information in the published correspondence.[12] Darwin's account books, and a nearly complete set of Emma Darwin's account books, are now preserved and displayed at Down House in Kent, Darwin's home from 1842 to 1882, the year of his death. They are owned by English Heritage, the government body that runs the house as a museum for the nation.[13] The documents are invaluable, not only for Darwin studies but as a major untapped resource for the history of Victorian family life. Plans to digitize the accounts and make them available online are being discussed.

For bookkeeping is not a neutral activity. Well before Foucault's day, historians have attended to accountancy and recordkeeping as a significant factor in the process of manufacturing knowledge and power. All nation states have used the authority of numbers to understand, confine, and control, and keeping accounts has long been recognized as a potent force in the emergence of capitalism, the bureaucratization of human affairs during the eighteenth century, and the rise of science.[14] The establishment of a numerical paper record is a vital stage in systematizing otherwise unintelligible entities; or, as important new work in the history of science would characterize it, the practices of measurement and number are increasingly understood as the hallmark of disinterested knowledge, the terrain on which science is built, closely associated with the consolidation of objectivity and authority.[15]

Further than this, there are deep-rooted theological echoes in bookkeeping practices that tell of the divine "book of life," of recording credits and debits in God's great ledger, of being called to account on the final day and experiencing divine judgment. Bookkeeping, in fact, presents some of the most powerful continuing metaphors for the moral life in Western culture. At root Hebraic in origin, possibly even Babylonian, and then successively naturalized by the Catholic and Protestant traditions, the language of accountancy has always expressed combined moral and political meaning.[16] The Psalmist speaks of God's book in which only the names of the righteous are written "and from which the unrighteous are blotted out."[17] People are still exhorted to keep their books, like their consciences, in a constant state of preparedness and the word "account" variously describes the written pages of a ledger, a debit-credit relationship with God, a person's economic responsibility, and those things that are to be remembered or recorded about a significant event. These epistemic continuities run through the study of "natural economy," as succinctly discussed by Margaret Schabas in relation to the work of Adam Smith, where the "book of nature" displayed moral meaning to *savants*.[18] To Linnaeus, too, as Tore Frängsmyr and Lisbet Koerner have indicated, the eighteenth-century concept of a polity of nature was critical both for his understanding of the natural world and his pious belief in religious retribution.[19] Emma Spary and Hal Cook similarly emphasize, in different spheres, how naturalists made explicit comparisons between the productivity of the earth and commercial productivity.[20] By the time of the Victorians, as Ted Porter comprehensively describes it, the vocabulary of bookkeeping was elevated into a distinctive voice of the times, a rhetoric of orderliness and probity that came to a peak in industrialized England, a voice that was simultaneously moral, methodical, and numerical.[21]

Darwin began keeping accounts early in 1839, ten days after his marriage to his cousin Emma Wedgwood, and continued to the day before his death in April 1882. As far as is known, he did not keep similar records while on board the *Beagle*, although it can safely be presumed that he kept track in some alternative manner of the money provided by his father.[22] The thing to note is that the responsibility of marriage inspired in him the need to make a particular kind of record. These account books deal with three aspects of financial activity, namely household expenses, investments, and the receipt and issue of cheques. His wife, Emma Darwin, also kept a set of domestic accounts, as was customary for women of her social status. The couple's books differed in content and function. It is worth remarking that Emma Darwin made her own entries, and at times kept her husband's books when he was unwell, and after his death. Though no great mathematician, she was competent enough. Indeed, this competence will form part of my argument later.

Darwin used the traditional accountancy system of double entry bookkeeping that he learned from his father.[23] Daily transactions were written down, in date order, in a cash book so as not to be forgotten, credits on one side, debits on the other.

Items were transferred on a regular basis to ledgers that were divided into categories, or classed, as Darwin called it, according to various categories. These categories encompassed topics such as Science, Gardens, Personal, Household (including man servants' and governesses' salaries), and so forth. Every September, the expenditure under each heading was totalled and compared with that of previous years and with his overall income.[24] Under the double entry system, credits and debits should exactly match (or to be precise, add up to zero) at the end of each accounting period. Year by year, he could track defined categories of goods and services.

Another ledger was dedicated to recording income against the cheques and cash sums he paid out to his wife, shopkeepers, salaried staff, investment companies, and suchlike, the historical equivalent of modern bank statements. When Francis Darwin came to compile his father's *Life and Letters* (1887), he recollected that "In money and business matters he [Darwin] was remarkably careful and

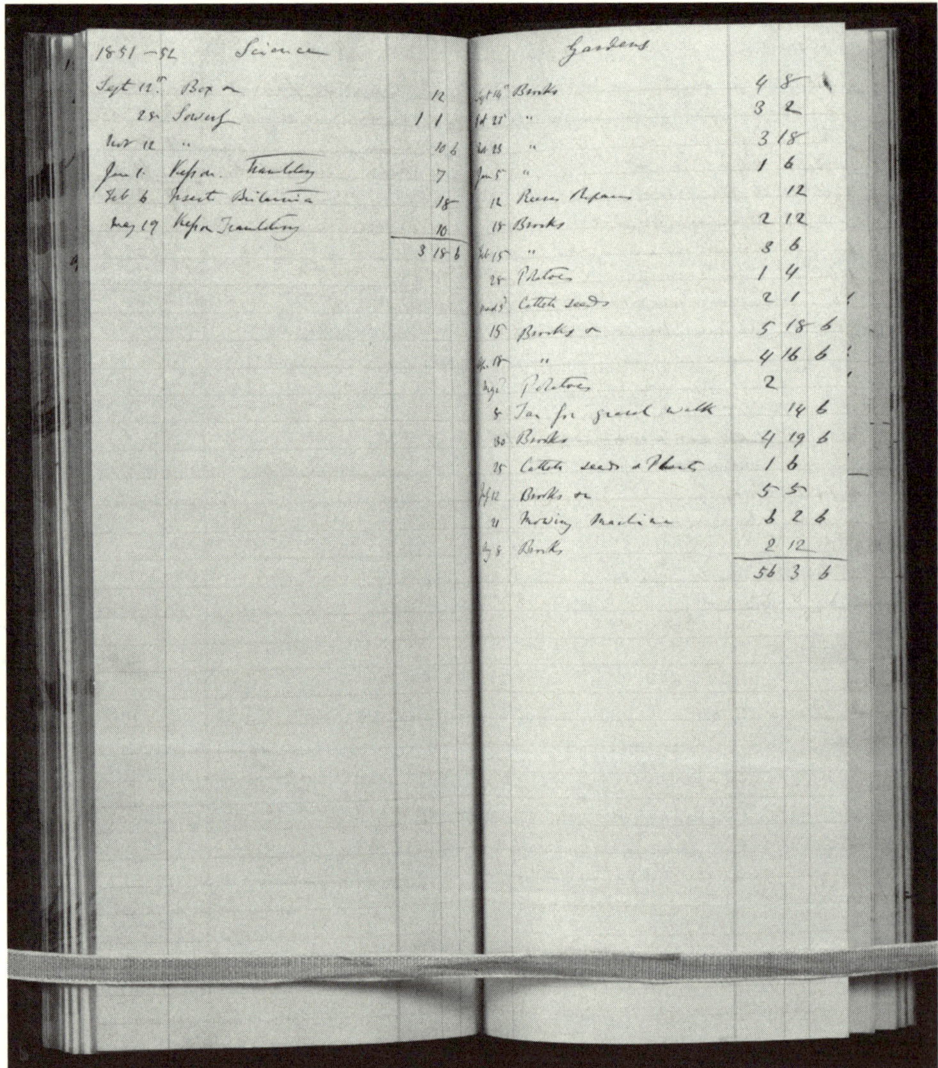

FIGURE 2.   A page from Darwin's Classed Accounts, 1851–1852, giving the accumulated expenditure in categories for the year, Science (to the left) and Gardens (to the right). Courtesy of English Heritage, Down House MS.

exact. He kept accounts with great care, classifying them, and balancing at the end of the year like a merchant. I remember the quick way in which he would reach out for his account-book to enter each cheque paid, as though he were in a hurry to get it entered before he had forgotten it."[25]

Darwin's investment books had a slightly different function in that they recorded the movement of paper wealth. They reveal that the household principally lived off returns from capital. Every year Darwin made a considerable income from invest-

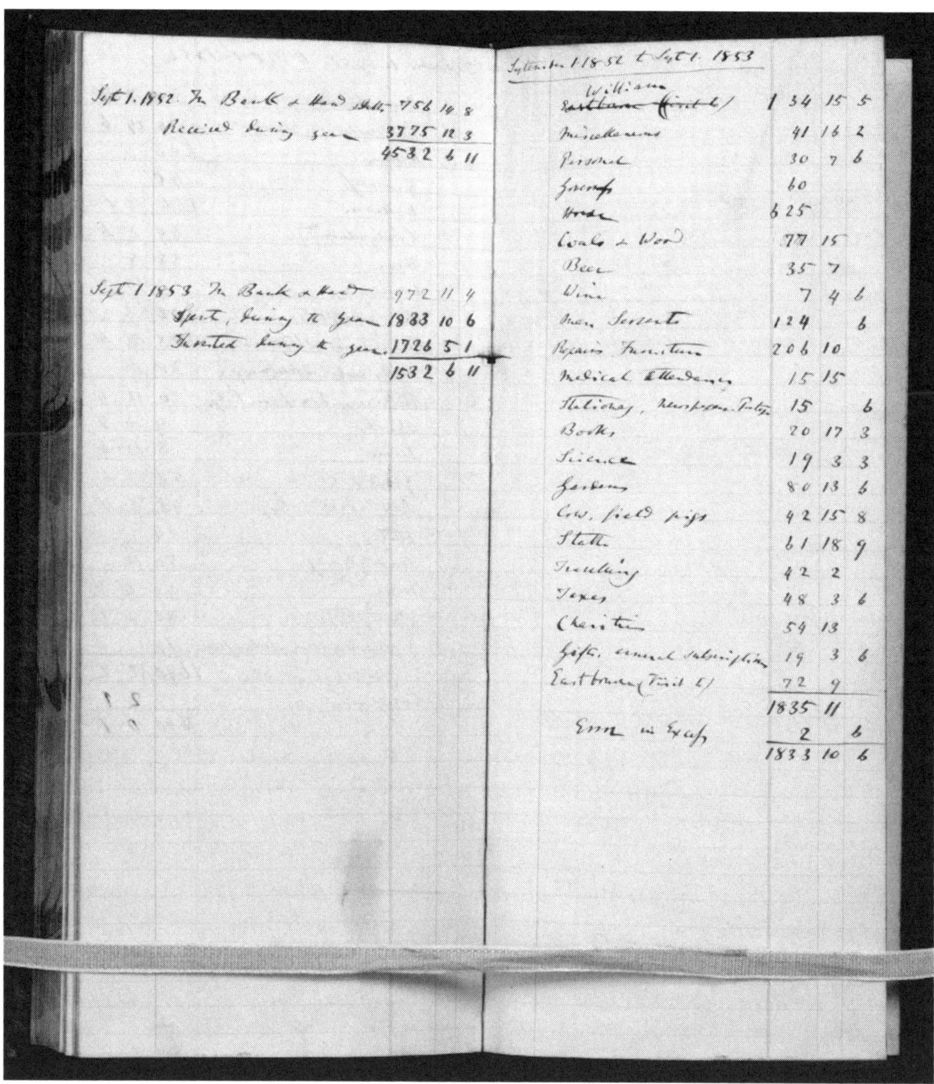

FIGURE 3.　A page from Darwin's Annual Ledger, 1852–1853, indicating his total expenditure for the year. Income on the left, expenditure on the right. English Heritage, Down House MS

ments (such as dividends from shares, maturation of bonds, rent, and interest on loans). Out of this he paid his bills and reinvested the rest. After his father's death in 1848, when he inherited substantial assets, Darwin's annual income began to rise, occasionally reaching £5–6,000, of which he usually reinvested around two to four thousand pounds according to his current situation. He found expenditure on education was high while the children were growing up, and cut back on investment. Later on, when he encountered the pleasant experience of rising surplus, he began to distribute some of this as a cash bonus among his grown children, which they remembered with gratitude.

The investment books show that in the first twenty years of his marriage, Darwin owned shares exclusively in railway companies, canals, and dockyard schemes. The preponderance of transport in his portfolio indicates an appreciation of the role of infrastructure in economic advance. He also owned farmland in East Anglia bought in 1845 as a rental concern (this purchase was probably financed by drawing in advance on money due to him from his mother's Wedgwood inheritance) and ran several loans in the form of mortgages on Shropshire property. These loans were acquired after the decease of his father at an interest rate of just over 3 percent. Railway stock was volatile in the second quarter of the nineteenth century and Darwin was caught once or twice by company crashes. Mostly, however, it was a good time to invest in the rapidly expanding railway network and he benefited greatly from share-issues (preference shares), company buy-outs, and commercial take-overs, especially those carried out by the voracious North Eastern Railway Company. Although Herbert Spencer was to criticize the rapacity of the age, dubbing it the era of "railway morals," Darwin apparently had no such qualms.[26] He disposed of most of his railway stock by the mid-1860s and bought into government debentures and consolidateds (consols), which guaranteed a fixed return, no doubt an indication of his growing family responsibilities.

Because of this diversity of assets, it is not always easy for historians to work out Darwin's net worth at any one point. Nevertheless, Atkins gives an extremely useful chart of Darwin's annual income from 1854 to 1880.[27] The account books record that Darwin began married life in 1839 with a marriage bond of £10,000 from his father, £573 in the bank, and £36 in his pocket. Living in central London, soon with young children and staff, his income of about £600 per annum did not stretch far. He complained it was hard to make ends meet. By 1881, the year before his death, his capital had increased to £282,000, excluding earnings from his books, as calculated by his banking son William. Much of this had accrued from the death of Darwin's older brother Erasmus in 1881, the occasion for the reckoning by William, and the moment for Charles Darwin to draw up a new will. "Did you ever expect to be worth over a $\frac{1}{4}$ of a million?" William Darwin asked with some amusement.[28]

On the other side of the family, Emma Darwin's marriage bond, given by her father, Josiah Wedgwood, was initially worth £5000.[29] Neither of these marriage bonds (trust funds) given by the two fathers were strictly realizable assets. They were intended to provide income during the holder's lifetime, followed by distribution of the capital to the holder's children on death. Darwin and two other trustees (brothers and cousins) managed Emma Darwin's money on her behalf, a customary responsibility in the moneyed classes because married women were not legally able to own property or investments. In this case, income from Emma's bond was reinvested for her by Darwin. On her death the accumulated capital, some £24,000 by 1881, was divided among her children as a direct inheritance.

FIGURE 4. "The Great Land Serpent," an image that satirized the voracity of the owners and shareholders of early Victorian railway companies. Bags of money are being swallowed and converted into smoke labelled "Dividend." From *Puppet-Show*, ii (1848), p. 87. Courtesy British Library.

Alive to every penny spent, Darwin recorded each and every financial transaction, however small, for the whole of his married life. To be sure, this was a necessity not a choice. Until 1844, when the Bank Charter Act was passed in the United Kingdom, banks were private businesses drawing on capital from prominent local backers and myriad small savers, and could easily crash if there were an extended run on resources. Many small rural banks operated on independent lines, including the issue of their own banknotes. The Bank of England, founded in 1694, was as precarious as the rest. Caricatured in *Punch* as the Old Lady of Threadneedle Street in need of medical attention, the bank was tight-laced into a corset. Any restriction of credit in one quarter, symbolized by the drawn-in waistline, would lead to dramatic inflation in another.

Elsewhere in the world of British finance, limited companies did not yet exist, nor were investors or stockholders protected by law in any meaningful way.[30] Several notorious upsets in the early decades of the nineteenth century were caused by immoderate speculation on railway companies, private bank failures, insurance scams, and high-profile court cases for civil bankruptcy; and always the specter loomed of the "South Sea Bubble" scandal, in which the government and many

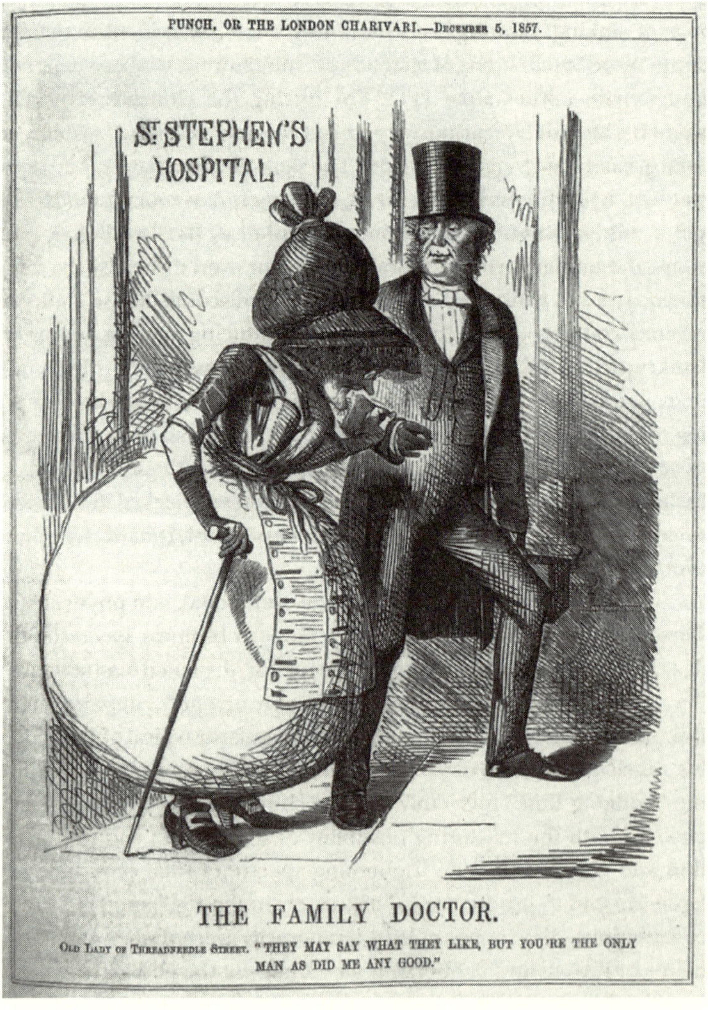

FIGURE 5.   Inflation at the Bank of England, caricatured in *Punch* as an old woman in need of medical attention from the nation's treasurer, Lord Palmerston. "The Family Doctor," *Punch or the London Charivari*, 5 December 1857, p. 231. Courtesy Wellcome Library London.

thousands of investors lost everything in a sudden market crash, some hundred years before. From the 1840s, the British government therefore began to take tougher measures to control speculation and risk, first by introducing the Joint Stock Companies Act, followed by a Limited Liability Act in 1855, and then a Royal Inquiry into Assurance Associations to deal with increasing evidence of insurance company fraud. Sadly, these mostly highlighted wrongdoing rather than introducing remedies. Famously, Charles Dickens put a double-dealing insurance company at the heart of *Martin Chuzzlewit*. In 1857, Darwin's own bank, the Union Bank in Maidstone, Kent, suffered from fraud. So Darwin needed to keep accounts. If middle-class money was not managed in the home then it was not managed at all.

These account books also reveal something of Darwin's character. The language of arithmetic is a language of distance, bound by strict rules. It exacts a discipline that is very nearly uniform all over the globe. It endorses the view that numbers are objective and universal. For Darwin, this constant attention to financial matters, the regular entering of figures, the adding up of credits and debits, was a way to order his personal life, to create rigor and certainty out of a ceaseless flow of information and activity. Entries were not neat: Darwin's handwriting was just as bad here as elsewhere. But the figures were accurate, comprehensive, and regular. The columns boxed up Darwin's life in the same way as his health diary counted out the number of sick periods,[31] or his "journal" described the precise hours that he was able to work each month,[32] or the backgammon book that recorded the running total of the games he and his wife played every night for forty years. One well-known, yet still engaging, instance of Darwin's balance sheet mentality is the sheet of paper he drew up when contemplating marriage, headed "Marry, Not Marry. This is the Question."

On each side of the page, he listed the advantages and disadvantages as they occurred to him. If he remained unmarried, he would be free to go to gentlemen's clubs and societies, and have plenty of time to work. On the other hand, a wife offered the charms of music and "female chit-chat," a companion for his old age. "Better than a dog anyhow," he stated.[33] To a large degree his social interactions with the working villagers in Down were similarly structured by the written record of cash-flow. With the aid of the Down vicar, John Brodie Innes, Darwin set up a small savings club for the local poor, called The Down Friendly Society, which encouraged, as he saw it, the Malthusian virtues of thrift and foresight. Subscribers to another small society, the Coal and Clothing Club, paid a weekly sum over to Darwin (who was registered with the civic authorities as the club treasurer) and were then able to purchase food and coal in the village shop against the account. Emma Darwin took responsibility for the issue of bread "tickets" for redemption in the shop.[34]

The point here is a methodological one. The act of writing and classifying was something that materially helped Darwin arrange his thoughts. Once written, he could process the information, systematize, compare and collate, plan and reflect. He is known today as one of the most prolific note-takers in the history of science,

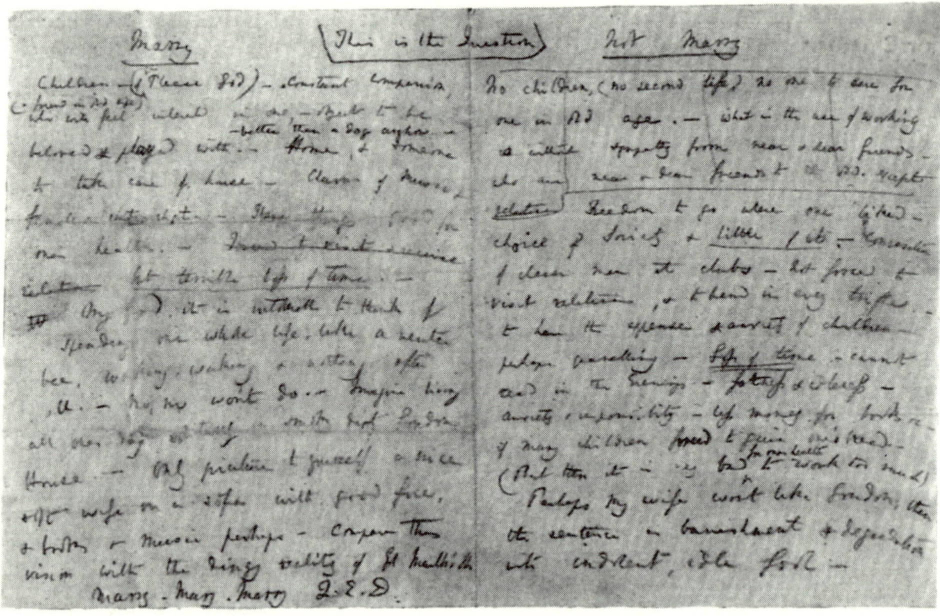

FIGURE 6. "Marry or Not Marry." Charles Darwin's notes to himself, probably July 1838. Here Darwin drew up the benefits and disadvantages of marriage in the form of a balance sheet. Cambridge University Library, DAR 210.10. Courtesy Syndics of Cambridge University Library.

the creator of an archive abounding in detail and variety of sources. Writing things down freed his mind from carrying vast quantities of information—an understandable habit of using the pen as a substitute for memory. And there was an element of compulsion, an urge to convert daily events into records, a realization that the first step in the coming together of a scientific fact is the act of inscription. Naturally enough, in the years on the *Beagle*, writing and recording were Darwin's primary research tools. Darwin quickly learned on his travels that an observation meant nothing unless it was captured in some form of permanent record. He continued the methodology afterwards, even intensifying its scope. As Larry Holmes once noted about Lavoisier, there is considerable kinship between the field or laboratory notebook, the diary, and an account book.[35] We could add the astronomer's record book; no doubt many more such inventories of facts. This is the process of constructing stable information described by Latour and Woolgar.[36]

The point can moreover be extended. Darwin insistently recorded his scientific thoughts for his future use. Most famous are his notes on transmutation in the so-called Species notebooks, composed from 1837 to 1842.[37] He also made observations on his children,[38] listed specimens, and kept careful records of botanical and animal breeding projects. He itemized books he wished to read and marked each item when done. He kept letters that he thought contained useful information, or cut out the useful portions, in portfolios, each catalogued by subject matter. It is as if data had no meaning for Darwin until it was captured on paper and distributed into larger categories—the very essence of the scientific classifying spirit. Numerous items in Darwin's archive reflect this classifying practice, whether the subject matter was cash, backgammon, a new mineralogical specimen, ill health, a useful fact, or time. In sum, he elevated the act of making records into a methodology. Darwin's techniques therefore repay study in the light of fresh interest among historians in the manual practices of science—the doing, the calculating, the writing— and what they signify about the changing structure of science from the seventeenth century onwards. Steven Shapin, for example, describes the cultural authority conferred by an increasing emphasis on the discipline and hard work of science.[39] Characterized by Darwin as a mundane slog (although clearly involving intense self-control and perseverance), and described by him with the catchphrase derived from one of Anthony Trollope's Victorian novels "it's dogged as does it," this ideology of hard work was more elaborately formulated by Samuel Smiles in the mid-nineteenth century and Darwin seemingly identified very closely with the attributes that Smiles made popular.[40] For Darwin, the theme of rigorous perseverance and record-keeping was central to his self-identity as a naturalist. This sense of inhabiting a life dominated by writing might gainfully be addressed.

Emma Darwin's books were also double entry, although the goods and services recorded there are different in purpose from her husband's. Her books documented kitchen expenditure and domestic accounts, including wages for the female servants, excluding the governess. Entries dealt with the business of feeding the family, entertaining, and maintaining the domestic space.[41] Emma's classed accounts for 1867 covered meat, candles, groceries, soap, loaf-sugar, tea, eggs, bread, and so forth, to a total of 20 headings. Meat was the largest expenditure at an annual total of £250. Dress came next at £213. Servants cost her £71. Bread for the same year was £63. She also included incidentals like piano tuning, concert tickets, presents, cab fares, and sheet music. The total for 1870, for example, was £1089.[42] On the incoming side of the book, Emma recorded the cash sums given to her by Darwin to cover these expenses and made a note of the cheques he paid to local suppliers on her behalf. As in her husband's accounts, these figures were meant to match up at the end of the year. At the end of 1867, Emma Darwin discovered she had made arithmetic mistakes to a total of £7-19s-8d, which were recorded as "Errors."

A comparison of the two sets of books, male and female, confirm the roles generally taken by men and women in middle-class Victorian homes. Such a sharing out of the household accounts was typical for many British households and reflected a gendered division of labor where the home was envisaged by both partners as an efficiently run business enterprise in which separate and clearly defined activities were pursued. While a man was the nominal head of a household in legal terms, the domestic sphere was the main place in which a woman displayed her autonomy and authority.[43] This has long been recognized for the early modern period, during which women expected to keep accounts, whether aristocratic ladies controlling their estates, or female servants schooled in elementary arithmetic to ensure they were not cheated by shopkeepers. From the earliest times in Europe, independent women ran retail businesses, participated in the skilled clothing trade, were self-employed artisans, hairdressers, or milliners, or became active as *rentiers*, handling significant sums of money. In a small family concern, women usually kept the books. Single women were also a notable source of cash in the lending markets of northern England in the early eighteenth century. Each and every one of them had good reason to keep close track of what they possessed.

With the onset of Victorian sensibilities and class differentiation, it is usual to claim that much of this female business acumen disappeared. But a number of careful local studies show that a woman's authority inside the Victorian domestic domain was not lost but merely hidden from public view. Middle-class wives may indeed have withdrawn from productive work in the national economy but nonetheless moved toward more exclusive control of the household accounts and childrearing. From this power base of domestic control, as Prochaska and others argue, women progressively expanded their authority through the later years of the century to more public activities, social reform, and major charitable movements.[44] And despite the difficulty that married women were not entitled to own or dispose of property, Victorian women were often the means by which families recruited new business partners, strengthened wealth, or reaped investment profits. At the very least, most middle-class husbands hoped to find in their wives a prudent domestic manager. Hence, in Emma Darwin's hands, account books were a symbol of female knowledge, her mastery of many strands of specialized information over which her husband had no claim, a visual manifestation of her skill in managing a home.

Nevertheless, as already indicated, the source of the money was her husband. Emma's own money (the marriage bond) was not used in any way for domestic expenditure. More than this, Darwin exerted masculine appropriation of his wife's accounts at the end of each year. Her categories of meat, fish, and bread were incorporated into his annual summary but reclassified and redistributed into his own categories, categories of expenditure that were based on different criteria and served a different function. Despite considerable independence in running the household, it can be seen that Emma Darwin's accounts were at root ancillary to those of her husband. The budget was his business not hers.

Here in these ledgers lies a gendered natural economy that can be directly related to Darwin's views on masculine responsibility.[45] First of all, Darwin felt an overwhelming duty to maintain his share of the Darwin-Wedgwood fortune. His marriage to his cousin had brought prosperity by consolidating family funds and ensuring that money did not move out of the Wedgwood line through marriage with a non-family member. Feelings, loyalties, kinship, and obligations were for him profoundly united. To Darwin, then, Emma was an investment as well as a wife. He believed that his highest responsibility as a husband was to provide a secure financial environment for her and the children. Indeed, according to Galton's questionnaire, he apparently came to judge himself as a man almost entirely on his ability to manage money. In this he was evidently not alone. Invited by Galton to evaluate their business abilities, many of the scientific respondents to the 1873 questionnaire felt that they possessed business aptitude, by which they seemingly meant neatness, orderliness, and self-control.[46] Darwin's ledger pages became a literal indication of his respectable place in society, an annual reassurance that he was fulfilling his role as head of the household. With no direct belief in God to guide his conduct, good management of the family finances came to signify deep moral commitment and virtuous behavior.

The Malthusian doctrine of moral responsibility, especially the proper management of money as an indicator of social merit, consequently had a place in Darwin's domestic sphere as surely as it was a leading feature of his theory of natural selection. As is well known, Thomas Robert Malthus stipulated that mankind was part of the natural world, subject to natural law just as much as any animal or plant.[47] The laws were God's laws: they were designed by God for mankind and should be accepted stoically as such. In his *Essay on the Principle of Population* (1798), Malthus described the checks and balances that he thought kept human populations at a level broadly equivalent to the available food. Population numbers would soar far beyond any possible means of sustenance without the God-given and necessary checks of war, famine, disease, and pestilence, or the inevitable consequences of sinful behavior (by which Malthus meant infanticide, homosexuality, abortion, and contraception). The only way voluntarily to ameliorate the deadly struggle for limited food resources was to exercise what Malthus called "moral restraint." In other words, people should not have children that they could not afford to feed.

Malthus laced these stern views into an equally stern view of social behavior. He was convinced that only the respectable, educated classes were capable of moral restraint. The poor, he argued, were unable to control their passions and produced far too many children. These individuals were regarded as indigent wastrels, a class of person increasingly obvious to British intellectuals and politicians who saw, with horrified concern, the urban crowding, rural deprivation, drunkenness, disease, high death rates, and abject poverty accompanying the nation's commercial and industrial advance. If such people were given money to relieve their situation, stated Malthus, they would merely spend it on drink and sin, and beget more children, adding to the misery. Patricia James, among others, describes how in the 1830s Malthus's writings formed the background to parliamentary reforms brought in by the Whigs, led by William Pitt the younger. Whereas earlier legislation stipulated that the support of the destitute was the responsibility of town authorities and parish churches, Pitt's government passed the Poor Law Amendment Act of 1834 that introduced the work-house. In the work-house, a cash economy prevailed. Poor people were required to labor for their bread. By contrast, Malthus valorized the middle-classes as respectable, morally-responsible individuals, thrifty with their money—the very people who should be reproducing their kind. Malthus thus amalgamated the economy of work with the economy of morals, all set in a world operating according to God's beneficent design.

It is indisputable that Darwin associated himself with this same group of responsible people. Several members of Darwin's family circle knew Malthus before his death in 1834, and Darwin personally knew the prominent writer Harriet Martineau, who at one point published exhortatory Malthusian tales.[48] In September 1838, Darwin picked up Malthus's *Essay* and read it, as he said, "for amusement." It provided the source of his principle of natural selection and the key to the rest of his life. Though Darwin soon dropped any notion of natural theology—the belief in God's laws that permeated Malthus's writings—he retained a lasting commitment to the combined natural and moral economy that he found there.

So Darwin was entirely Malthusian in focusing on his ability to support his wife and family. Although he possessed a considerable inheritance and increasing earnings from the sale of his books, his was not a self-satisfied existence, and he identified very strongly with the ethos of hard work and thrift spelled out by Smiles in his *Self-Help* (1859). The old ideas of the intellectual as a supremely talented individual, the genius, the hero of Thomas Carlyle's writings, were at that time being replaced by images of disciplined application and modest tastes, an image with which Darwin empathized.[49] He did not indulge in an arriviste taste for opulence. At the end of each financial year, he would add up his accounts, and tell the children that "at least they would have bread and cheese."[50]

Something of these close interrelations between natural economy and moral economy, between money and social values, can be discerned in Darwin's later work on the *Descent of Man* (1871). In this book, Darwin finally tackled the question of human ancestry and cultural development, issues that he deliberately avoided in the *Origin of Species* (1859) but which played a central role in the controversies following publication. In the *Descent of Man*, Darwin reasserted his belief in Malthusian social philosophy by boldly stating that "all ought to refrain from marriage who cannot avoid abject poverty for their children." If the poor members of society were given too much charitable relief, Darwin continued, there would be a lessened struggle for survival and humanity would sink into indolence. It was vital, in Darwin's view, that the "more highly-gifted men," the more successful men, should have children and pass their attributes on to the next generation.[51]

Moreover Darwin declared in the *Descent of Man* that wealth was a factor in the emergence of a civilized society.

For without the accumulation of capital the arts could not progress; and it is chiefly through their power that the civilized races have extended, and are now everywhere extending, their range, so as to take the place of the lower races. Nor does the moderate accumulation of wealth interfere with the process of selection. When a poor man becomes rich, his children enter trades or professions in which there is struggle enough, so that the able in body and mind succeed best. The presence of a body of well-instructed men, who have not to labor for their daily bread, is important to a degree which cannot be over-estimated; as all high intellectual work is carried on by them, and on such work material progress of all kinds mainly depends, not to mention other and higher advantages. No doubt wealth when very great tends to convert men into useless drones, but their number is never large; and some degree of elimination here occurs, as we daily see rich men, who happen to be fools or profligate, squandering away all their wealth.[52]

In these sentences Darwin indicated that he believed, like many of his era, that the accumulation of wealth, freedom from physical labor, along with appropriate values of thrift and responsibility, would allow intellectuals the necessary opportunities to foster civilization and national progress. Typical enough for its time, Darwin obviously held the prejudices and assumptions of his class, ethnic identity, sex, and nationality, as Desmond and Moore so vigorously describe in their *Darwin*.[53]

Since Desmond and Moore, perhaps even since the heyday of Young's influential writings, it is scarcely contentious to see Darwin as a member of the elite manufacturing classes, familiar with commercial concerns, well-versed in the entrepreneurial ethos of an expanding marketplace for British goods, subscribing to liberal reformist social principles, and adopting the values of the new industrialized society, including what Perkin called the middle-classes' transition into paper wealth. So it is no surprise to see him endorsing the accumulation of financial assets and endowing the process with a civilizing, moralistic status. These sentiments held real meaning for him, in that he felt the acquisition and preservation of capital was a worthy endeavor in itself. Furthermore, to live off interest—as he did—is presented in the *Descent of Man* as a valuable contribution to the progress of nations. Self-serving though this many well have been, to a large degree Darwin probably refers here to the intellectual freedom that financial independence conferred. Much debate in his own day, ranging from Charles Babbage's *Reflections on the Decline of Science* (1830), to Thomas Henry Huxley's clamor for paid employment for experts, focused exactly on this tension between salary and obligations. In the words of the 21st century, intellectual property was at stake. Independent financial resources—a private income—allowed Victorian men or women to pursue whatever line of thought appeared important without fear of punishment or loss of ownership. Conversely, to accept a salary from (say) the Geological Survey of Great Britain, or the Natural History Museum, obliged a talented young naturalist to give all his results to the government and to publish under the institution's name.[54] Darwin sincerely believed that private income was a vital element in fostering an independent life of the mind. Over and above these special concerns, a civilized society, as he described it, was clearly one that emphasized the rule of natural economic law, the balance between those who labor to make wealth and those who spend it, a more or less harmonious design in the minds of Victorians, carrying moral significance and cultural advantage for nations. The natural theological framework of mid-Victorian Britain was here secularized into patterns of financial creditworthiness.

In the larger sense, these views may have had some real impact on the theory of natural selection. It is likely that the act of recording and classifying money was a transferable skill that generated fresh insights for Darwin when pursued in other domains. For I want to propose that Darwin's daily records of the accumulation of capital, his search for profit, and careful attention to reinvestment, provided a model for animal and plant evolution. An animal or plant species could—in this light—be regarded as possessing debits and credits in the natural economy. Each favorable variation added to an individual's capital, each profitable adaptation led to a consolidation of market share.

The parallel is not completely fanciful. The word "specie," the old financial term for money or coinage, derives from the Greek word "species" as used by Aristotle in the classification of animals. It is also used in the legal profession to indicate a fact. It is not a large step to imagine that the word conveys not only the idea of "kind" but also "commodity," as Gordan McOuat proposes in relation to the process of cataloguing species at the British Museum. Hal Cook alludes to the same merging of categories in his study of the golden age of Dutch commercial enterprise.[55] Darwin's financial activities can be seen to generate a practical framework, a way of thinking about data, an activity combining hand and head that allowed him to link a plethora of minute local variations with higher categories and an estimation of future profit or loss. As Ted Porter reminds us, account books keep track of past performance, extrapolate trends, and forecast profitability. The balance sheets of accountancy are intended to make objective, generalized statements out of many individualized transactions, and ledgers such as Darwin's stabilized the variability of a long series of financial proceedings. The numbers provided distance, imposed a rhetoric of objectivity. Surely the structured framework of accounts made it possible for Darwin, when ruminating about natural processes, to calculate the probability of a small number of favorable variants maintaining their place in a larger population? Once at a viable level these variants, so to speak, might reinvest their biological capital until a stage was reached where profit and dividends could be reaped. Such symbolic biological capital was based on the accumulation of favorable adaptive traits and, like monetary capital, could be converted into all kinds of resources necessary for continued activity and profitability.[56] Modern biologists speak broadly in these terms when discussing evolutionary strategy.[57]

Darwin's attention to financial issues therefore made available a technique that allowed him to distance himself from the individual variant and to think of animal and plant populations in an objective, statistical way. His view of money emphasized a natural economy in which acquisition, probability, and accumulation were closely linked to success. Here in a nutshell was the central metaphor of natural selection.

To conclude: in the history and philosophy of science, we are at ease with the idea that Darwin's science rested in the industrialized economy of his day. As a member of the elite manufacturing classes, embedded in commercial concerns, well-versed in the entrepreneurial ethos of an expanding marketplace for goods and cultural achievements, and committed to liberal reformist social principles, Darwin was just the kind of man who might be expected to come up with a system of biological change based on economic metaphors of competition, struggle, and survival.[58] And Darwin's dependence on Malthusian modes of thought is well established, not least through Darwin's own statements in his autobiography and *On the Origin of Species*.[59] There is no pressing need for historians to ransack his remaining documents to find further evidence of dependence on such metaphors. For these reasons, it is useful to remember that Darwin's account books, begun in 1839, had no direct causal role in his formulation of the principle of natural selection in September 1838. Moreover, it is entirely feasible that the archives of other notable scientific thinkers, for instance the business accounts of Sadi Carnot or William Thomson, or the tax records of Antoine Laurent Lavoisier, might be equally full of notions of competition, profit, and loss; and yet have no discernable place in those individuals' scientific achievements. If Herbert Spencer's account books existed, or Alfred Russel Wallace's, would they supply a technique for thinking about biology? With Darwin, there is always the danger of seeking evolutionary meaning in documents that might have no such meaning at all.

However, a study of Darwin's account books raises interesting issues relating to the control of information and the classification of data. Instead of asking whether there are causal links between Darwin's accounts and his theories, it is more relevant here to think of these ordering principles and habits of work as a common resource that were applied across different intellectual domains. Darwin used numbers to reduce the many variables of his financial existence to consistent categories that could be compared, standardized, and turned into strategy, a process that Mary Poovey describes as an important element in the creation of a modern fact.[60] His ledger books gathered individual expenses into generalizations that helped him visualize his overall financial position in terms of moral credit. This credit was for him devoid of the old theological sense of divine judgment, yet fully echoed the ethics of duty and responsibility held by middle-class Victorian gentlemen steeped in Malthusian economic policy and the Smilesean work ethic. He made this notion explicit in the *Descent of Man*, where he endorsed the creation of wealth by men, men he imagined to be like himself, members of the newly industrialized, culturally advanced, commercial, and geographically expanding British nation, the scientific figures that Galton suggested were the nation's rightful leaders.

I have pushed this further to suggest that quantification and classification were for Darwin a technology of linkage. The activity of adding up numbers transformed small things into big things, facilitating the construction of categories of a different order. This conversion of small into large through the accumulation of many tiny instances lay at the heart of Darwin's approach to nature, a methodological commitment initially derived from Charles Lyell that can be seen throughout his researches, ranging from his earliest work on coral reefs, through studies of geological elevation and subsidence, animal and plant variation, domestic breeding experiments, plant movements, and earthworm activity.[61] Many small instances, for Darwin, added up to large effects. This way of processing data can be expanded beyond the circulation of money to propose a symbolic capital circulating in Darwin's evolutionary biology, where the accumulation of individual adaptations leads to the increased productivity and profitability of species. Natural selection theory, as well as drawing on Victorian economic policy and Malthusian ideas, can perhaps be regarded as part of a larger history of bureaucracy, accountancy, and quantification, in which the notion of natural economy went hand in hand with the natural theological underpinnings, gender relations, and moral commitments of nineteenth-century life.

Notes

1. Darwin's Account Books are owned by English Heritage, and displayed at Down House, Kent. The curator, Ms. Tori Reeve, has generously facilitated my research. I am also grateful to Paul White, of the Darwin Correspondence Project, Cambridge University Library, England; Margaret Schabas, of the Philosophy Department, University of British Columbia, Vancouver; and Steven Shapin, Department of the History of Science, Harvard University, USA, for their helpful comments on this article which have deepened my argument considerably. Participants at the 2007 Lancaster conference on Science and Religion offered some robust opinions. I warmly acknowledge the assistance of the Darwin Correspondence Project. I am particularly grateful to Ted Porter's work, see especially Theodore M. Porter, "Quantification and the Accounting Ideal in Science," *Social Studies of Science* 22 (1992), 633–651.

2. The original questionnaire sent to Darwin and filled in by him does not seem to have survived. A collection of manuscript answers to the same questionnaire is the focus of Victor L. Hilts's study, *A Guide to Francis Galton's English Men of Science* (Philadelphia: American Philosophical Society, 1975). However a printed version of Darwin's response was given by his son Francis Darwin, in F. Darwin, ed., *The Life and Letters of Charles Darwin* (London, 1887), vol. 3, pp. 177–180. See also Karl Pearson, ed., *Life, Letters and Labours of Francis Galton* (Cambridge: Cambridge University Press, 1924), vol. 2, pp. 177–179. Hilts discusses the case in Hilts, pp.10–11, 44–45.

3. Hilts 1975, pp. 9, 14, and Francis Galton, *English Men of Science: Their Nature and Nurture* (London: Macmillan, 1874), pp. 261–266. The survey was a key part of Galton's study of hereditary genius. Galton was familiar with statistical surveys from his acquaintance with the Statistical Society, with William Farr, commissioner for the census, and with ethnographical surveys, and went on to transform statistical procedures. For recent studies of Galton, see Nicholas Wright Gillham, *A life of Sir Francis Galton: From African exploration to the birth of Eugenics* (Oxford: Oxford University Press, 2001), and Milo Keynes, ed., *Sir Francis Galton, FRS: The legacy of his ideas. Proceedings of the twenty-eighth annual symposium of the Galton Institute, London* (Basingstoke: Macmillan in association with the Galton Institute, 1993). The history of the questionnaire as a means of gathering information is mostly covered through individual case studies and can be linked to a long tradition of "instructions" to travelers. As a vehicle for attaining trustworthy information at a distance and a means for capitalizing on the presumed objectivity of numbers, the rise of Victorian questionnaires would well repay study. Darwin issued several printed questionnaires, see Richard B. Freeman and Peter J. Gautry, "Charles Darwin's Queries about Expression," *Bulletin of the British Museum (Natural History) Historical Series* 4 (1975), pp. 205–219.

4. In a covering letter Darwin dwelled on his early love of natural history and stated that "my education really began on board the *Beagle*," Pearson, 1924, vol. 2, p. 179 (28 May 1873).

5. Quoted from F. Darwin, 1887, vol. 3, p. 179.

6. Robert M. Young, *Darwin's Metaphor: Nature's place in Victorian culture* (Cambridge: Cambridge University Press, 1985). See also his "Darwinism is Social," in *The Darwinian Heritage*, ed. David Kohn (Princeton: Princeton University Press, 1985), pp. 609–638.

7. It was not always so, see Steven Shapin and Barry Barnes, "Darwin and social Darwinism: Purity and history," in B. Barnes and S. Shapin, eds., *Natural order: Historical studies of scientific culture* (Beverly Hills, Calif.: Sage, 1979), pp.125–142. A good summary of recent work in this area is given by Evelleen Richards, "Democratizing Darwin," *Annals of Science* 52 (1995), 509–517. For biographical contextualization of Darwin, see Janet Browne, *Charles Darwin: Voyaging* (New York: Knopf, 1995), and *Charles Darwin: The Power of Place* (New York: Knopf, 2002). Pathbreaking work in this area is found in Adrian Desmond and James R. Moore, *Darwin* (London: Michael Joseph, 1991).

8. Barbara Wedgwood, *The Wedgwood Circle 1730–1897: Four Generations of a Family and their Friends* (London: Studio Vista, 1980).

9. Harold James Perkin, *The Origins of Modern English Society*, 2nd ed. (London: Routledge, 2002).

10. Browne, 1995, pp. 6–10. In Darwin's autobiography, he recalled his father as "a cautious and good man of business, so that he hardly ever lost money by any investment, and left to his children a very large property," *The Autobiography of Charles Darwin 1809–1882: With original omissions Restored*, ed. Nora Barlow (London: Collins, 1958), p. 40.

11. Hedley Atkins, *Down, the Home of the Darwins: The Story of a House and the People who lived there*, rev. ed. (London: Royal College of Surgeons of England, 1974).

12. Frederick H. Burkhardt, Sydney Smith, et al., eds., *The Correspondence of Charles Darwin*. Vols. 1–16 (1821–68) (Cambridge: Cambridge University Press, 1983–2007).

13. Down House, Beckenham, Kent. I thank English Heritage for permission to cite these manuscripts.

14. Most famously, Max Weber, *The Protestant Ethic and the Spirit of Capitalism*, trans. T. Parsons (New York: Scribner's, 1958).

15. Theodore Porter, *Trust in Numbers: The Pursuit of Objectivity in Science and Public Life* (Princeton: Princeton University Press, 1995), and Lorraine Daston and Peter Galison, *Objectivity* (Zone Books/MIT Press, 2007). See also Tore Frängsmyr, John Heilbron, and Robin E. Rider, eds., *The Quantifying Spirit in the Eighteenth Century* (Berkeley: University of California Press, 1990). For pioneering work on inscriptions, see Bruno Latour and Steve Woolgar, *Laboratory Life: The Construction of Scientific Facts*, 2nd ed. (Princeton: Princeton University Press, 1986). I am grateful to Steven Shapin for letting me read parts of his forthcoming book, of which Chapter 2 examines truth, method, and vocation, and distinguishes the practices of science from the truth claims of science.

16. James Alfred Aho, *Confession and Bookkeeping: The Religious, Moral, and Rhetorical Roots of Modern Accounting* (Albany: State University of New York Press, 2005).

17. *Psalms*, lxix. 28.

18. Margaret Schabas, *The Natural Origins of Economics* (Chicago: University of Chicago Press, 2005).

19. Tore Frängsmyr, ed., *Linnaeus: The Man and His Work*, rev. ed. (Canton, MA: Science History Publications, 1994), and "Linnaeus and the classification tradition in Sweden," in T. Frängsmyr, ed., *The Structure of Knowledge: Classifications of Science and Learning since the Renaissance* (University of California, Berkeley: Office for History of Science and Technology, 2001), pp. 77–91. See also Lisbet Koerner, *Linnaeus: Nature and Nation* (Cambridge, MA: Harvard University Press, 1999), and Wolf Lepenies, "Linnaeus's 'Nemesis divina' and the concept of divine retaliation," *Isis* 73 (1982), pp. 11–27.

20. Emma Spary, "Political, natural and bodily economies," in *Cultures of Natural History*, ed. N. Jardine, J.A. Secord and E.C. Spary (Cambridge: Cambridge University Press, 1996), pp. 178–195, and Harold J. Cook, *Matters of Exchange: Commerce, Medicine, and Science in the Dutch Golden Age* (New Haven: Yale University Press, 2007), especially pp. 42–81.

21. Theodore Porter, *The Rise of Statistical Thinking, 1820–1900* (Princeton: Princeton University Press, 1986). Lorraine Daston illuminates the important roles that quantification, empiricism,

and objectivity play in moral economy in "The Moral Economy of Science," *Osiris* 10 (1995), pp. 3–24.

22. Browne, 1995, pp. 228–229.

23. Robert Waring Darwin's medical practice books and financial books are in Cambridge University Library, DAR 227. For background on medical fees, see George Rosen, *Fees and Bills: Some Economic Aspects of Medical Practice in Nineteenth Century America* (Baltimore: Johns Hopkins Press, 1946). The development of accountancy practices is described in Herbert J. Eldridge, *The Evolution of the Science of Book-Keeping*, 2nd ed. (London: Gee, 1954).

24. In 1868, Darwin shifted his financial year to coincide with the calendar year. Charles Darwin's Account Books, Down House MS.

25. F. Darwin, 1887, vol. 1, p. 120.

26. Herbert Spencer, "Railway morals and railway policy," *Edinburgh Review* 116 (1854), pp. 420–461. See also Geoffrey Channon, "The business morals of British Railway Companies in the mid-Nineteenth century," *Business and Economic History* 28 (1999), pp. 69–79, and James Taylor, "Business in Pictures: Representations of Railway Enterprise in the Satirical Press in Britain, 1845–1870," *Past and Present* 189 (2005), pp. 111–145.

27. Atkins, 1974, p. 97.

28. Letter dated 8 September 1881, printed in Atkins, 1974, p. 100. For a discussion of Darwin's accounts, see pp. 95–100. See also Browne, 2002, pp. 461–465, 491. A draft of Darwin's intentions for his new will is in Cambridge University Library, DAR 202. Another letter dated 16 September 1881, addressed to his children, Cambridge University Library DAR 210.6, indicates the division of his estate. Further letters were written later in 1881 and early 1882. See Frederick Burkhardt and Sydney Smith, eds., *A Calendar of the Correspondence of Charles Darwin, 1821–1882: With supplement* (Cambridge: Cambridge University Press, 1994). Throughout 1881, Darwin discussed his assets carefully with William Erasmus Darwin, his eldest son, who was a partner in a bank in Southampton.

29. Charles Darwin's Investment Book, Down House MS, p. 33.

30. A helpful account is in Timothy Alborn, *Conceiving Companies: Joint Stock Politics in Victorian England* (London: Routledge, 1998).

31. Ralph Colp, *To be an Invalid: The Illness of Charles Darwin* (Chicago: University of Chicago Press, 1977), pp. 43–53. Down House MS.

32. Gavin De Beer, ed., "Darwin's Journal," *Bulletin of the British Museum (Natural History) Historical Series* 2, no. 1 (1959), pp. 1–21, and "The Complete Works of Charles Darwin Online," darwin-online.org.uk.

33. Cambridge University Library, DAR 210.10. Transcribed in Burkhardt and Smith *Correspondence*, vol. 2, pp. 444–445, and discussed in Browne, 1995, pp. 378–379.

34. Darwin's philanthropic activities in Down village are described in Browne, 2002, pp. 452–456. See also James R. Moore, "Darwin of Down: The Evolutionist as Squarson-naturalist," in D. Kohn, ed., *The Darwinian Heritage* (Princeton: Princeton University Press, 1985), pp. 435–481, and R. M. Stecher, "The Darwin-Innes letters: The Correspondence of an Evolutionist with his Vicar," *Annals of Science* 17 (1961), pp. 201–258.

35. F. L. Holmes, "Scientific writing and scientific discovery," *Isis* 78 (1987), pp. 220–235; Daston, "Moral Economy," p. 23.

36. Latour and Woolgar, 1986, pp. 45–53, in which they observed scientists filling in forms, punching in numbers, and taking observations. For Darwin's writing work on the *Beagle* see Browne, 1995, pp. 193–194.

37. Paul H. Barrett, et al., eds, *Charles Darwin's notebooks, 1836–1844: Geology, Transmutation of species, Metaphysical enquiries* (Cambridge: Cambridge University Press, 1987).

38. Cambridge University Library, DAR 210.17, transcribed in Burkhardt and Smith, et al., *Correspondence*, vol. 4, pp. 410–433. The notebook is briefly discussed in Janet Browne, "Darwin and

Expression," in D. Kohn, ed., *The Darwinian Heritage* (Princeton: Princeton University Press, 1985), pp. 307–326, and more directly in Randal Keynes, *Annie's box: Charles Darwin, his Daughter and Human evolution* (London: Fourth Estate, 2001), pp. 55–58.

39. Forthcoming, see note 15. See also Theodore M. Porter, "Objectivity as standardization: The rhetoric of impersonality in measurement, statistics, and cost-benefit analysis," *Annals of Scholarship: Metastudies of the Humanities and Social Sciences* 9 (1992), pp. 19–59; Lorraine Daston, "Objectivity and the escape from perspective," *Social Studies of Science* 22 (1992), 597–618, and Peter Dear, "From truth to disinterestedness in the seventeenth century," *idem*, pp. 619–631.

40. Samuel Smiles, *Self-Help* (London, 1859). Smiles' impact on Victorian ideology is discussed in a slightly different context by Anne Secord, "Be What You Would Seem to Be: Samuel Smiles, Thomas Edward, and the Making of a Working-Class Scientific Hero," *Science in Context* 16 (2003), pp. 147–173. See Browne, 2002, pp. 396–400.

41. Henrietta Litchfield, ed., *Emma Darwin: A Century of Family letters, 1792–1896* (London: J. Murray, 1915); Edna Healey, *Emma Darwin: The inspirational Wife of a genius* (London: Headline, 2001). Emma Darwin is discussed by Evelleen Richards, "Darwin and the Descent of Woman," in *The Wider Domain of Evolutionary Thought*, ed. David Oldroyd and Ian Langham (Dordrecht: D. Reidel, 1983), pp. 57–112 and Keynes 2001. For a more general account of Victorian married life, see Phyllis Rose, *Parallel lives: Five Victorian marriages* (New York: Knopf, 1983). For women and bookkeeping, see Gail Wilson, *Money in the Family: Financial Organisation and Women's Responsibility* (Aldershot: Avebury, 1987); Mary S. Hartman, *The Household and the Making of History: A Subversive View of the Western Past* (Cambridge: Cambridge University Press, 2004); and Andrew August, *Poor Women's Lives: Gender, Work, and Poverty in late Victorian London* (Madison: Fairleigh Dickinson University Press, 1999). I have specifically drawn on Amanda Vickery, *The Gentleman's Daughter: Women's Lives in Georgian England* (New Haven: Yale University Press, 1998).

42. Emma Darwin's Classed Account Books, Down House MS.

43. Leonore Davidoff and Catherine Hall, *Family Fortunes: Men and Women of the English Middle Class, 1780–1850*, rev. ed. (London: Routledge, 2002).

44. Frank K. Prochaska, *Women and Philanthropy in Nineteenth-century England* (Oxford: Clarendon Press, 1980).

45. John Tosh, *A Man's Place: Masculinity and the Middle-class Home in Victorian England* (New Haven: Yale University Press, 1999). A good lead in this direction has been taken by Paul White, *Thomas Huxley: Making the "Man of Science"* (Cambridge: Cambridge University Press, 2003). See also his "Science at home: the space between Henrietta Heathorn and Thomas Huxley," *History of Science* 34 (1996), pp. 33–56.

46. Hilts, 1975, p. 27.

47. Patricia James, *Population Malthus: His Life and Times* (London: Routledge & Kegan Paul, 1979); and Brian Dolan, ed., *Malthus, Medicine and Morality: Malthusianism after 1798* (Amsterdam: Rodopi, 2000).

48. Desmond and Moore make this connection a central plank in their *Darwin*. See Desmond and Moore, 1990, especially pp. 256–279.

49. Janet Browne, "Making Darwin: Biography and Character," unpublished, Distinguished Lecture, History of Science Society, November 2005.

50. Burkhardt and Smith, et al., *Correspondence*, vol. 2, especially pp. 81, 123.

51. Charles Darwin, *The Descent of Man, and Selection in Relation to Sex* (London: John Murray, 1871), vol. 1, pp. 167–182.

52. Darwin *Descent*, vol 1, p.169.

53. Desmond and Moore, *Darwin*, pp. xviii–xxi. Out of an extremely large literature on Darwin as a social Darwinist, see Richard Weikart, "A recently discovered Darwin letter on social Darwinism," *Isis* 86 (1995), 609–611; and Evelleen Richards, "Redrawing the boundaries: Darwinian

science and Victorian women intellectuals," in *Victorian Science in Context*, ed. Bernard Lightman (Chicago: University of Chicago Press, 1997), pp. 119–142. A general review of current research is given by Mike Hawkins, *Social Darwinism in European and American thought, 1860–1945: Nature as model and nature as threat* (Cambridge: Cambridge University Press, 1997). See also Peter Bowler, "Social metaphors in evolutionary biology, 1870–1930: The wider dimension of Social Darwinism," in Sabine Maasen et al., eds., *Biology as society, society as biology: Metaphors* (Dordrecht: Kluwer Academic, 1995), pp. 107–126.

54. James A. Secord, "The Geological Survey of Great Britain as a research school, 1839–1855," *History of Science* 24 (1986), pp. 223–275.

55. Gordan McOuat, "From Cutting Nature at its Joints to Measuring it: New Kinds and New Kinds of People in Biology," *Studies in the History and Philosophy of Science A* 32A (2001), pp. 613–645, on p. 618; and Cook, 2007, pp. 49–57, 141–142, 207–209, 414–416. See also Browne, 1995, pp. 206–210. Michael O'Malley compares the language of race to the language of money in "Specie and species: Race and the money question in 19th-century America" *American Historical Review* 99 (1994), pp. 369–395.

56. These analogies were echoed in Herbert Spencer's argument that the social body is constituted like the human body. For Spencer, coin and currency mimicked the blood in providing the economy's nutrition. See Herbert Spencer, "The Social organism," in *Essays, scientific, political, and speculative* (New York: D. Appleton, 1899), p. 294. In a different domain, Hillard Kaplan speaks of evolutionary capital. See H. Kaplan, J. B. Lancaster, and A. Robinson, "Embodied capital and the evolutionary economics of the human lifespan," *Population and Development Review* 29, Suppl. (2003), pp. 152–182. Charles Blinderman discusses economic metaphors in "Mudlarks of the Thames: Victorian oeconomy as scientific metaphor," *Journal of American Culture* 16 (1993), pp. 55–65.

57. For example, a search on Web of Science (biology) for English-language use of the word "investment" in the title brings forward 445 articles published in the period 2000-2008. A typical example is RM Evans, "Parental investment and quality of insurance offspring in an obligate brood-reducing species, the American white pelican" *Behavioral Ecology* (1997), 8(4): 378–383.

58. An excellent framework for analysis was given by Silvan S. Schweber, "Darwin and the political economists: Divergence of character," *Journal of the History of Biology* 13 (1980), 195–289.

59. Charles Darwin, *On the Origin of Species by Means of Natural Selection, or the Preservation of Favoured Races in the Struggle for Life* (London: John Murray, 1859), pp. 5, 80–95. See also Darwin, *Autobiography*, p. 120.

60. Mary Poovey, *A History of the Modern Fact: Problems of Knowledge in the Sciences of Wealth and Society* (Chicago: University of Chicago Press, 1998).

61. Browne, 2002, pp. 489–490.

The Persistent Use and Abuse of Wilhelm von Humboldt in History and Politics— noch einmal

THORSTEN NYBOM

Introduction

The ongoing, steadily more intense, European debate on the need for rapid and fundamental changes in higher education, research funding and education policy planning has certainly not been an exception from the traditional rule of systematic ideological abuse. On the contrary, many of the most frequent arguments in these deliberations—both on the national and supra-national level—have had clear political or ideological connotations. This is perhaps most obvious when different protagonists have resorted to historical arguments to promote their own particular reform agenda and recommendations.[1] Most notably this has been the case in regard to the perpetual and sometimes heated discussions on the relevance/ irrelevance, impact—negative or positive, actual significance and, indeed even existence of the so-called Humboldt University both in the history of European higher learning during the last two centuries and regarding its present-day role and repercussions in European higher education organization and policy-making.[2] In view of this persistent but nevertheless rather uninformed ideological obsession, it seems reasonable and indeed necessary to, once again, return to questions concerning the actual significance and impact of the Humboldtian ideas and ideals.

Traditionally, the rise of the systematic pursuit of original knowledge—subsumed under the label science/*Wissenschaft*—is viewed as the most significant and important outcome of the profound European intellectual and institutional transformation process in the 17th and 18th centuries called *the* Scientific Revolution and *the* Enlightenment. Paradoxically, perhaps, the existing European universities, with the possible exception of the two untypical examples of Leiden and Göttingen,[3] were not only side-stepped but to a high degree the victims of this dual intellectual and ideological revolution, which, eventually, also led to what probably could be described as the most serious institutional crisis that the European University has hitherto gone through in its almost millennial-long history. Thus, it is my deep-seated personal conviction that the thorough-going transformation process, which took place in European higher learning during the "long 19th century,"[4] was not primarily a period of reform of the *existing* institutions but should rather be seen as an era of *reorganization* and restoration of all three major European university "systems"[5]: the French—where the university in everything but name was side-stepped by the Napoleonic Reforms, the German with the "Humboldtian" revolutionary reforms, and the English with what Sheldon Rothblatt has described as the "revolution of the dons."[6] The only place where the existing universities actually flourished during this particular era was probably Scotland.[7]

The "Humboldt Revolution": Part One—"Professionalization" and "Verwissenschaftlichung," 1810–1850

The first part of these seminal shifts started in 1810, in the then not particularly illustrious Prussian *Krähwinkel* of Berlin. The immediate driving forces behind this truly revolutionary break in the history of the University, which has gone down in history, as the establishment of the so-called Humboldt University, can be found in a combination of a number of integrated and random historical factors:

- *Ideology*: An almost unique combination of aggressive neo-humanism (classicism), "*Spät-Aufklärung*" and pre-romantic German idealism. (The most important intellectual point of departure probably being Immanuel Kant's satirical pamphlet *Streit der Fakultäten*, 1798).
- *Politics*: Prussia's national catastrophe after the Napoleonic wars and the ensuing political and institutional reconstruction of central functions of the state (vom Stein–Hardenberg reform era).
  Mentality: A historically—possibly unique—concentration of creative intelligence and a general interest in—almost obsession with—education in general.[8]
- *Institutional*: The total external and internal intellectual and institutional decline of the German university system.

I am prepared to state that the seminal and even revolutionary importance of these events did *not* primarily concern the institutional level but the *ideological* level. Thus, the main and enduring achievement of Wilhelm—*Freiherr*—von Humboldt was that he, out of the almost innumerable philosophical and pedagogical ideas on knowledge and learning floating around, was able to deduce and articulate and produce a consistent *Idea* of the University. Traditionally, the defining properties and basis of the "idea" of the Humboldt's University/vision have rightfully been described as:

- Knowledge as a unified indivisible entity.
- *Einheit von Forschung und Lehre* (Unity of research and teaching).
- Primacy of *Wissenschaft* and research, which also presupposed a new institutional order and cognitive hierarchy.
- The individual and common pursuit of "truth" in "*Einsamkeit und Freiheit*" (Solitude and freedom).[9]
- *Lehr- und Lernfreiheit* (Freedom of teaching and learning).
- The creation of a unified national culture with *Wissenschaft* and University as the centerpiece: "*Bildung,*" *Wissenschaft* and (higher) education as the second categorical imperative of the central state beside national defense: as the basis of a modern "*Kulturstaat.*"[10]

The Humboldtian reforms, nevertheless, also had far-reaching institutional consequences. As regards Wilhelm von Humboldt himself, his main institutional dilemma and concern could be formulated as follows: How is it possible to establish a *socially integrated yet autonomous institutional order for qualified scientific training?* An institutional order, which, at the same time, could guarantee an optimal and perpetual growth in knowledge but also provide a dimension of "*Sittlichkeit*" (virtue) to the individual?

Wilhelm von Humboldt's pragmatic solution or even functioning historical compromise was: *The regally (state) protected and fully endowed Ivory Tower combined with an elitist and gate-keeping Gymnasium/Abitur.* And even if the label Ivory Tower has gradually nowadays become one of the most frequently used degrading metaphors for the supposed societal and even cultural irrelevance of the Humboldtian University, it certainly had no derogatory connotations for Wilhelm von Humboldt. On the contrary! The creation of an Ivory Tower was precisely what he ultimately was striving to achieve. Accordingly, the state must be persuaded that it was in its own well-founded, long-term interest to optimally promote the expansion of scientific knowledge, and this could only be accomplished by securing the individual freedom of the scholar. Reciprocally, the king should keep the prerogative of appointing professors—not primarily as a means of control but in order to protect the institutions from succumbing to the vice of internal strife and nepotism. Furthermore, to be worthy of enjoying this extended freedom and autonomy, the professors should refrain from political and other "external" strife, and hence the delimitation of the university from society at large should be clear.

Wilhelm von Humboldt presented two main arguments for why the king/state should play the role of a more or less passive guardian angel to an institutional order with unparalleled autonomy in an absolute monarchy. First, there was the "philosophical/moral" argument that new and original knowledge could only be pursued and produced in "*Einsamkeit und Freiheit*." Second, and logically following the first, he also presented the purely "utilitarian" argument, which is usually forgotten in the present deliberations on the Humboldt university: Since the intellectual *and* economic prosperity, and even physical existence of a modern *Kulturstaat* was directly and inexorably linked to the optimal pursuit and production of qualified knowledge, the principle of individual autonomy was not only a *desirable* institutional solution but a *utilitarian necessity* and central *moral obligation* of the central state—*sine non qua*. This "utilitarian" side of Wilhelm von Humboldt's argumentation—sometimes said to have been originally invented by his brilliant younger brother, Alexander[11]—may perhaps surprise his present-day academic admirers, who tend to believe that Wilhelm von Humboldt did not care about such "worldly things" as instrumental usefulness. He certainly did.

Contrary to Wilhelm von Humboldt's original proposal and repeated pleas for a fully endowed and thus economically autonomous university, the Prussian *Staatsrat* decided to treat the new University as just an ordinary state agency with annual and hence politically controlled short-term allowances. This decision was a major disappointment to von Humboldt, who considered economic autonomy as a necessary precondition also for academic autonomy. The governmental decision was probably the main reason why he quite suddenly and promptly decided to leave his office as Secretary of State in the *Section des Cultus und Unterricht* after only 16 months in office and officially declared that he wished to have nothing to do with the further planning arrangements of the new institution of higher learning.[12] This definite demission has been reason enough to ask how much "Humboldt" the "Humboldt" university actually contained—all the more so as it was not until almost a century later that the Berlin University was hailed as the outcome of Wilhelm von Humboldt's genius.[13]

Last, but not least, a strict state-control of student admission (*Abitur*) should be established. Modern academic "politicians" tend to forget that Wilhelm von Humboldt's greatest and lasting *institutional* achievement in education was probably not the reorganization of the Berlin University, but laying the foundations of the German *Humanistisches Gymnasium*.[14] The European secondary school thus became totally integrated in, and dominated by, higher education, and subsequently also the real gatekeeper and guardian of excellence. From now on the *Abitur* became the only but also powerful selection process for the comprehensive, "open" Humboldtian but nevertheless highly elitist university system.[15] The illusion that the present-day European secondary school systems are actually still performing this crucial gatekeeping task is very much alive in many parts of continental Europe, which has contributed to aggravate some of the structural dysfunctions in European higher education.[16]

In the initial five decades (1810–1860) of its existence, the "new" German university underwent a gradual institutional and professional transformation, which eventually and in different degrees, would permeate and influence almost all Western university systems.[17] At the institutional level, the modern organizational and hierarchical triad of *Fakultäten—Disziplinen—Lehrstühle* (Chairs) was formally established and cemented, where the actual power rested with the full professors (*die Ordinarien*). The European university then became a rule-governed community of scholars—a loosely coupled institutional framework without an administrative center of gravity within which individual professors remained more or less autonomous. The *Rektor* remained a purely representative position, and the *Kanzler*, as administrative head, did not even formally belong to the university but to the ministry. In due course, this institutional fragmentation would turn out to be one of more decisive institutional difference between the European universities and its rapidly expanding North American sisters and competitors. When it comes to *pedagogical* change, the introduction of the Seminar could be seen as an ambition to establish an ideal-typical form of free, discursive and common scientific inquiry of professors and students.

From having been regarded as "*Trivia*," the Philosophical Faculty was elevated to the indispensable core of the "new" University. A revolutionary transformation, which although it had deep-set institutional consequences, primarily reflected the epistemological and ideological cornerstones in German Neo-humanist thinking. The unity of knowledge was not only a cognitive and epistemological pillar of German idealistic philosophy; it also constitutes, in some respects, its basic philosophical and moral foundation. This unity should primarily be achieved and secured through the reign of philosophy.[18]

This did not just mean that the natural and cultural sciences could be merged on the higher philosophical level. Philosophy—together eventually with history—were also given the central task or duty to supervise the so-called "*Brotwissenschaften*," i.e. Medicine, Technology, and Law. These fields of study should not be able to corrupt, or even influence, the institutional order and the intellectual content of higher education, since those disciplines—to quote Wilhelm von Humboldt: "don't have their immediate, spiritual home in *Wissenschaft* but in qualified handicraft." This Kantian idea actually meant that the existing medieval university was turned on its head when the traditionally "lower" Philosophical Faculty suddenly became top dog. In reality, one could say that this faculty, from now on, constituted the genuine and "real" new university, since, according to Kant, von Humboldt and others, the philosophical faculty was the only one immediately connected with "truth," while the other three had their rationale in instrumental "usefulness."[19] Eventually, this also quite early led to an institutional differentiation.[20] "Thus, the dual identity of the modern European University became established: it was supposed to be at the same time a place for research and an educational institution for *civil servants*. [my italics]"[21]

On the *professional* level, it has been argued that this period signified the emergence of the modern competitive academic career system and consequently also the establishment of an informal but nevertheless obvious institutional hierarchy. Until the mid 19th century, the recruitment of professors had in the German realm been extremely local—and to some extent even a family affair.[22] In the second half of the century, Germany had become a national academic labor market where professors pursued highly competitive academic jobs and careers. It was also now that the Berlin University gradually established itself as the pinnacle of academic excellence and fame.[23] Simultaneously, the individuals devoted to the noble task of perpetual "truth-seekers"—i.e. University professors—advanced markedly in social and economic status until they, eventually, in the imperial era attained a mandarin-like position—or in the words of the German professor of philosophy, Jürgen Mittelstrass: "What God was among the angels, the learned man should be among his fellow men."[24]

Finally, it must be pointed out, however, that the driving-force behind this broad and massive international impact of the German University in the second half of the 19th century was not primarily a matter of any formal organization or institution building but rather an effect of an almost instant and exceptional expansion of scientific knowledge in Germany in practically all scholarly fields, which moreover only seemed to accelerate over time. And since nothing succeeds like success in academia as elsewhere, in less than half a century the *Friedrich-Wilhelms-Universität zu Berlin* became the undisputed model institution for practically all university systems in the world.[25]

The explanation of scientific productivity has long been a central concern of the history and the sociology of science. Should the undisputable success of German science and scholarship, in the 19th century and onwards, be explained by specific or generalizable cultural or/and economic factors?[26] Although there are many different theories accounting for successful scientific performances, social scientists seem to agree on at least this one factor: "advance was dependent upon the number of talented individuals who select science as a career."[27] But even if one certainly can find a fair number of scientific geniuses in German 19th-century intellectual history, such geniuses, nevertheless, are in need of milieus where their genius can thrive and where their achievements can be duly acknowledged. So, the question remains: what are the factors that seemingly influence and possibly increase the pool of scientific talents? In the history and sociology of science, there are at least three different types of theories: cultural, organizational, and wealth-oriented.

One classic theory is the *cultural theory*, which Robert K. Merton advanced in the 1930s.[28] His study of 17th-century English science showed that after the Reformation in England, the rate of scientists increased considerably. Implicit in Merton's theory was the hypothesis that Protestant societies place a higher value on scientific activity, and hence, these societies will profit from that greater activity. But before applying this handy theory modeled on Max Weber's study of the affinity between the Protestant Ethic and Capitalism, we need to consider the fact that many European countries had been Protestant for centuries without showing a similar development, at least not in the 19th century.[29] Nevertheless, even if it is no longer Protestantism and related religious and moral values that are significant in explaining scientific success, I maintain that Merton's theory has a certain historical significance for 19th-century Prussia, and indeed the Northwestern parts of

Europe in the sense that *education* became a *central national state priority* in most countries during the first half of the 19th century.[30]

In Prussia, furthermore, after the above-mentioned defeat in the Napoleonic war it became a deep conviction among the reformers around *Freiherr* vom Stein and *Fürst* Hardenberg that the state must be reformed and rebuilt from *within*, or in the words attributed to King Friedrich Wilhelm III himself, Prussia had to "[. . .] make up in spiritual strength for the physical strength it has lost,"[31] which certainly included the notion or concept of national education as an absolute centerpiece.[32] Thus, to summarize, one could quote a fellow German scholar:

> The Prussian imperial desire to strengthen the "spiritual strength," humanist-idealist demand for "national education," and the reformers' aim of having a tertiary educational institution in the service of civilian society all came together and formed the amalgam, which ran like a red tread through the university success story of the 19th century [. . .][33]

However, we need to be highly cautious as to "motivational" factors that may operate in the case of young talents choosing a science career simply because of "higher" idealistic reasons. So it is most probable that apart from inner driving-forces external, material stimuli were also in operation: good research facilities, good salaries, good career opportunities, etc., and such a (materialistic) motivation structure is linked to some other kind of theories. In the late 1950s, Joseph Ben-David advanced a theory of scientific success linked to *structural-organizational factors* prior to motivational-cultural ones.[34] In order to increase the pool of talented scientists, the crucial mechanism is institutional/educational reform. When more universities are created in a country, competition between these universities increase, and talented youth are offered richer opportunities. Hence, the pool of talents expands and intensifies; a motivational structure of high performance is the result.[35]

Ben-David's theory seems easily applicable to the development of the Prussian and the German higher education system in general in the 19th century.[36] Likewise, many have also pointed to a growing tendency to expand and reform the higher education and research systems in response to rising and changing demands and requirements of a rapidly growing and innovative industry, which would also be consistent with the rapid German industrialization process in the second half of the 19th century. And even more important, in this second phase of the industrial revolution, the new electro- and chemical industries became the "cycle leaders," i.e. industrial undertakings that not only craved sophisticated skills but even scientific knowledge to flourish and expand.[37]

Taking all these factors or driving forces into account, it actually doesn't matter if the different international followers often had a less well-grounded or even non-existing knowledge of the actual Humboldtian ideas and their implications and significance.[38] Thus, it would be quite possible to make the argument that the next flash of genius in institutional, University history—the establishment of the North American Graduate School—is, at the same time, both absolutely inconceivable without and fundamentally at odds with the *Friedrich-Wilhelms-Universität*.[39] To make the argument a little more provocative, I believe one could, quite convincingly, argue that the Graduate School, and subsequently the great U.S. research university, was founded on one of the most successful and productive "misunderstandings" in modern intellectual history! Hence, Daniel Coit Gilman, Abraham Flexner and other U.S. reformers could serve as instructive illustrations to Friedrich Nietzsche's warning of the dangers of knowing too much history if you wish to be an active and successful political and social actor in our own time.

"Bildung": A Dubious (?) Ideological Blessing— A Necessary Exursus

An almost endlessly discussed key word in Humboldt's thinking and reform plans is the concept of *Bildung* which, eventually, would have such a powerful, but at the same time imprecise, impact not only in German culture and public debate for almost two centuries. The Humboldtian concept of *Bildung* was not only a matter of understanding the rational features of knowledge and *Wissenschaft*, but also the possibilities of developing a person's natural abilities through an unlimited, spontaneous, spiritual process of self-cultivation guided from within.[40] *Bildung* involved more than the narrow learning process; it was also related to a particular concept of the human being that emerged in the closing decades of the 18th century and the first decade of the 19th century. As a matter of fact, *Bildung* became the catchword for a whole philosophy of pedagogy, and indeed national culture, spreading from German-speaking cultural circles to the Nordic countries and Russia. The original power of the *Bildung* concept was that it referred to the objective, as well as subjective, aspects of knowledge. Thus, on the one hand, the subjective aspect of knowledge was emphasized, but at the same time *Bildung* would also serve as a barrier against arbitrariness.

Inspired by the Swiss educationalist Johann Heinrich Pestalozzi, the central aim in Humboldt's "*Bildungs-vision*" was the establishment of a national three-level educational system (*Nationalerziehung*) where the university was the third and final level.[41] In this comprehensive system the first or elementary level should only be concerned with *Menschenbildung*. The secondary schools should, through the intense study of languages (classic), history, and mathematics, primarily teach the students *how* to learn, since mastery of the learning process was absolutely necessary for the kind of university education Humboldt wanted to establish, in which the student is striving to attain "pure knowledge" in "*Einsamkeit und Freiheit*." At this highest level, the most a teacher could do was to awaken the student's natural will to learn and act as an experienced counselor and *Meister*.

The Humboldtian ideas and the ensuing German *Bildungsideal* never created a unitary national culture—if there ever was such a thing anywhere in the world. But it did contribute to the rise of a specific national "super-ideology." Research and higher education became integrated in, and were a central component of, a well-structured societal status and power brokering hierarchy.[42] In this extraordinary ideological brew, it is possible to find, at least partly, the roots of the peculiar German ideological "*Sonderweg*," which in the second half of the 19th century was condensed into the conviction—or illusion—of a unique German road to modernity—*Kultur*. The allegedly unique German development was not only and in many and fundamental ways supposed to be different from, but also superior to the "normal," Anglo-French process of "*Civilization*." This hierarchical, not to say aristocratic, national ideology got its perhaps most ideal typical expression, in 1918, in Thomas Mann's equally brilliant and chilling treatise *Betrachtungen eines Unpolitischen*.[43]

The "Humboldt" Revolution, 1860–1920: Part Two—
The Rise of the Modern Research University and
the Comeback of Wilhelm von Humboldt as Myth

The second "Humboldtian revolution," the emergence of the modern research university, which in reality brought about a gradual restructuring and reorganization of all university systems, at least in the so-called Western world, took place in the period between 1850 and the outbreak of the First World War in 1914. The driving forces behind these fundamental and simultaneous changes came not least from *within* science and scientific theory itself. With the emergence of the modern—and post-Newtonian—natural sciences and its soon-demonstrated industrial potential it became virtually impossible to define the scientific endeavor and the academic profession as "the pursuit of curious individual gentlemen of ingenious minds." After Justus Liebing and, subsequently, Herman von Helmholz, Robert Koch, Louis Pasteur, *et al.* (laboratory), Albert Einstein, Max Planck, Niels Bohr, Ernest Rutherford, *et al.* (theory), and also Wilhelm Röntgen, Carl Bosch, Fritz Haber, *et al.* (application), the pursuit of knowledge had become a central concern for almost every sector of modern society.

Hence, the combined effects of the fundamental breakthroughs and revolution on the scientific cognitive level and the demonstrated and potential impact on the macro-economic and eventually also political level, had deeply penetrating ideological, professional, institutional and policy consequences, which in many ways collided with the basic Humboldtian ideas and ideals.

- First, science had turned into a collective task or "intellectual industry," which demanded scale, organization and, perhaps above all, money and in which the notion of "*Einsamkeit und Freiheit*" seemed to be utterly obsolete.

- Second, and for more or less the same reasons, the ambition to amalgamate "*Forschung und Lehre*" gradually became almost impossible.[44] The most striking illustration and manifestation of this fact became the establishment of the *Kaiser-Wilhelm-Gesellschaft* and its string of more or less autonomous research institutes in 1911. It was, perhaps also, the ultimate indication of the deplorable fact that "excellence" had actually started its gradual exodus from the Humboldt University.
- Third, the steadily growing costs and societal impact of research did not only lead to institutional changes but also to innovations in research policy and (targeted) funding, which had consequences for the institutional autonomy.[45]
- Fourth, and perhaps, even more seminal, modern science finally and irrevocably crushed the illusion of the "unity of knowledge under benevolent aegis of philosophy" and was gradually superseded by the idea of two distinct scientific "cultures." Significantly enough, it was in Germany that this distinction between "*Natur- und Geisteswissenschaften*" was first discussed and philosophically codified in the second half of the 19th century by scholars such as Wilhelm Dilthey, Heinrich Rickert and Max Weber, but was also discussed by intellectual industrialists like Werner von Siemens.[46]

This process of cognitive disintegration and specialization was, furthermore, institutionally manifested by the foundation of the modern *Technische Hochschulen*, which responded to the rapidly growing demand for a new type of qualified professional training and skills.[47]

In our context, it is, however, equally interesting and remarkable that this process of cognitive and institutional disintegration, which in many respects signified a fundamental break with the original Humboldtian ideals, was not only explicitly presented as the ultimate fulfillment of Humboldtian dreams, it also, ironically enough, marked the reinvention and even canonization of Wilhelm von Humboldt himself as the spiritual *and* practical founding-father of German (European) University.[48] Accordingly, it is typical that when the prime intellectual and bureaucratic movers, the theologian Adolf von Harnack and the almighty *Ministerial-Direktor* Friedrich Althoff, instigated the institutional revolution of the *Kaiser-Wilhem-Institute*,[49] they were nevertheless very keen to use and stress all the supportive arguments they could possibly find in Wilhelm von Humboldt's rediscovered and immediately canonized *Denkschrift*. Luckily enough, in his deliberations Humboldt had indicated that a complete science organization should have three major institutional components or levels: beside the free Academy and the University, there should also be "*Hilfs-Institute*." But with these "*leblose* ('lifeless') Institute*," Humboldt had hardly meant the powerful centers of excellence that were now established.[50]

It is also at this point in time, especially in connection with the centennial anniversary in 1910 that Wilhelm von Humboldt's ideas and ghost is transformed into some kind of "universal weapon" ("*Allzweckwaffe*")[51] in the German and gradually also the international debate on higher education institution building and higher education policy. It is perhaps interesting that this ideological innovation process or transfer was already from the start, and continued to be, driven and promoted not in scholarly works by professional historians but primarily in interventions and pamphlets by academics with a "university political cause" or education politicians on the national level. And in the German context, this becomes particularly true in the reoccurring times of national or institutional euphoria, deep crisis and ongoing restructuring.[52]

Even if technological innovation *per se* cannot be said to have played an important role in the process of restructuring of university life between 1860 and 1920, the short- and long-term *technological consequences* of the internal scientific revolution were to become almost "cosmic." From now on, and increasingly so, "Big science" did not only become heavily dependent on modern, sophisticated technology, it also became the absolute necessary prerequisite for, and powerhouse of, this path-breaking new tool, soon to be called "high tech." Or to put it differently, in more socio-political terms: As the English crystallographer and historian/politician of science, John Desmond Bernal *hoped* in the 1930s, and M.I.T. President and Scientific Advisor to the President, Vannevar Bush *feared* in the 1940s,[53] the demonstrated tremendous impact or obvious and immediate "social function of science" had ultimately made science and scientific training too important a matter to be left to the scientists and so it was eventually turned into a separate sector of national policy making.

At least in the German case, the most frequent explanation for the obvious lack of structural reforms has been attributed to the lasting and overpowering impact of the "Humboldtian ideals."[54] And regardless of the *actual* presence of any form of "Humboldtianism" in present-day European higher education, it must nevertheless be stated that the undisputed success of the North American research universities in the last century and particularly in the last 30 years (the same period in which their European sisters declined) could, at least to a certain extent, be explained by their readiness and superior ability to react to social, economic, scientific, and political changes.[55] The European university, on the other hand, *has not changed* in the last 50 years—it has *been changed*. Paradoxically enough, this has been achieved rather by systematic negligence than by bold intervention on the part of the politicians, but the end result is, nevertheless, that the European university has become a seemingly helpless political football.

Concluding Remarks

Coming back to the Humboldt revolutions, I would, in this connection, first like to point to the fact that successful transformations in higher education are not always—and have even seldom been—to expand the number of tasks, duties, and obligations performed by the University. I have the slightly worrying impression, that we, being caught in a curious type of a-historic and simplistic analogy think-ing, have a tendency to believe that the developments since the 1960s and 1970s are forever true and relevant. In short, when, and if, the university has to respond to "new challenges" or is asked to "reformulate its agenda" or "mission," the univer-sities tend to conclude that they must take on any new task or responsibility "soci-ety," on an almost daily basis, suggests or demands. *This is not true*, simply because, when it comes to knowledge, "*society*" very seldom actually knows what it really "needs" in 15 years time!

The dual Berlin-based "revolution" discussed above, which thoroughly reorga-nized and rejuvenated the Euro-American universities and turned them into the real intellectual *and* industrial powerhouses of their societies, for almost two cen-turies, had nothing to do with expansion. On the contrary! Wilhelm von Hum-boldt's exceptionally successful institutional reforms of 1810 in Berlin meant retraction and "purification." The establishment of the modern European and North American research university at the turn of the previous century also meant that the universities defined their core mission in a much more restricted way than they had previously done. So, when we, today, are discussing how to respond to the "new challenges and demands" and to "redefine our new role/mission" in society, we should also perhaps remember that all great universities always have, *at the same time*, been institutionally adaptive, intellectually creative, *and* ideologically conservative institutions.

Sometimes the impact of intellectual or mental transformation is so powerful that actual reality is more or less superseded by this projection and thus becomes myth. Certain concepts, and ideas may acquire an "afterlife" that makes them sig-nificant far beyond the times in which they were created and sometimes for reasons far different from those the original creator probably envisioned. This is most cer-tainly the case with Wilhelm von Humboldt's *Denkschrift* and his final proposal, *Antrag auf Einrichtung der Universität Berlin Juli 1808*. These few and scattered pages, written in clear and beautifully un-bureaucratic German, have triggered off an almost innumerable number of more or less qualified scholarly, political, and other kinds of reflections during the last 200 years. In these two centuries, there has probably not been delivered one academic *Festrede*—at least on the European con-tinent—that did not mention either Wilhelm von Humboldt or the "Humboldtian Idea of the University."

But despite the never-ending deluge of speeches, essays, and books from both Humboldt's friends and foes, I do, nevertheless, believe that it is necessary for the European academic community to discuss and confront this overpowering and nebulous image repeatedly, simply because of its continued presence in almost every European discussion on the mission and future of higher education and re-search. In some curious way, the central question then is perhaps not Wilhelm von Humboldt's *actual thoughts* but rather why these ideas have come to play such an exceptional role during two centuries almost regardless of how far from his origi-nal intentions the European university systems have moved.

One tentative answer to this fundamental question would be that Humboldt was not only able to formulate a comprehensive idea of what institutionalized higher learning and the systematic pursuit of knowledge should be, but he was also able to convincingly argue why it must be considered as one of the central interests and indeed obligations of the nation state to support such an undisputed public good. In an era when almost everybody in politics, business, civil service, academia, and so on are almost incapable of delivering a single speech without referring to the alleged strategic importance of such things as "research, education, knowledge, competence, excellence" in the present and future "knowledge society," and when most European universities neither seem to have a formative idea nor are they adequately supported or trusted by their formal political owners and masters, then ideological references to the "noble Humboldtian ideas" can either be used as an eternal source of moral and intellectual legitimation or be dismissed as an obsolete and detrimental institutional European heritage, which is hampering the necessary restructuring of European higher education and research.

As we have seen, this is nothing new in the history of German and European university politics, it has been going on for more than a century, or at least since the turn of the previous century when the then-existing Berlin University almost officially was declared to be the institutional and physical embodiment of Wilhelm von Humboldt's *corpus* of ideas and ideals. One of the reasons why this ideological traffic has persisted is, in my view, precisely because Humboldt was primarily interested in *pursuing and realizing a coherent but nevertheless imprecise body of neo-humanist ideas*. His actual interest in institution building was secondary or at least not concretely and precisely articulated. And, however brilliant, a slightly nebulous set of ideas can readily and steadily by used and abused in ideologically infested conflicts.

So, if the other important university ideologue of the 19th century, Cardinal John Henry Newman, who incidentally formulated his vision of the University in direct opposition to the German/Humboldtian "*Wissenschafts-Universität*," could be said to have taken an existing formal *institutional order*, the Oxford College, and transformed into an *Idea of a University*,[56] then Wilhelm—*Freiherr*—von Humboldt's major achievement was to *synthesize a number of ideas on science, Bildung, and learning*, which 100 years later were transformed, or elevated, or perhaps even perverted into *an institution* soon to be decreed as *the* University, and which, another 100 years later, is freely and indiscriminately used in the European debate on higher education; either hailed as an eternally valid ideal-type or disdained as a suitable scapegoat that is responsible for nearly all our alleged present miseries. From this saga we may thus learn that not only "institutions matter." This is equally true of ideas.

Notes

1. Recently, see the contributions in Johan P. Olsen and Peter Maassen, eds., *University Dynamics and European Integration* (Dordrecht: Springer, 2007).

2. The Humboldt literature has, not least in the last few years, become almost boundless. For an overview, see Thorsten Nybom, "The Humboldt Legacy," in *Higher Education Policy*, 16 (2003), pp. 141–159; Olaf Bartz, "Bundesrepublikanische Universitätsbilder: Blüte und Zerfall des Humboldtianismus," in *die hochschule. journal für wissenschaft und bildung* 2005:2, pp. 99–133, and not least Rainer Christoph Schwinges, ed., *Humboldt International: Der Export des deutschen Universitätsmodells im 19. und 20. Jahrhundert* (Basel: Schwabe & Co., 2001), and Michael G. Ash, ed., *Mythos Humboldt: Vergagenheit und Zukunft der deutschen Universitäten* (Vienna: Böhlau verlag, 1999). Also the contributions by Th. Nybom, "Creative Intellectual Destruction Destructive Political Creativity?"; Inge Jonsson, "Universities, Research and Politics in Historical Perspective"; Bernd Henningsen, "A Joyful Good-Bye to Wilhelm von Humboldt"; and Björn Wittrock, "The Legacy of Wilhelm von Humboldt and the Future of the European University" in Guy Neave, Kjell Blückert & Thorsten Nybom, eds., *The European Research University—An Historical Parenthesis?* (New York: Palgrave, 2006).

3. And possibly Uppsala simply because of the presence of Carl Linnaeus, Anders Celsius, Torben Bergman, not otherwise, for an overview and further references, see Tore Frängsmyr, *Svensk idéhistoria, I* (Stockholm: Natur & Kultur, 2000), chap. 4.

4. The label was coined by Eric J. Hobsbawm in the 3rd volume of his brilliant exposé of world history since 1780, *The Age of Empire* (London: Weidenfield & Nicholson, 1987).

5. Thorsten Nybom, in Olsen and Maassen, 2007.

6. Sheldon Rothblatt, *The Revolution of the Dons: Cambridge and Society in Victorian England* (Cambridge: Cambridge University Press, 1981).

7. On the particular Scottish University and the Scottish Enlightenment, see Douglas Sloan, *The Scottish Enlightenment and the American College Ideal* (New York: Teachers College Press, 1971). For a comparative dimension see Sheldon Rothblatt, *The Modern University and its Discontents* (Cambridge: Cambridge University Press, 1997).

8. A complete list of the participants in this intense and extensive debate would almost be equal to a comprehensive "Who is who in German *Wissenschaft & Kultur.*" For an overview see Ernst Müller, ed., *Gelegentliche Gedanken über Universitäten: J.J. Engel, J.B. Erhard, F.A. Wolf, J.G. Fichte, F.D.E. Schleiermacher, K.F. Savignyv, W.v. Humboldt, G.W.F. Hegel* (Leipzig: Reclam, 1990) and Ludwig Fertig, ed., *Bildungsgang und Lebensplan: Briefe über Erziehung von 1750 bis 1900* (Darmstadt: Wissenschaftliche Buchgesellschaft, 1991).

9. It should be pointed out that by *Einsamkeit* von Humboldt certainly did mean individual *intellectual isolation*. On the contrary, *Wissenschaft* = the never-ending search for truth, was indeed seen by both Humboldt brothers as a common enterprise of the republic of scholars/students. The claim for *Einsamkeit* entailed the right to devote oneself to scholarly work without any intervention from external forces.

10. Apart from the argument of a superior "critical mass," the decision to locate the new University in the capital also reflected the central strategic position of the University in the nation state, Rüdiger von Bruch, "Die Gründung der Berliner Universität," in Schwinges, 2001, p. 59. This pattern was soon to be followed in other German and European states.

11. See for instance, Alexander von Humboldt, "Einleitende Betrachtungen über die Verschiedenartigkeit des Naturgenusses," *Kosmos, Bd1.* (1845), pp. 3–40; Alexander von Humboldt, *Kosmische Naturbetrachtung,* ed. Rudolph Zaunick (Stuttgart: Alfred Körner, 1958), p. 344. For a brilliant condensed introduction to Alexander von Humboldt, see Wolf Lepenies, "Alexander von Humboldt: His Past and His Present." Keynote Speech, Jahrestagung der Alexander von Humboldt-Stiftung, May 31, 1999.

12. Heinz Steinberg, *Wilhelm von Humboldt, Preußische Köpfe, 32* (Berlin: Stapp, 2001), p. 81.

13. The actual Humboldt heritage becomes even more ironic and dubious when the Professor of Pedagogy Eduard Spranger in *Über das Wesen der Universität* (Lepzig: Dürr'schen Buchhandlung, 1910)—one of the many "Gedenkschriften/Reden" published at the centennial anniver-

sary—boldly declares: "The great achievement of Wilhelm von Humboldt was that he was able to cog (Verzahnung) *Wissenschaft* and state together into an organic whole." (p. XLI), and even more so in the light of one of Humboldt's most frequently quoted statements: "The state must always be aware that it [. . .] is always a hindrance as soon as it becomes involved in things that would go so much better without it." (*Werke Band 4*, p. 257)—and one of those "things" was precisely the University/*Wissenschaft*. For an early classic study on the "Berlin-type" University, see Friedrich Paulsen, *Die deutschen Universitäten und das Universitätsstudium* (Berlin: Asher & Co., 1902).

14. From 1810, all German Gymnasium teachers had to get their degree from the university, see Peter Lundgreen in Ash, 1999, p. 148; see also Rudolf Vierhaus, "Wilhelm von Humboldt," in Wolfgang Treue and Karlfried Gründer, ed., *Berlinische Lebensbilder: Wissenschaftspolitik in Berlin. Minister, Beamte, Ratgeber* (Berlin, Colloquium Verlag, 2004), pp. 63–76 and Jürgen Mittelstrass, *Die unzeitgemässe Universität* (Frankfurt am Main: Suhrkamp, 1994), pp. 149–174.

15. Nachwort in Müller, ed., 1990, p. 306.

16. One interesting case of the lack of serious consequential, long-term analysis would be how Sweden from the 1970s gradually changed its secondary school system from a "German" into a U.S.-type high school system.

17. The modern statutes of the Uppsala University from 1852 could serve as a typical example; see Göran Blomquist, *Elfenbenstorn och statsskepp: Stat, universitet och akademisk frihet, 1820–1920* (Lund: Lund University Press, 1992).

18. Vierhaus in Treue and Gründer, 1987, p. 69.

19. Mittelstrass, 1994, pp. 22 and 43. Also Nachwort in Müller, 1990, pp. 294 and 306.

20. But the dream of an indivisible body of knowledge lived on and, accordingly, when the soon much envied German *Technische Hochschulen* were given the right to grant doctorates, in the second half of the 19th century, they did so only on the condition that they established chairs in philosophy and/or history "to secure their scientific quality." These chairs are, by the way, still with us today, see Lundgreen, in Ash, 1999, p. 157.

21. Henningsen in Neave et al., eds., 2006, p. 99, also Franz Schnabel, *Deutsche Geschichte im 19. Jahrhundert, 2* (Freiburg, 1964), p. 207.

22. Martina Baumgarten, *Professoren und Universitäten im 19. Jahrhundert. Zur Sozialgeschichte deutscher Geistes- und Naturwissenschaftler* (Göttingen: V&R, 1997).

23. Martina Baumgartner, Professoren- und Universitätsprofile im "Humboldt'schen Modell" 1810–1914, in Schwinges, 2001, pp. 105–129.

24. Mittelstrass 1994, p. 83. On the German Mandarin, see Fritz K. Ringer's interesting and controversial, *The Decline of the German Mandarins: The German Academic Community, 1890–1933* (Harvard University Press, Cambridge, MA, 1969) and the ensuing debate, Jürgen Habermas, "Die deutsche Mandarine," in Habermas, *Philosophisch-politische Profile* (Frankfurt am Main: Suhrkamp, 1971), pp. 239–251. Also Wolfgang J. Mommsen, *Bürgerliche Kultur und künstlerische Avantgarde: Kultur und Politik im deutschen Kaiserreich 1870–1918* (Frankfurt am Main: Suhrkamp, 1994).

25. To illustrate the self-understanding and the almost unbounded self-confidence of the German professoriate already in 1869 one can quote from a speech, "Über Universitätseinrichtungen," by the Rector of the Friedrich-Wilhelms-Universität, Emil Du Bois-Reymond: "It is reasonable to maintain that in the field of higher learning the German universities are superior to those of any other country. Indeed, given the fact that none of man's works is perfect, the German universities have such an organizational strength that they could only have been created by an act of the most fundamental legislative wisdom" (my trans.), in Bois-Reymond, *Biographie, Wissenschaft, Ansprachen*, 2. Folge (Leipzig, 1887), p. 337. As an illustration of the long-term international impact, one can quote Abraham Flexner's "self evident" introduction to the German chapter in his famous book *Universities American—English—German*, from 1930(!): "Of the countries dealt

with in this volume, Germany has in theory and practice come nearest to giving higher education its due position," p. 305.

26. The same question is applicable in the case of Ireland showing a staggering R&D growth and publication rates in the last couple ofdecades —or for that matter also the cases of Finland and the Netherlands, see Thora Margareta Bertilsson, *Researchers in Europe: A Scarce Resource?* (Athens: Muscipol Workshop, 2002-10-10–11).

27. Stephen Cole and Thomas J. Phelan, "The Scientific Productivity of Nations," in *Minerva* 1999:37, pp. 1–23.

28. Robert K. Merton, *Science, Technology and Society in Seventeenth Century England* [1970] (New York: Howard Fertig Press, 1938).

29. As a negating case as regards the continued relevance of Merton's thesis of religious connection, one could point to the contemporary German case, in which the Roman Catholic south is considered to be far more successful in science (e.g., the outcome of the recently carried through national "*Exzellenz-Initiative*" in which *seven* of the ten selected universities are located in Baden-Württtenberg and Bayern!).

30. Henningsen in Neave, Blückert, Nybom, 2006, pp. 94 and Hans-Ulrich Wehler, *Deutsche Gesellschaftsgeschichte. Vol. 1: 1700–1815* (München: C.H. Beck, 1987), pp. 405–485 and for the Swedish case, Jonsson, in Neave, Blückert, Nybom, 2006, pp. 51–60.

31. Cit. from, p. 473.

32. Wehler, 1987, pp. 405–485 and also Helmut Schelsky, *Einsamkeit und Freiheit: Idee und Gestalt der deutschen Universität und ihrer Reformer* (Reinbek, 1963).

33. Henningsen in Neave, Blückert, Nybom, 2006, p. 95

34. Joseph Ben-David, "Scientific Productivity and Academic Organisation in Nineteenth-Century Medicine," in *American Sociological Review* XXV (1960), pp. 828–843.

35. This was not only how Ben-David explained the lead of United States from the 1920s and forward but also how he explained the success of German universities in the 19th century: a federal structure that promoted competition, which in turn promoted adequate funding and innovation, Ben-David, "Rivalität und Kooperation. Wettwerbbedingungen an amerikanischen und deutschen Universitäten," in *Hochschulpolitische Information* 14 (1983), pp. 3–6.

36. Sylvia Paletschek, "Verbreitete sich 'ein Humbldt'sches Modell' an den deutschen Universitäten im 19. Jahrhundert?" in Schwinges, 2001, pp. 75–114.

37. Hans-Ulrich Wehler, *Deutsche Gesellschaftsgeschichte, 1849–1918. Vol. 3* (München: C.H. Beck, 1987). In this connection one should neither underestimate the constant impact of war and armament as an "ultimate" driving force also in the development of national higher education and research systems; for the U.S. case, see Roger Geiger, *Research and Relevant Knowledge: American Research Universities since World War II* (Oxford: Oxford University Press, 1993).

38. For a systematic discussion on the impact of the German university in different parts of Europe and overseas, see the contributions in Schwinges, 2001.

39. Talcott Parsons and Gerald Platt, *The American University* (Cambridge, MA: Harvard University Press, 1973), pp. 304–45 and Steven Muller, "Deutsche und amerikanische Universitäten im Zeitalter der Kalkulation," in Ash, 1999, p. 199.

40. Fritz K. Ringer, *Fields of Knowledge* (Cambridge: Cambridge University Press, 1992), pp. 95–108.

41. The central elements in Humboldt's educational thinking were presented in the "white paper" *Der Köningsberger und litauische Schulplan* from 1809, see Sven-Eric Liedman, *I skuggan av framtiden: Modernitetens idéhistoria* (Stockholm: Bonnier Alba, 1997), p. 227 and idem. "Hur formas en människa? Om bildning och utbildning," in A. Björnsson, M. Kylhammar, Å. Linderborg, eds., *Ord i rättan tid* (Stockholm: Carlssons, 2005), p. 217.

42. Henningsen in Neave, Blückert, Nybom, 2006, p. 101. For a uncompromising and negative evaluation of the impact of the "Humboldtian Bildung-Ideal," see Theodor Litt, *Das Bildungsideal der deutschen Klassik und die moderne Arbeitswelt* (Bonn, 1955).

43. Thomas Mann, *Betrachtungen eines Unpolitischen* (Berlin, 1918). For a penetrating discussion, see Wolf Lepenies, *The Seduction of Culture* (Princeton: Princeton University Press, 2006), also Henningsen, in Neave, Blückert, Nybom, 2006, p. 101.
44. For instance, when Albert Einstein was called to Berlin in 1913, he had no teaching obligations, and he was not the only one, see Vierhaus in Treue, Gründer, 1987, p. 73.
45. Bernhard vom Brocke, "Von der Wissenschaftsverwaltung zur Wissenschaftspolitik: Friedrich Althoff," in *Berichte zur Wissenschaftsgeschichte* 1 (1988), pp. 1–26.
46. Rüdiger von Bruch, Langsamer Abschied von Humboldt? Etappen deutscher Universitätsgeschichte 1810–1945, in Ash, 1999, p. 46.
47. Lundgreen in Ash, 1997, p. 157.
48. During the entire 19th century, Wilhelm von Humboldt was hardly a reference point, or even mentioned, in the University policy discussion. The Humboldt that indeed was often referred to was his brother *Alexander*, whose crucial importance regarding the development of sciences in Germany was frequently emphasized, see Paletschek in Schwinges, 2001, pp. 98–104. (It is in this connection perhaps significant to note that even if the two brothers have remained almost equally illustrious and constantly referred to, each epoch of German political history has crafted its very own Alexander—and sometimes (1949–1989) even more than one—while Wilhelm, on the other hand, seems to have always remained the unchangeable "neo-humanist genius and university-builder"! On "The many lives of (A) von Humboldt," see Nicolaas Rupke, in *Humboldtkosmos* 87 (2006), pp. 34–39.
49. On the KWG, see Rudolf Vierhaus and Bernhard vom Brocke, eds., *Forschung im Spannungsfeld von Politik und Gesellschaft: Geschichte der Kaiser-Wilhelm/Max-Planck-Gesellschaft* (Stuttgart, 1990).
50. See Vierhaus in Treue, Gründer, 1987, p. 72. On Althoff's central position in research and university policy-making, see Bernhard vom Brocke, Friedrich Althoff, in Treue, Gründer, 1987, pp. 195–214.
51. Paletschek in Schwinges, 2001, p. 103.
52. As a starting point, one could choose the above-mentioned *Gedenkschrift* by Eduard Spranger from 1910. In 1919 the Prussian *Kultusminister* Carl Heinrich Becker published *Gedanken zur Hochschulreform*. In 1946 the philosopher and university ideologue Karl Jaspers published his important *Die Idee der Universität*. In 1963 the sociologist and University reformer Helmut Schelsky published his equally seminal book *Einsamkeit und Freiheit*.
53. Bernal's influential book *The Social Function of Science*, appeared in London, 1939 and Vannevar Bush's equally important *Science—The Endless Frontier* in Washington, 1945.
54. Helmut Schelsky, who still in 1963 (*Einsamkeit und Freiheit*) had invoked and endorsed the Humboldt Legacy as the basis for future university reforms, was, only four years later in his *Festrede* at the bicentennial anniversary of Wilhelm von Humboldt's birth, inclined to warn against the tradition to make Humboldt into the eternal litmus test for higher education policy: "In our considerations on education (*Bildung*) we have elevated Humboldt to the rank of Church Father, and subsequently, every attempt or suggestion to change anything in what is held to be the founding elements in his University structure, is condemned as blasphemy," "Die Universitätsidee Wilhelm von Humboldts und die gegenwärtige Universitätsreform," in Schelsky, *Abschied von der Hochschulpolitik. Oder: Die Universität im Fadenkreuz des Versagens* (Bielefeld: Bertelsmann, 1969), p. 152. For a discussion, see Bartz in *die hochschule* 2 (2005), pp. 105–110. For the Humboldt "heritage" in the GDR, see John Connelly, "Humboldt im Staatsdienst. Ostdeutsche Universitäten 1945–1989," in Ash, 1999, pp. 80–104, and Wittrock, in Neave, Blückert, Nybom, 2006, pp. 119–123.
55. Kerr, 1991 and Kerr, *Troubled Times for American Higher Education: The 1990s and Beyond* (Albany: SUNY Press, 1994). Even if Clark Kerr sometimes has argued that U.S. universities also have changed mainly due to external pressures, I do, nevertheless, humbly maintain that the

North American research universities and central university actors have shown a relatively remarkable ability to act and reform. Not least Clark Kerr himself (the California Master Plan) must be considered to be an almost ideal typical example of this capacity. See also Martin Trow, "The Exceptionalism of American Higher Education," in Martin Trow and Thorsten Nybom, eds., *University and Society: Essays on the Social Role of Research and Higher Education* (London: Jessica Kingsley Publishers, 1991), pp. 156–172 and Morton and Phyllis Keller, *Making Harvard Modern: The Rise of the American University* (New York: Oxford University Press, 2000).

56. For a penetrating analysis of Cardinal Newman and his important *The Idea of a University* from 1852, see Rothblatt, 1997.

Fruit Nationalism

Horticulture in the United States—From the Revolution to the First Centennial

DANIEL J. KEVLES

*I*n the eighteenth century, the North American continent provided its native inhabitants and European colonizers with more than two hundred species of tree, bush, vine, and small fruits as well as at least fifty varieties of nuts. It was a "natural orchard," to use the phrase of U. P. Hedrick, long the leading historian of horticulture in the United States. But like the Native Americans, the European settlers neither improved nor even cultivated the indigenous small fruits—the black and red raspberries, blackberries, strawberries, grapes, plums, and crab apples. They grew abundantly in the wild, and the colonists were content to pick them from forest clearings and waterways, where they were readily available.[1]

From the first Spanish arrivals, European conquerors and settlers supplemented the native fruits by introducing stalwarts from the Old World—notably apple, pear, peach, and cherry trees—in the form of seeds and seedlings. The imported fruit trees often fell victim to American soils, climate, and insects. Thomas Jefferson attempted to develop orchards of diverse fruits at Monticello and nearby Colle, thinking that the region resembled a fruit-rich area of Italy, but early spring frost did in most of the trees. Pear trees were particularly notorious for failing, and in the early days of settlement were associated with aristocrats, because only those who owned opulent gardens could afford to plant and nurse them.[2] Still, farmers and orchardists sowed seeds and seedlings by the thousands, with the result that some hardy growths survived to be selected and nurtured, and to produce seed for new plantings. Encouraged through such successive selections, fruit trees proliferated in the colonies that lacked the distinction to merit a varietal name but that provided serviceable fruit and were adapted to the various local environments of settlement.[3]

The most common were apple trees, essential in the American farm economy. Apple orchards were mainly devoted to the production of cider and apple brandy— known as applejack—and peach orchards were primarily given over to the making of peach brandy. Even unappetizing fruits served well for such purposes, and the fruits could be preserved longer in fermented liquid form than in their original whole state. The fruit from a single apple tree would yield about six barrels of cider, some as many as seven. About one in ten farmers in New England had a cider mill. Farmers close to towns and cities sold their cider and brandy in urban markets. Farmers beyond reach of markets paid directly in cider the cobbler, tailor, lawyer, or doctor as well as their children's school board. They also regularly served cider to guests, and families, including children, daily consumed it at their tables. They used the excess production of their apple and peach orchards to feed hogs, permitting the animals to forage for fallen fruit. A southern farmer noted that "one acre of well-grown sweet apple trees will fatten more hogs than five acres of the average farm crops; and the expense to the farmer is nothing in comparison to the latter method of fattening."[4]

Surplus apples and peaches were also dried in slices in the kitchen, the sun, or the dry house, a common outbuilding. The dried slices remained edible for at least a year and were used in apple pies and apple dumplings, both daily staples of the American diet. Dried peaches and plums were kept for holidays and parties. Farmers also dried the peels and cores of apples, using them to brew a popular type of beer and to harvest the yeast that grew in the brewing process for baking bread.[5] American farmers thus valued their apples as essential parts of their domestic economy, but, since for most sugar was rarely to be had, they treasured their apples whether consumed in liquid or solid form as a convenient source of sweetness directly on the palate and in cooking.

Few fruits were available to urban dwellers. At the end of the War of 1812, there was hardly any fruit in city markets. When New York City organized a banquet to celebrate the Treaty of Ghent, no raisins were to be had for plum puddings nor could dried plums, cherries, or currants be purchased. The cooks had to rely on dried apples, which were difficult to get.[6] In general, when fruits were available in towns and cities they were a luxury. Citrus fruits came at a particularly high premium, especially in the north. They were provided by specialty suppliers who grew dwarf orange and lemon trees under glass. Fetching especially high prices in the winter off-season, they were fare for well-to-do Americans, some of whom raised them on their estates in heated greenhouses that they called orangeries, after the French, planting them in tubs that they put out in gardens during the summer.[7]

The principal reason for the urban scarcity was that the commercial market for fruits was limited to what could be reached by wagon or water and limited to the distance that could be covered before the fruits went bad. Few orchards, gardens, and hothouses operated within practical transportation range of cities and towns. Withal, at the end of the eighteenth century, most fruits, as with apples and peaches, were grown for local rather than market consumption.

Under the circumstances, there were few commercial incentives to provide fruit trees or to invest in the development of new varieties. Only a handful of nurseries existed in the new United States, most of them in the Northeast and operating on a total of perhaps a few hundred acres.[8] The largest and most important was the Prince Nursery, which had been founded in 1737 by Robert Prince at Flushing Landing, Long Island, New York, and which had been taken over by the time of the Revolution by his son, William. The Princes, who in 1771 published the first nursery catalogue in the colonies, offered a great variety of fruit trees, including apple, plum, peach, nectarine, cherry, apricot, pear, and pecan, and sold to farmers and orchardists in a number of colonies.[9]

Like nurserymen elsewhere in the colonies, the Princes offered few of these fruits in varieties of American origin. Most colonists lacked the knowledge and know-how to devise new fruits. Hardly any literature on the science or practice of fruit breeding—"pomology," in the technical term—was available except a handful of European works, almost all of them in foreign languages. Nor were there any American journals of horticulture to consult.[10] Still, American plantings yielded some new and favorable fruits worthy of naming—for example, the Seckel pear, which came from the farm of a Mr. Seckel, about four miles from Philadelphia, or the Tyson pear, which originated on the farm of one Jonathan Tyson, also near Philadelphia, both fruits appearing in the late eighteenth century.[11] Like these pears, most of the young nation's indigenous new varieties were the products of fortunate natural cross-pollinations that farmers chanced upon in their orchards and fields.

Apples told the general tale. Americans later celebrated John Chapman, commonly known as "Johnny Appleseed," who in the early nineteenth century populated the Ohio Valley and parts of the upper Midwest with apple trees by planting seeds, starting a nursery, then moving on to repeat the process. Chapman's elevation to mythic hero may be merited by the mystic and selfless nature of his character, but it is not by the consequence of his plantings. Apple trees grown from seed yield notoriously degenerate fruit, and most grown from Chapman's seeds were good only for cider.[12] Still, the thousands of apple-seed plantings, many of which grew up to be cross-pollinated with each other and with the common wild crab apple, generated a small but increasing number of trees with tasty fruits. Among them were the Baldwin, which appeared on a farm near Lowell, Massachusetts, in the 1740s, and was brought to wider notice by one Colonel Baldwin of Woburn; the Jonathan, which was found on a farm in Kingston, New York, and made known by Jonathan Hasbrouck to a local judge, who returned the favor with an eponymous compliment; and the Newton Pippin, which popped up in an orchard in Newton, on Long Island, New York, and which a later authority on American fruits called "the finest apple of our country, and probably of the world."[13]

New and desirable varieties could be multiplied by planting their seeds, but fruit from the products of such sexual reproduction tended, like John Chapman's apples, to be qualitatively impoverished compared with the parental fruit. By the late eighteenth century, orchardists were increasingly propagating valuable fruit trees by grafting scions or buds onto common stock, a method of asexual reproduction that yielded the same quality fruit on the new trees. Such trees were spread through their regions by sale or, in the case of gentlemen farmers, gift or exchange. For example, George Washington, an avid horticulturalist, obtained from friends in Virginia scions of new varieties of pear, apple, and plum trees that he grafted and propagated in his orchards.[14] Some of the new American varieties were even propagated to the Old World. Benjamin Franklin's British friends in and out of Court so liked the Green Newton apples—one of the first varieties native to and named in the New World—that he could hardly keep up with the demand for cuttings from the tree that the botanist John Bartram sent him for distribution.[15]

Yet for all the accidental innovations from the fields, at the time of the Revolution, most distinctive fruits of quality had come from abroad, especially England and France. Some—among them the Bartlett pear—rapidly became staples in American fruit baskets. That pear was first brought to the United States in the late eighteenth century under a different name. Cultivated on a farm in Dorchester, Massachusetts by Enoch Bartlett, it soon spread under his name via asexual reproduction throughout the colonies.[16] The imports were obtained by nurserymen— the Princes relied on them heavily—and by men of wealth with estates. Typical of them was Eleuthère-Irénée du Pont de Nemours, who practiced horticulture on his estate in Delaware and in 1805 introduced the European chestnut to the United States.[17] During the first years of the American republic, the young United States remained a colonial dependency in the arena of quality fruits.

§ § § §

During the first half of the nineteenth century, a kind of fruit nationalism developed among horticulturalists in the Northeastern United States. Centered in greater Boston, it was led by members of the area's mercantile and then incipient industrial elite. Their fortunes made or accumulating, they embraced farming as a commitment to republican virtue removed from mere money-making. They bought and cultivated large farms, turning some of them into opulently improved estates. To the end of agricultural improvement, they imported fine breeds of animals and plants from Europe, built vast greenhouses for fruits and flowers, planted huge orchards with scores of varieties of the same trees. An exemplar of the trend was John Lowell, a wealthy Boston lawyer and political writer who had retired to an estate that he called Bromley Vale, where he grew fruits, some in a greenhouse, and pursued scientific farming.[18]

At first gentlemen like Lowell declared themselves devoted to agricultural innovation for the sake of their region's economy, which was losing migrants to the rapidly developing Midwest. Later they became acolytes of ornament and display, pointing proudly to their luscious fruits and flowers (and fine breeds of cattle, too). They were connoisseurs, praising varieties, that, "though not of sufficient excellence for extensive cultivation," as one of their circle later commented, "were yet so good that a single tree should be in every large collection."[19] In 1792, they founded the Massachusetts Society for the Promotion of Agriculture, whose rules declared that "the members of the Society have no other interest than the benefit of the human species at large."[20] They felt that horticulture especially offered a devotion to beauty and a retreat from materialism. Throughout the period they regarded and ran their operations in a spirit of noblesse oblige, divorced from considerations of personal profit.

Their enthusiasm for horticulture was, however, shared by commercial nurserymen. They were eager to enlarge the abundance and diversity of American fruits not only for nationalist reasons but also for commercial ones. Some straddled the world of gentlemanly amateurs and profit-minded nurserymen. The historian Francis Parkman, who proudly offered ornamental plants for sale from his nursery in Jamaica Plain, Massachusetts, near Boston, had a foot in both camps.[21] So did Marshall P. Wilder, who was prospering in the dry goods business in Boston when, in 1832, following the death of his wife, he bought a thirteen-acre farm in nearby Dorchester, Massachusetts, that he named Hawthorne Grove. Wilder went on to make a small fortune selling wool and cotton as well as uniforms to the Union Army during the Civil War, but all the while he developed his Dorchester acreage into a showplace nursery, importing and growing hundreds of flowers, shrubs, and trees, including prize-winning camellias and 900 varieties of pears. He was a prominent participant in Boston's civic affairs, and his horticultural collection was used in 1839 to create the first plantings of the Boston Public Garden.[22]

A number of the era's leading nurserymen embraced horticultural service and science even as they attended to profit. William Prince's two sons divided and continued the family nursery business. One of them, also named William, called his portion the Linnaean Botanic Garden and in his catalogue for 1820 identified the plants and trees he offered for sale solely by their Latin names. Prince published his later catalogues in English, offering in the early 1840s more fruits in number and variety than all others. During the first third of the century, the proprietors of the largest nursery in New England were John Kenrick, who had founded it in 1790 and grown wealthy from the business, and his son William, who in 1833 published *The New American Orchardist*, valuable for the information it provided on the growing number of new fruits in the United States.[23]

Together the gentlemen farmers and nurserymen, joined by gardeners, and horticultural hobbyists, established societies devoted to horticulture, notably in New York, in 1818, in Pennsylvania, in 1828, and, in 1829, the Massachusetts Horticultural Society. By mid-century, thirty-six such societies had been founded, almost half of them in Massachusetts, New York, and Pennsylvania, but most of the rest in the Midwest and border states.[24] In 1848, several of the leaders in the eastern state societies, Marshall Wilder notable among them, initiated the establishment of what became the first national organization of fruitmen—the American Pomological Society, as it was formally named in 1852.[25] At regular meetings of the Massachusetts society, and no doubt others, members proudly exhibited their fruits, flowers, and vegetables, offered prizes for the best fruits, and shared seeds and scions with their fellow cultivators.[26]

The fruit nationalists benefited from the proceedings published by these societies as well as from a growing number of broad-gauged agricultural journals, notably the *American Farmer*, begun in 1819, the *New England Farmer*, in 1822, the *Genesee Farmer*, in 1831, and the *Albany Cultivator*, in 1835. New treatises on pomology were appearing in Europe and they found a receptive audience on the western side of the Atlantic. In 1817, William Coxe, a prosperous nurserymen in Burlington, New Jersey, published *A View of the Cultivation of Fruit Trees*, the first work devoted to pomology to appear in the United States. In the next twenty years it was followed by a spate of similar books, including works by a number of the period's leading figures in American horticulture.[27]

Significant contributions to these developments came from Charles M. Hovey, a Boston nurseryman. A native of the city, he had taken to horticulture as a teenager, keeping detailed notes on pears, apples, and peaches. In 1832, at age 22, he founded a nursery with his older brother, Phineas, on one acre in Cambridgeport, Massachusetts. They sold their products through Hovey and Company, a store on Cornhill Street in Boston that was just below the meeting hall of the Massachusetts Horticultural Society, which Charles joined, in 1833. Charles was an avid reader of horticultural books and articles from abroad. In 1835, the Hovey brothers founded *The American Gardener's Magazine and Register of Useful Discoveries and Improvements in Horticulture and Rural Affairs*. Renamed two years later *The Magazine of Horticulture* and edited thereafter by Charles alone, the magazine published contributions from the leading practitioners at the multiple levels of horticulture. It was the first national magazine in the subject as well as one of the most influential journals in the field for a half century.[28]

Hovey directly advanced American fruits in the field, notably with his Seedling Strawberry. A longtime collector of strawberry plants, he knew from experience that some of the finest specimens from abroad would die off in the New England environment, burned by summer heat or killed by winter cold. The difficulty of cultivating strawberries made the fruit a luxury in the market and compelled most Americans to harvest them laboriously from the wild. In the summer of 1832, Hovey turned to developing improved strawberries by crossing different varieties, aiming for the merits of hardiness, productivity, size, flavor, and beauty. The next year, he selected seed from promising plants, planted the seeds in February 1834, and, on scrutinizing the crop when it came into full bearing in the summer of 1836, plucked the Seedling Strawberry from among some thirty plants of note. Hovey recalled that in 1838, the runners from this single plant produced "one of the most remarkable crops of remarkable strawberries we ever saw." The fruits were very large, often 5.5 inches in circumference, with firm, shining red flesh and yielding to the palate "a most agreeable acid, and exceedingly delicious and high-flavored juice." Hovey offered his new strawberry for $5, then later for $2 per dozen plants, advertising them in his *Magazine of Horticulture*.[29]

Hovey may have reduced the price at least in part because the first purchasers of his Seedling found that it produced paltry fruit. Nicholas Longworth, a fruit grower in Cincinnatti, Ohio, learned from the son of a local—and successful— strawberry grower that strawberry plants fruited little if they were left to fertilize themselves. Different flowered varieties had to be planted in alternate rows to obtain good fruit. Longworth studied strawberry blossoms and concluded that strawberry plants have sex—that is, some, having mainly stamens, are male, and others, having mainly pistils, are female, and one must be fertilized by the other to produce vital fruit. Longworth published his finding in an essay, in 1842, thus initiating a fierce debate on the issue—Hedrick calls it the "Strawberry War"—among horticulturalists that persisted for at least two decades.[30]

Hovey, to whom Longworth sent his essay, waxed hot and cold on Longworth's theory, but eventually conceded its truth, noting in 1848 that if the plants were not set near a "staminate variety" for the completion of fertilization, "a very small crop will be the result." Hovey had arranged for the completion of fertilization by planting his Seedling Strawberries in rows that alternated with the Boston Pine variety, one of the descendants of the thirty plants among which he had found the Seedling, choosing it in a crop of 1841 for its "earliness, size, beauty, exquisite flavor, abundant product, and hardiness."[31] Between 1838 and 1854, his Seedling received the Massachusetts Horticultural Society's annual premium for strawberries at least twelve times. Strawberry growers around the country soon understood that the vines had to be planted in configurations that encouraged sexual fertilization, and Hovey's Seedling was so widely cultivated as to become the principal market and table strawberry for several decades.[32]

Nicholas Longworth was a man of expansive interests and accomplishments. A migrant from New Jersey, where his Loyalist family's property had been confiscated during the Revolution, he had arrived in Cincinnati, in 1804, at age twenty-two with a handsome wardrobe and little else. He became a lawyer and parlayed two copper whiskey stills—he accepted them from a client, an accused horse thief, in lieu of a fee—eventually into a fortune in Cincinnati real estate. He befriended artists, dressed, in a slovenly fashion, pinning reminders of the day's appointments to his clothing, and he gave generously and regularly to the indigent, insisting that he would provide charitable assistance only to "the idle, drunken, worthless vagabonds that nobody else will help." Like his wealthy counterparts in the Northeast, Longworth acquired an estate and devoted twenty acres of it to horticulture. A catalogue from his nursery in 1825 offered numerous varieties of fruits from trees and vines, but he took a special interest in wine grapes, partly with an eye to the gathering temperance movement. A number of wines at the time had an alcohol content as high as that in whiskey. Longworth thought to produce a wine whose alcohol content was much lower.[33]

For some two centuries would-be American vintners had been importing and attempting to grow European grapes, but the vines almost always died after a few years. Some growers turned to cultivating native grapes for wine, succeeding with a few, including the Scuppernong, found by the earliest settlers in Virginia and the Carolinas; the Alexander, the product of a natural cross-pollination between an imported and a native grape that James Alexander, a gardener for a son of William Penn, noticed around 1740 in the woods along the Schuylkill, near Philadelphia; and the Isabella, another chance foreign-native hybrid that had reportedly appeared in South Carolina and was brought to the attention of northern cultivators about 1818 by one Isabella Gibbs, in whose honor it was named.[34]

None, however, achieved Longworth's success with the Catawba grape, which was said to have originated as a chance native-foreign crossing in the Catawba Valley in North Carolina. Longworth had obtained the Catawba from John Adlum, who, encouraged by Thomas Jefferson, was an energetic promoter of American grapes and wines. In 1823, Adlum published *A Memoir on the Cultivation of the Vine in America and the Best Mode of Making Wine*, the first book to tout the production of wine from American grapes. He cultivated the Catawba grape in a vineyard he had established around 1816 in what is now Rock Creek Park, in Washington, D.C., having obtained the vine from a family in Western Maryland in 1819. By the early 1820s, the wine he produced from it was capturing widespread attention and was held to be superior to any other from native grapes. Adlum complained that he could not keep up with the demand for cuttings from his Catawba vines.[35]

Longworth, who introduced the grape to the Cincinnati area in the mid-1820s, contracted with tenant farmers, most of them recent German immigrants, to grow Catawba vines for him on south-facing acres that sloped upward from the Ohio River. The grape was very dry and spicily aromatic. Longworth at first produced a dry white Catawba wine for local consumption, mainly among the town's German-Americans. In the 1840s, however, he developed a sparkling white Catawba that soon impressed palates in Britain and Italy, gained considerable popularity in the United States, and prompted Henry Wadsworth Longfellow to compose a poem in tribute to it: "For richest and best/Is the wine of the West,/That grows by the Beautiful River;/Whose sweet perfume/Fills all the room/With a benison on the giver." By 1852, Longworth was producing 75,000 bottles of sparkling Catawba a year. At the time, many other proprietors were also cultivating wine grapes, mainly the Catawba, on some 1,200 acres in or near Cincinnati. The region was the nation's leading wine producer and people were calling the Ohio River "the Rhine of America."[36]

❧ ❧ ❧ ❧

Demand for fruiting vines and trees greatly expanded during the middle third of the century as Americans moved west into the Ohio and Mississippi valleys, then out to the Great Plains and the Far West of California, Oregon, and Washington. Land grants made under the Northwest Ordinance, which the U.S. Congress passed in 1787 to establish rules for settlement of the Northwest Territory (now the states of Ohio, Indiana, Illinois, Michigan, and Wisconsin) required that every homesteader plant at least fifty apple or pear trees. New means of transportation both accompanied and stimulated the westward migration.[37] The settlers established steamboat services on the major rivers and Great Lakes, built roads and canals beginning in the 1820s, and then railroads starting in the 1830s that linked their settlements to distant towns and cities across regions and, after the first transcontinental railroad was completed, in 1869, across the Continent.

The proliferating transportation networks vastly enlarged the reach of horticultural products and oriented fruit growers increasingly away from production for local consumption and towards production for market. The Long Island Railroad, which began service in 1836, enabled gardeners on the island to send to New York City in a few hours produce that had previously taken as many days to get there by boat. In June 1847, a single train on the Erie Railroad brought 80,000 bushels of strawberries to the city from the western areas of the state. With the opening of the Erie Canal in 1825, peaches grown in the Finger Lakes region that had previously gone to brandy now were sold as whole fruit in the numerous towns and cities that sprang up along the route. The new means of transportation similarly enabled peach growers in eastern Maryland to supply cities and towns in the region and growers along the eastern shore of Lake Michigan to provide peaches in abundance to the Midwest.[38]

The railroads also advantaged interregional sellers by, for example, bringing fruit to northern cities from the south, where it ripened weeks before northern crops were ready for harvest. In Illinois, strawberries were shipped from the southern part of the state in early May, when strawberry plants grown near Chicago were not yet even in bloom. Georgians in the American Pomological Society noted that railways brought "the fruits of the Southern sunshine to our doors almost before the frost has left the soil in our gardens" and thus extended the strawberry season in Chicago from a couple of weeks or so to almost a month and a half. The state could "supply large quantities of fruit to all the Northern markets, from six to eight weeks in advance of the Middle States."[39]

Fruit was often damaged or spoiled when transported over long distances. It tended to be shipped before it was mature so that it ripened in transit, giving off heat and gases. To get their fruit to distant markets in good condition, nurserymen called for well-ventilated freight cars mounted on decent springs. Beginning in the late 1860s, an increasing number of railway lines were equipped with such cars, and grapes and pears were said to be reaching the East Coast from California in "perfect condition."[40] By then, too, canning or large-scale drying allowed excess fruits that had been previously wasted to be preserved and sold. And the drying process, by removing the water in fruit, reduced the weight to be transported by three quarters or more. California growers now sent carloads of dried figs and grapes to Eastern markets.[41]

Reviewing such changes, Marshall Wilder noted with evident pleasure that while fruits had once been considered "as only a luxury for the opulent, they have now become not only a sanitary condiment, but a daily necessity of the meal." The onetime "aristocratic" pear tree was now "planted in the orchards of five or more thousands of a single variety" and is "enjoyed by the western pioneers as well as by the eastern magistrates." Now "the humblest cottager could sit in the shade of a grape arbor and pluck the fruit for his table." Not even Nicholas Longworth, Wilder reflected, could have predicted that manufacture of juice from the grape would in a single state exceed more than five million gallons a year.[42]

Americans had been compelled a generation earlier to harvest strawberries from the woods and fields, Wilder remembering them as the "small, dry, seedy, red and white, wood strawberries of our youth." Now luscious varieties were abundantly available and thousands of bushels of strawberries were sent to the northeastern cities from the Western States and New Jersey. In 1869, Norfolk, Virginia, a non-producer in the early 1850s, shipped three million quarts to northern markets. Thirty years earlier, Americans counted themselves fortunate if they could gather a small saucer full of raspberries, currants, or blackberries. Now farmers sent all three fruits to market by the hundreds or thousands of bushels.[43] The railroads not only brought fruits to urban markets earlier than local farmers might supply them. They also forced strawberry prices down dramatically, from one to two dollars per quart to ten or twelve cents a quart. To Wilder, it was clear that the huge numbers of baskets of fruits now being produced meant that "the time is fast approaching when all classes of society may enjoy this health-preserving condiment as a portion of their daily food."[44]

§ § § §

In good years, fruits brought high returns to their producers. Peach profits could run as high as $16,000 a season. In the 1850s, Wilder estimated that a single fine pear tree—for example, the White Doyenne in Western New York State—might produce $50 to $60 worth of fine fruit each year and as much as $3,750 over forty years. A well-connected nurseryman in western New York noted in 1853 that pears sold in Philadelphia were fetching prices "calculated to give an impetus to their culture. The Duchess d'Angouleme variety sold at Isaac Newton's Fruit and Ice Cream Store, on Chestnut Street, for one dollar each, and smaller specimens at seventy-five cents each .... Our correspondent says he immediately sat down and ordered pear trees for all the vacant spots in his garden. We only add that we think him a sensible man .... Bartlett pears have been selling on the New York markets at wholesale for $9.00 a barrel. One cultivator of this delicious fruit realized [a profit] at the rate of $9,200 per acre."[45]

All the while foreign demand for American fruits, both fresh and dried, increased steadily. In 1845, Boston exported some 10,000 barrels and in 1846, about 11,000. The English, who were said to have a liking for red apples, took from twelve to fifteen thousand barrels of the fruit each week. Wilder predicted that peaches and pears would find a huge market abroad once refrigerated ships were perfected. As it was, most of the crop went to London, where in some years they fetched as much as $21 a barrel, in contrast to the $6 a barrel they brought in New York City. Between 1872 and 1877, exports of fruits increased fivefold, reaching almost $3 million in value, with American producers shipping abroad in the latter year almost 400,000 barrels of apples.[46]

Americans responded to the rising domestic and foreign demand by producing fruit in greater quantities and kinds. Marshall Wilder noted in 1869 that two decades earlier trees and vines were purchased by the dozen or hundred. Now they were being bought by the thousands and nursery stocks numbered in the millions of trees and vines. In western New York during the 1840s, people started to speak of fruit kings—for example the "Apple King" Robert L. Pell, of Esopus, in Ulster County, whose orchard comprised 200 acres of Green Newtons and was touted as the world's largest apple orchard, with a crop in 1847 of 10,000 barrels. By 1855, the seven counties around the Finger Lakes and along the shores of Lake Erie were producing more than a half-million bushels of fruits from trees, mainly apples and peaches.[47]

Wilder found the progress of grape culture in the United States particularly exhilarating, declaring in 1867, "At last, the vine which has been so much neglected or persecuted, from fear of producing an intoxicating beverage, is becoming the great object of attraction." Grape vines were said now to cover thousands of hillsides from New York to Missouri and on out to California. According to an expert of Wilder's acquaintance, between the late 1850s and the late 1860s the number of acres devoted to the grape had multiplied from a few thousand to more than two million. In 1865, a region in Northern Ohio produced 279,000 gallons of wine, worth at wholesale prices between $500,000 and $600,000. Wilder took pleasure in the report that in a recent exposition in Paris the better wines of America compared favorably with foreign wines of similar character. He declared, "We are now more confident than ever, that America can and will be a great wine-producing country." Indeed, California was already exporting wines and brandies to Europe, which prompted Wilder to note that the "vintage of the golden western slope promises ere long to rival in value the riches of its mines."[48]

Together, the expanding market opportunities and returns for fruit growers stimulated an exploding demand for fruit trees and vines and in turn, from 1840 onward, vast growth in the nursery industry. In the mid-nineteenth century, according to a knowledgeable observer, between seventy and eighty nurseries were operating in the United States that were rightly celebrated for their abundant variety of plants, and many others were newly under way. The nurseries tended to be clustered in at least three major horticultural centers—greater Boston, Western New York, and, until disease killed off its vineyards in the 1860s, Cincinnati, all of them situated within practical reach of the rapidly growing transportation network.

In Massachusetts, some twenty nurseries issued catalogues, while nearly 200 comprised an acre or more of land.[49] In 1840, drawing on the profits from the Seedling Strawberry, the Hovey brothers enlarged their nursery, buying forty acres of woodland in Cambridge. They planted pear trees as well as other fruit trees and by 1848 had built four greenhouses, one of them the largest such structure in the country at 96 feet by 30 feet that they used chiefly for the growth of specimen plants.

The nursery's open specimen grounds included some 2,000 dwarf fruit trees. Its nursery stock comprised all told nearly 60,000 pear trees and thousands of young trees of other fruit, not to mention 20,000 geraniums and, according to Hovey, the "best collection in the country of Japan lilies [*Lilium speciosum*]." By the mid-1860s, the R.G. Dunn and Company, then emerging as one of the country's leading assessors of creditworthiness, rated Hovey as "rich and growing richer all the time."[50]

Nurseries had begun springing up in Western New York shortly after the completion of the Erie Canal. By the mid-to-late 1850s, Monroe County was home to 30 nurseries, including 18 in its principal city, Rochester. In 1856, the *Genesee Farmer* declared that "more nursery trees were grown in Monroe County than in all the United States besides." In 1854, the dollar total of the produce from Rochester nurseries alone amounted to more than a half million dollars.[51]

The flagship nursery in Rochester was Ellwanger & Barry, which was also one of the leading firms in the country. The firm had been established in 1840 as a partnership of George Ellwanger and Patrick Barry, both then age twenty-four—the former an enterprising native of Germany who had the year before bought out his employer's business on five acres, which he and Barry named The Mount Hope Botanical and Pomological Garden; the latter an immigrant from Belfast, Ireland, who had spent the previous four years with the Prince Nursery, in Flushing. Situated at the nexus of the Erie Canal, the Great Lakes, and railroads running east and west, the firm sold to a national market. It provided California and the rest of the Far West most of its cultivated trees until long after the Civil War. Many nurseries opened elsewhere were founded by graduates of Ellwanger & Barry's employ.[52]

By the 1850s, Mount Hope had expanded to 500 acres, where its proprietors cultivated thousands of fruit trees and vines as well as ornamental trees and shrubs, more than any other two nurseries combined. Its fruit stock included pears, plums, cherries, apples, peaches, quinces, currants, and more than one million grape vines. Patrick Barry, a prominent and respected leader among American fruit breeders, accumulated a comprehensive working library of the major pomological works in several languages, edited the *Genesee Farmer* from 1844 to 1852, and in 1851 published his own *Treatise on the Fruit Garden*, long one of the best books on pomology in the United States.[53]

After the Civil War, horticulture, especially the cultivation of citrus fruits, began taking strong hold in the South. Citrus fruits had been grown since colonial days in the semi-tropical South, having been introduced there by the Spanish, but southerners had been disadvantaged in developing their cultivation into an industry by the very limited miles of canals, turnpikes, and railroads that had marked the region before the Civil War. Then, too, as with all of horticulture, investment in citrus culture was dampened by the South's overall devotion to cotton and tobacco. Wild orange trees were cut down and burned over to clear land for cotton. It was estimated that not more than 75 acres, many of them on the Indian River, were planted in Florida for the commercial production of oranges.[54]

But in 1875 a report to the American Pomological Society announced that orange growing was beginning to emerge "as an American industry." Bereft of a slave labor force, an increasing number of land owners turned to the cultivation of oranges, drawn to the business also by the multiplying miles of roads and railroads in the region available to transport the fruit to local urban and northern markets. One tree might now yield fruit valued at $4. Between 1867 and 1875, wild orange land had risen 1,000 percent in value. Florida oranges were said to be absolutely without rival in American markets, selling in New York and Philadelphia among other places for one dollar a dozen, four times as much as oranges from the West Indies and the Mediterranean. In the mid-1870s, the income from easily accessible groves in full bearing came to about $1,000 a year. One could expect a net income of $500 per acre—100 trees to the acre, 500 oranges per tree, one dollar per hundred oranges net income above production costs.[55]

The national expansion of the nursery business generated increasing competition and encouraged investment in the creation of improved fruit varieties. Two methods were well known and debated by American pomologists. One, advanced by the English horticulturalist Thomas Andrew Knight, called for hybridization across different varieties. The other, championed by the Belgian pear breeder J. B. Van Mons, stressed the repeated selection of seeds from delicate specimens.[56] Hybridization, however, was laborious, and in the mid-nineteenth century American horticulturalists seemed to rely on seed selection from promising varieties that were often the product of natural hydbridization.

The combined pomological and financial success of the Seedling Strawberry evidently encouraged Hovey to try similar crossings and seed selection programs with other fruits, a venture that led to the Hovey Cherry, in 1855.[57] A similar program led Ephraim Bull, a nurseryman in Concord, Massachusetts, to produce the Concord grape by planting the seeds of a chance volunteer that he found in the corner of his garden, then after it fruited, in 1843—Bull thought the fruit resulted from fertilization by a Catawba—planting the seeds from the fruit and using Van Mons' rules to select the resulting plants for propagation. Bull's new variety ripened two weeks earlier than the earliest then known and yielded large clusters of delicious fruit, good for the table and for wine. It created a sensation in the horticultural world after it was first exhibited at the Massachusetts Horticultural Society, in 1853. Hovey marketed Bull's grape nationally for the premium price of $5 a cutting.[58]

Hovey's Seedling Strawberry was the foundation of his fortune, but he knew that he had been lucky in obtaining such a successful new variety from only two seasons of seed selection. In 1837, the year before the Seedling appeared, he had estimated that "the chance of raising a very superior fruit may be considered as one to five hundred."[59] The low odds may have been the reason seventeen years passed before Hovey devised his new cherry. The breeding of new varieties of trees was even more challenging. Whether done by seed selection or hybridization, many seedlings had to be cultivated in the hope that at least one of them would produce fruit with desirable qualities. Then, too, while vinous plants such as the strawberry and grape would fruit within a year or two, fruit did not appear on trees for at least several years—in the case of apple trees, for as many as seven to nine years. The breeding of new fruit-tree varieties thus required both sizable experimental orchards and a taxing combination of resources and patience.

Hovey, with his extensive nursery in Cambridgeport, had plenty of both. He bred a number of new varieties of pear trees. Elllwanger & Barry, in Rochester, possessed experimental capacities comparable to Hovey's. At a pomological gathering in Philadelphia, in 1858, the firm exhibited 233 kinds of pears, 80 varieties of apples, and 50 types of plums.[60] But few other nurseries cultivated experimental grounds of comparable size.

Under the circumstances, the principal sources of new fruit varieties, especially from trees, continued to be the result of favorable cross-pollinations in the fields and numerous imports. The importation of plants and trees had long been practiced by officials of the U.S. government, especially naval officers, whose duties took them to foreign ports. In 1838, Commander Charles Wilkes sailed for the Pacific Ocean, instructed by the Navy to obtain new plants, and in 1839 the U.S. Patent Office was given funds to collect and distribute seeds and plants from abroad. Gentlemen farmers obtained huge numbers of trees from Europe, and so did nurseries like Hovey and Co. and Ellwanger & Barry.[61]

The fruit nationalists took increasing satisfaction in what they were achieving. In 1856, Hovey published *The Fruits of America*, which helped pioneer in the United States the presentation of fruits in the form of chromolithographs, a recently developed technique of color illustration. Hovey's book included one hundred ten plates, each handsomely depicting the fruit, its stem, and its leaves. Hovey declared that he felt "a national pride" in "beautifully" portraying the "delicious fruits in our own country, many of them surpassed by none of foreign growth" and thus demonstrating the rapidly developing "skill of our Pomologists" to the "cultivators of the world."[62]

As the nation approached its bicentennial, in 1876, the American Pomological Society's members came from forty-three states, including California, Oregon, and Washington, an indicator that pomology was becoming a national occupation. Their skills increasingly included artificial hybridization, a strategy for pomology whose advantages Wilder persistently emphasized in his presidential addresses.[63] Natural or artificial crossings between European and native grapes had yielded several additional new varieties, including the Diana, Hartford Prolific, Creveling, and Delaware grapes. Wilder enthusiastically predicted that "we may expect the appearance, every year, of many new varieties, from which we shall be able to select those possessing every desirable quality."[64] He predicted the same for pears and strawberries, new varieties of which many cultivators were creating by selections from seedlings.[65]

Exhibits at the biannual meetings of the Pomological Society comprised as many as 6,000 dishes of fruits, and prizes were now offered for exotics such as figs grown in the open air. When at the opening of the 1870s the Society published its biannual Catalogue of Fruits, more than half the apples, peaches, plums, quince, raspberries, and strawberries were identified as of native origin and so were almost half the grapes. Of a total 151 varieties of fruits in the Catalogue for 1873, all but 37 were of American origin. Marshall Wilder happily declared, "Formerly we were obliged to rely mostly on imported kinds for our best fruits, but as time progresses these are gradually disappearing, and their places are being filled by those of American origins."[66]

Notes

1. Ulysses P. Hedrick, *Horticulture in America* (New York: Oxford University Press, 1950), pp. 4, 5, 29; Philip J. Pauly, *Fruits and Plains: The Horticultural Transformation of America* (Cambridge: Harvard University Press, 2008), pp. 13, 54. I acknowledge with thanks the assistance in the research for this paper of Karin Matchett and Jenny Smith, the support of the Andrew W. Mellon Foundation, and the help of Maureen Horn at the Library of the Massachusetts Horticultural Society as well as that of Stephanie Clark and her staff at the Society's archival depository.
2. Marshall P. Wilder, "President's Address," *American Pomological Society Proceedings* (Hereafter, *APS Proc.*), 1869, pp. 23–24; Pauly, *Fruits and Plains*, pp. 19, 24–26, 69.
3. Hedrick, *Horticulture in America*, pp. 19, 30–31, 34, 155.
4. Michael Pollan, *The Botany of Desire: A Plant's-Eye View of the World* (New York: Random House, 2001), pp. 9–13, 22; Hedrick, *Horticulture in America*, pp. 32–33, 38–39, 59, 61, 105, 184.
5. Hedrick, *Horticulture in America*, p. 60.
6. Ibid., pp. 148–49, 241; M. L. Dunlap, "On Packing and Marketing Fruit," *APS Proc.* (1867), pp. 72–73.
7. Hedrick, *Horticulture in America*, pp. 90–91, 221, 223, 265.
8. H.A.S. Dearborn, "Character, History and Culture of the Pear," *Transactions of the Massachusetts Horticultural Society*, Vol. I, (1847), p. 25; Marshall P. Wilder, "President's Address," *APS Proc.* (1869), p. 28
9. Hedrick, *Horticulture in America*, pp. 71, 163; a photo of the catalogue is reproduced between pp. 48 and 49.
10. Wilder, "President's Address," *APS Proc.* (1877), pp. 22–23.
11. Hedrick, *Horticulture in America*, pp. 40, 83, 153–57; A. J. Downing, *The Fruits and Fruit Trees of America* (New York: Wiley and Putnam, 1846), p. 415; Charles M. Hovey, *The Fruits of America, Containing Richly Colored Figures, and Descriptions of all the Choicest Varieties Cultivated in the United States* (New York: D. Appleton & Co., 1853), p. 33.
12. Hedrick, *Horticulture in America*, p. 311; Downing, *The Fruits and Fruit Trees of America*, pp. 113–14.
13. Pollan, *The Botany of Desire*, pp. 9–10; Hovey, *The Fruits of America*, p. 11; Downing, *The Fruits and Fruit Trees of America*, p. 118; Hedrick, *Horticulture in America*, p. 57; Pauly, *Fruits and Plains*, p. 67.
14. Hedrick, *Horticulture in America*, p. 171.
15. Ibid., pp. 40, 83, 153–57.
16. Marshall P. Wilder, "President's Address," *APS Proc.* (1858), pp. 15–18; "President's Address," *APS Proc.* (1873), p. 21; Downing, *Fruits and Fruit Trees of America*, p. 334. In England, the pear was originally Williams' Bonchretien.
17. Hedrick, *Horticulture in America*, pp. 145–47.

18. See Tamara Plakins Thornton, *Cultivating Gentlemen: The Meaning of Country Life among the Boston Elite, 1785–1860* (New Haven: Yale University Press, 1989), passim; B. June Hutchinson, "A Taste for Horticulture," *Arnoldia*, 40 (Jan./Feb. 1980), p. 37.

19. Wilder, "President's Address," *APS Proc.* (1867), p. 45.

20. Thornton, *Cultivating Gentlemen*, p. 67.

21. Francis Parkman, *Hardy Ornamental Plants*, [1861?], Catalogue Collection, Massachusetts Horticultural Society, Wellesley, Mass., Box 592.

22. Richard Heath, "From Pear Orchard to Multi-Family Housing: A History of the Marshall P. Wilder Estate," *Dorchester Athenaeum*, Nov. 20, 2003, accessed at http://www.dorchesterathenaeum.org/page.php?id=603, Aug. 4, 2007.

23. Catalogue of American Indigenous Trees, Plants, and Seeds Cultivated and for Sale at the Linnaean Botanic Garden, William Prince, Proprietor, 1820, Catalogue Collection, Massachusetts Horticultural Society, Box 621; Hedrick, *Horticulture in America*, pp. 208–10.

24. [H. A. S. Dearborn], "Historical Sketch," *Transactions of the Massachusetts Horticultural Society*, vol. I (1847–1851), p. 84; Hedrick, *Horticulture in America*, p. 146. The sixteen founders of the Massachusetts Horticultural Society included both gentlemen farmers such as Lowell and nurserymen such as William Kenrick and Robert Manning, of Salem, one of the leading pomologists of the time. Hutchinson, "A Taste for Horticulture," p. 35, n. 3. In establishing these societies, horticulturalists in America were emulating Englishmen, who had founded the Horticultural Society of London, in 1804. Pauly, *Fruits and Plains*, p. 53.

25. The Society was formed from two groups, the North American Pomological Convention and The American Congress of Fruit-Growers, both founded in 1848. In 1849, the two agreed to consolidate as the American Pomological Congress, which in 1852 renamed itself, agreed to meet biannually, and adopted a constitution and by-laws. Wilder, "President's Address," *APS Proc.* (1867), pp. 29–32.

26. [H. A. S. Dearborn], "Historical Sketch," *Transactions of the Massachusetts Horticultural Society*, vol. I (1847–1851), p. 84. In 1851, the Massachusetts Horticultural Society offered $750 in prizes ranging from $30 to $60 each. "Preface to the Third Number," *Transactions of the Massachusetts Horticultural Society*, vol. I (1847–1851), pp. iii–iv. The mixture of amateurs and commercial nurserymen in these societies is evident from the list of people who exhibited fruits at the annual meetings. See "Report of Committee on Fruits Exhibited," *APS Proc.* (1860), pp. 112–14. Through the auspices of the American Pomological Society, various other state societies, nurseries, and individuals offered medals and/or cash prizes ranging from ten dollars to $150 for prize fruits, including apples, pears, peaches, grapes, figs, and pomegranates and also for the best bottles of wines made from particular grapes. See "Premiums," *APS Proc.* (1871), pp. 6–7.

27. William Coxe, *A view of the cultivation of fruit trees, and the management of orchards and cider; with accurate descriptions of the most estimable varieties of native and foreign apples, pears, peaches, plums, and cherries, cultivated in the middle states of America* (Philadelphia: M. Carey & Son, 1817). For other works as well as a broad introduction to the horticultural literature of the period, see Hedrick, *Horticulture in America*, pp. 467–498.

28. The magazine ceased publication in 1868, its cumulative pages comprising what the Cornell plant scientist Liberty Hyde Bailey later described as a record of "the budding stage of New World Horticulture." Its name, design, and layout were originally borrowed from *The Gardener's Magazine*, which was edited by J. C. Loudon and was probably the first such journal in England. Hutchinson, "A Taste for Horticulture," pp. 31–37 and p. 36, n. 7; Pauly, *Fruits and Plains*, p. 59.

29. Hutchinson, "A Taste for Horticulture," p. 41; Hovey, *The Fruits of America*, pp. 25–28; Pauly, *Fruits and Plains*, pp. 70–72.

30. Hedrick, *Horticulture in America*, p. 312.

31. Pauly, *Fruits, and Plains*, pp. 71–72; *Fruits of America*, pp. 25–28.

32. Hutchinson, "A Taste for Horticulture," p. 41.

33. Charles Boewe, "Nicholas Longworth," John A. Garraty and Mark C. Carnes, eds., *American National Biography* (Oxford University Press, 1999), vol. 13, pp. 898–99; Thomas Pinney, *A History of Wine in America: From the Beginnings to Prohibition* (Berkeley: University of California Press, 1989), p. 158.

34. Pinney, *A History of Wine in America*, pp. 7, 84–85, 128–29, 148; Downing, *The Fruits and Fruit Trees of America*, p. 255.

35. Hedrick, *Horticulture in America*, p. 124; Pauly, *Fruits and Plains*, pp. 12, 22, 26, 74. Pinney, *A History of Wine in America*, pp. 139–49. Adlum was for a time regarded as the country's leading authority on wine grapes and wine making, but he fell into poverty after 1830 and obscurity after his death, in 1836. According to one account, the Catawba was discovered in Buncombe County, North Carolina, in 1802. Robert Buchanan, *The Culture of the Grape and Wine-Making . . .* 5th ed. (Cincinnati: Moore, Wilstach, Keys, & Co., 1855), p. 71.

36. Boewe, "Nicholas Longworth," p. 898; Pinney, *A History of Wine in America*, pp. 160–65; Hedrick, *Horticulture in America*, pp. 208–9, 311. The full text of Henry Wadsworth Longfellow, *Catawba Wine*, is available at "Archive of Classical Poems," http://www.everypoet.com/Archive/ Poetry/Henry_Wadsworth_Longfellow/longfellow_birds_of_passage_catawba_wine.htm, accessed Dec. 8, 2007. Cincinnati's leadership in wine was shortlived. In the 1850s, mildew and black rot began to destroy the vines and by 1870 the vineyards were moribund. Pinney, *A History of Wine in America*, pp. 171–72.

37. Pollan, *The Botany of Desire*, p. 16.

38. Hedrick, *Horticulture in America*, pp. 232–33, 241.

39. M. L. Dunlap, "On Packing and Marketing Fruit," *APS Proc.* (1867), pp. 72–73; "Report from the State of Georgia," *APS Proc.* (1860), p. 183.

40. M. L. Dunlap, "On Packing and Marketing Fruit," *APS Proc.* (1867), pp. 73–74; Wilder, "President's. Address," *APS Proc.* (1869), p. 28; John H. White, Jr., *The American Railroad Freight Car: From the Wood-Car Era to the Coming of Steel* (Baltimore: The Johns Hopkins University Press, 1993), pp. 290–301.

41. Wilder, "President's Address," *APS Proc.* (1877), p. 20.

42. Wilder, "President's Address," *APS Proc.* (1869), pp. 16–17, 23; (1868), p. 25.

43. Wilder, "President's Address," *APS Proc.* (1869), pp. 25–26.

44. Hedrick, *Horticulture in America*, p. 241; Wilder, "President's Address," *APS Proc.* (1871), p. 27.

45. Wilder, "President's Address," *APS Proc.* (1858), p. 20; Hedrick, *Horticulture in America*, pp. 232–37.

46. "Preface," *Transactions of the Massachusetts Horticultural Society*, vol. I (1847), pp. vii–viii; Wilder, "President's Address," *APS Proc.* (1877), p. 20; (1858), p. 20; Hedrick, *Horticulture in America*, pp. 239–40.

47. Wilder, "President's Address," *APS Proc.* (1869), pp. 16–17; Hedrick, *Horticulture in America*, pp. 230–40. Wilder noted, "We hear no more of varieties which, though not of sufficient excellence for extensive cultivation, were yet so good that a single tree should be in every large collection." Wilder, "President's Address," *APS Proc.* (1867), p. 45.

48. Wilder, "President's Address," *APS Proc.* (1867), pp. 41–43.

49. [H. A. S. Dearborn], "Historical Sketch," *Transactions of the Massachusetts Horticultural Society*, vol. I (1847–1851), pp. 84–85; Hedrick, *Horticulture in America*, p. 208, 230. Wilder noted in 1858 that less than forty years earlier not a nursery in New England was entitled to the name but that now some "contain nearly as great a variety of fruit trees as the most extensive in England and France." Wilder, "President's Address," *APS Proc.* (1858), p. 20.

50. Hutchinson, "A Taste for Horticulture," pp. 39, 45; Pauly, *Fruits and Plains*, p. 59.

51. The first significant nursery in Western New York was established by Lewis F. Allen, in 1827, on Grand Island, in the Niagara River near Buffalo. Hedrick, *Horticulture in America*, pp. 238–39, 245.

52. Hedrick, *Horticulture in America*, pp. 243–45.

53. Ibid.

54. Ibid., pp. 119–20, 277–78; P. P. Bishop, "Orange Growing in Florida," *APS Proc.* (1875), pp. 50–51.

55. P. P. Bishop, "Orange Growing in Florida," *APS Proc.* (1875), pp. 48–51.

56. Pauly, *Fruits and Plains*, pp. 65–68.

57. Hutchinson, "A Taste for Horticulture," pp. 40–42.

58. Wilder, "President's Address," *APS Proc.* (1867), pp. 43–44; Albert Emerson Benson, *History of the Massachusetts Horticultural Society* (Norwood, Massachusetts: Plimpton Press, 1929), pp. 107–108; Pauly, *Fruits and Plains*, pp. 74–76.

59. Hutchinson, "A Taste for Horticulture," p. 40.

60. Ibid., p. 42; *APS Proc.* (1858), pp. 227–28. Other pear varieties worthy of note originated with other nurserymen in greater Boston—for example, the Dix, the Dearborn, and Clapp's Favorite—but it is unclear whether these pears were the products of breeding or of chance finds in the fields. Wilder, "President's Address," *APS Proc.* (1873), p. 21.

61. H. A. S. Dearborn, "Character, History and Culture of the Pear," *Transactions of the Massachusetts Horticultural Society*, vol. I (1847), pp. 25–26; Hutchinson, "A Taste for Horticulture," p. 39; Pauly, *Fruits and Plains*, pp. 112–13; Jack R. Kloppenburg, Jr., *First the Seed: The Political Economy of Plant Biotechnology*, 2nd ed. (Madison: University of Wisconsin Press, 2004), pp. 50–57. At mid-century, Charles Hovey rightly attributed the "rapid advancement" of the nation's pomology to the "accession of numerous foreign varieties of fruits" and the "introduction to notice of an immense number of native seedlings." Hovey, *The Fruits of America*, "Preface," p. v. The encouragement of plant importation continued through the nineteenth and twentieth centuries. See David Fairchild, *The World was my Garden: Travels of a Plant Explorer* (London: C. Scribner's Sons, 1938); Nelson Klose, *America's Crop Heritage: The History of Foreign Plant Introduction by the Federal Government* (Ames: Iowa State College Press, 1950); Ken Kraft, *Garden to Order* (New York: Doubleday & Company, 1962); John Fisher, *The Origins of Garden Plants* (London: Constable, 1982).

62. Charles M. Hovey, *The Fruits of America, Containing Richly Colored Figures, and Descriptions of all the Choicest Varieties Cultivated in the United States* (New York: D. Appleton & Co., 1856). The quotations are from the "Preface," pp. v–vi, to the first edition of the book, 1853, which contained forty-eight colored plates. At horticultural meetings exhibits of pears alone included hundreds of varieties. *APS Proc.* (1858), pp. 227–28; (1860), pp. 112–14. According to Hedrick, "The three decades from 1830–1860 showed greater progress in horticulture in America than in all the time before." Hedrick, *Horticulture in America to 1860*, p. 230.

63. Wilder, "President's Address," *APS Proc.* (1867), p. 32.

64. Ibid., pp. 41, 43–44.

65. Wilder, "President's Address," *APS Proc.* (1873), p. 26; Wilder, "President's Address," *APS Proc.* (1867), p. 32.

66. *APS Proc.* (1871), pp. 9–10, 15. Fifty percent or more of the apricots, cherries, currants, nectarines on exhibit were of foreign origin. See the "Catalogue of Fruits," *APS Proc.* (1871), pp. V–XXVII; Wilder, "President's Address," *APS Proc.* (1873), pp. 8, 16, 26. The meaning of "native" was ambiguous, encompassing both plants that were indigenous to North America or that were crosses between indigenous and imported varieties. Either way, however, they originated in the United States. For a discussion of this issue, see Pauly, *Fruits and Plains*, pp. 52, 64–65.

Jules Verne—The Divided Modernist

JEAN-FRANÇOIS BATTAIL

*J*ules Verne (1828–1905) is one of world literature's most read and most translated authors, and, according to UNESCO's annual *Index Translatorum*, his international influence has remained as widespread and powerful as ever in our post-modern times. He has been accorded the distinction of being the prophet of science and technology, an apologist for progress, one who creates exciting adventures to convey his didactic message. We have often been staring at everything associated with the advance of science and technology—at the aspirations and dreams of the West, along with its visions and predictions of the future—in a somewhat one-sided fashion. The question is, whether this age-related perspective provides adequate explanation of the fascination Verne has exerted, and still exerts, over his readers. There must exist something else behind the author's rare ability to withstand the forces of time and oblivion.

His calling as a writer of fiction was genuine beyond a shadow of a doubt. Growing up in an extremely middle-class environment, he had a hard struggle to escape the destiny his family had preordained for him, namely to become a lawyer and succeed his father as notary in Nantes. The escape took its time. However, he completed his law studies, if with reluctance, and supported himself for a number of years as a stockbroker; even then his literary pen was his most important tool: he wrote from 5 to 10 o'clock every morning, until he achieved success at the age of 35 with *Five Weeks in a Balloon* (1863). From that time on, he was able to devote himself entirely to writing. The success of his first publication ensured his future but also sealed his fate. Hetzel, his publisher and mentor, himself a novelist (under the pseudonym of P. J. Stahl), was a demanding employer, offering Verne a totally binding, if lucrative, contract, and keeping his star author on a tight rein. He was expected to write two novels a year, which from 1864 would first appear as serials in *Magasin d'éducation et de récréation*, a publication intended for the education of youth. The aim was to extol the achievements of science and give expression to a secularized bourgeois morality with courage, health and daring as cardinal virtues. Verne clearly struck the right chord, perhaps a little too well. The reading public could scarcely see the difference between the conventions of the genre and Verne's own philosophy of life. He was imprisoned, apparently non-controversially, in the ghetto of youth literature, which only provided meager symbolic capital. However popular he became, he suffered from the way he was branded. If he achieved recognition as an author, it was mainly as one who described the wonders of science, saw into the future and predicted technical innovations. Even if the French Academy did award him a prize, there was never any question of his being elected to that exclusive assembly.

It is worth noting, however, that he aroused interest and admiration among his fellow-authors—George Sand, Rimbaud, Mallarmé, Strindberg and Gorky, among others. Would it have been possible had his work not displayed specific literary merit? Range and ambition in his novels call to mind Balzac's in some way. While the latter strove to portray every facet of society ("the human comedy"), Verne aimed to chart every kind of achievement, describe the whole of the known world—and even extend its frontiers in searching for the unknown. But he was never canonized as Balzac was. Rather, he is compared with Alexandre Dumas as a skilled craftsman and entertainer—the one had plundered history, the other science and technology, and both were seen as moneymakers who sold a little too well to be taken really seriously. On top of that, Verne's original texts were not shown much respect—a large number of adaptations, retellings and slimmed-down versions have been in circulation. As in the case of all extremely productive writers (over a hundred titles), he naturally left behind a mixed bag of sweets, including some really poor works, but the best he produced deserve more than a mere shrug. Yet that is not what my French teacher thought, as I well remember from my school years in Paris. Undiplomatically, I chose Jules Verne when we were instructed to write an essay on "My Favorite Author," and the comment was along the lines: Why choose a mediocre writer when we have so many brilliant authors in our country?

After a great deal of wandering in the wilderness, Verne was at long last rediscovered, around 1960, by the French avant-garde, which in its turn stimulated academic research. Little by little new perspectives were constructed from different viewpoints—from those of literary studies, the history of ideas, psychoanalysis—at the same time as what had once been Verne's distinctive characteristic, viz. his scientific and technological visions, conspicuously paled and occasionally disappeared altogether from view. The pendulum had swung, for both good and ill. Some of the results of this research led to interesting reinterpretations, but these quite often turned out to be just as reductionist as the traditional clichés bandied about before.

It is all the more gratifying to come across studies that recognize the complexity of the writing—a body of work with many depths, a force-field where apparently conflicting elements create tension. Without doubt one of these is the study Tore

Frängsmyr devoted to the author of *The Mysterious Island* in *Framsteg eller förfall: Framtidsbilder och utopier i västerländsk tanketradition* [Progress or Decline: Visions of the future and utopias in western thought] (Stockholm, 1981). Verne is the subject of Chapter 6, "Naturvetenskapen som idealbild" [Natural Science as Ideal Image], which also contains a discussion of "The New Enlightenment" (science as an attitude to life in connection with the Industrial Revolution), "The Positive Stage" (Auguste Comte as philosopher of science, moralist and utopist), "Progress as Law of Nature" (Comte again, together with Hegel, Darwin, Buckle and others), and "The Evolution of Society" (with Spencer and Thomas Huxley heading the list). Within this extensive history-of-ideas framework Verne is accorded an analysis covering 20 or so pages, which, despite this limited amount of space, makes an amazingly large number of points. Tore Frängsmyr allows that "the idea of Jules Verne as the apostle of modern technology is not ill-conceived [. . .] He combines adventure and technology in a marvelous way in his novels," but he adds that "a description of this sort does not do Jules Verne justice." He plays down the importance of the technological fantasies on which his reputation is founded—technology is some kind of game to him, with a frequent streak of parody—and confirms that the best-seller is at bottom a moralist and critic of civilization with pessimistic undertones. Verne's biographers have asserted that he became more and more retiring and melancholic as the years went by, especially after 1886, but Tore Frängsmyr shows convincingly that pessimistic traits can be discerned even in his very first novel, *Five Weeks in a Balloon* (1863). I intend to take this re-evaluation, here drastically condensed, as my starting-point for a number of reflections on Verne's actual relationship to science and technology, as well as on his sincerely held view of life. And finally I shall sketch out a somewhat different perspective, beyond the pair of opposites progress/decline.

More than a hundred years after the author's death, it is legitimate to cast a critical eye over the traditional picture, still predominant, of Verne as the mouthpiece of science and the apostle of technology. In Sweden, for example, it is noteworthy that the only biographical reference in *Nationalencyclopedin* [The National Encyclopaedia] is *Jules Verne, Inventor of Science Fiction* (P. Costello, 1978), and that during recent years articles have been published with such significant titles as "The Creator of Science Fiction Predicted USA's Voyage to the Moon," "Jules Verne's Voyage to the Moon Almost Uncannily Correct," "Jules Verne Was Long Before His Time," "Technology Took 100 Years to Catch Up with Jules Verne: the Future Has Come True," etc. To begin with, one can object that such assessments are far too one-sided. Verne has also written a large number of stories, including superior works such as *Michael Strogoff* (*The Courier of the Czar*), in which science and technology are visible by their absence or play an insignificant role. Nor is it certain that the concept "science fiction" is especially relevant in characterizing his work. He sticks to the tried and tested laws of nature, with classical mechanics as his basis, even if at times he concocts totally impossible adventures, like *Hector Servadac* or the fantastic *Journey to the Centre of the Earth*. What cannot be doubted at all is that science and its applications were an important source of inspiration for him; he had the ambition of establishing a new genre, the scientific novel—in the same sense as one speaks of the historical novel. Even if science and technology were closely connected during the Industrial Revolution, theory and practice can be seen as separate, pure knowledge and might over matter. Verne was genuinely interested in the search for truth in itself, irrespective of any applications. Let us first look more closely at his relationship to science.

Jules Verne was a typical closet scholar, an indefatigable reader with a pen permanently in his hand. In Paris, he was a regular visitor to the Bibliothèque Nationale, and later he made full use of the Industry Society's extensive library in Amiens. Studying systematically, he acquired the scientific vocabulary he needed for his literary ventures. He made copious notes from popular science periodicals such as *La Nature, La Science illustrée, L'Astronomie,* and others. Merely to draft the details of *The Albatross* (the airplane in *Robur the Conqueror*), he collected data from over a hundred aeronautical journals. When he died, he left behind a huge card index comprising more than 20,000 items, organized in a complicated system.

He wasn't satisfied with simply summarizing numerous scientific articles. Sensibly enough, he also fell into the habit of consulting specialists around him before handing over his science-inspired narratives to the printer. And it has to be acknowledged that he managed well and on the whole achieved an accurate scientific discourse—even if, like so many others in those times, he went along with certain quasi-scientific theories, not least a naïve, ethnocentric anthropology that could easily degenerate into racism, anti-Semitism and even sexism (the female element rarely appears in his work, and when it does it is portrayed only in a highly conventional way). In the main, however, he provides accurate, well-documented descriptions of the encyclopaedic knowledge of his age. In its entirety, his work stands out as a handbook of handbooks, like a giant catalogue. Child of an age when people yearned for facts and reality, he gave himself the task of reflecting the achievements of science, often with the help of lists, classifications and enumerations. He embodies the dream of comprehensive, exhaustive knowledge, and his works give monumental expression to the imperialism of learning and the triumphant breakthrough of geography that characterize the nineteenth century.

In order not to drown in all the empirical diversity, he needed a theoretical framework such as he found in Auguste Comte's *Cours de philosophie positive* (I–IV, 1830–1842). The time was ripe for him to leave behind speculations about origins and purpose and devote himself to factual laws instead. Verne adopted Comte's classification of the sciences, from the most general and exact to the most concrete and individual: mathematics and mechanics, astronomy, physics, chemistry, biology, sociology. They portray therefore an ascending series whereby every successive science has the previous one as its basis and as its prerequisite. That part of Verne's work that has a didactic purpose can be regarded as a popular scientific representation of positivism, both as scientific theory and as ideology. Ever since Bacon and Descartes, the domination of nature has been the desired goal. But, according to Comte, human society is also part of nature, with conformity to laws existing there too. Hence the ambition to found a social physics with a view to ensuring both order and development by means of a technology for controlling society. Jules Verne is no stranger to such dreams: in a couple of stories he advocates a type of social engineering and outlines a number of socio-political programs, e.g. in *The 500 millions of the Begum*. Yet it isn't the dawning social sciences, which capture his interest more than anything else. It is, rather, at the so-called "hard sciences" that he directs his telescope, but he treats them differently, according to his own preferences.

There are, to be sure, many sciences present in his work—classical mechanics and triumphant chemistry as well as astronomy, ballistics, geodesy, geology, paleontology, crystallography, and so on. Thermodynamics as a background to Verne's portrayal of the development of the universe is particularly interesting. After Kant and Laplace, it is now Carnot and Clausius and their like who lead the way. Classical mechanics with the mechanism of the clock as its metaphor is no longer the dominant paradigm. A short story like "Master Zacharias, or the Clockmaker Who Lost His Soul" (1874), which, from a literary point of view, recalls E. T. A. Hoffmann, has the death of clockwork as its main theme, which can be metaphorically interpreted as the burial of classical learning—and the demise of the divine architect also in the bargain. Fire and furnace rule now, the universe having become a gigantic engine. The successful steam technology is daughter to the theory of heat. The machine needs no longer be fuelled from an outside source, be it horse, water or wind. And yet it has also been discovered that heat is not an indestructible substance. Work can be procured from both higher and lower temperatures, but this process is not reversible. The concept of entropy, a mathematically calculable state variable, triggered speculations about the world's thermal death, even if in a very distant future. This *Weltanschauung* is without doubt Verne's, and it is no accident that volcanoes and ice crop up again and again in his novels. "Our globe will end up like the moon," prophesizes Cyrus Smith in *The Mysterious Island*. Speculation as to the nature of the world's future thermal death raged in scientific circles at this time. In the long term, the extinction of the sun was inevitable, and with that the end of the earth. In the shorter term, however, humankind had a responsibility. We have learned to use fire, which both creates and destroys; we have succeeded in producing power—in short, we have become Zeus in human form and are now replacing the almighty clockmaker of the Classical Age, who has played out his role, with the Industrial Revolution. Did Jules Verne go the whole way? Here and there he acknowledges the Creator and alludes to Providence, but with the help of rhetorical formulas, which sound conventional and rather bloodless.

It is worth noting, however, that the science he pays greatest attention to is the softest of all, viz. geography, a synthetic discipline which isn't included in Comte's classification. Verne was even inclined to look upon cartography as the highest of all sciences. The mapping of newly discovered countries was without doubt an urgent undertaking, but hardly of the revolutionary kind. Jules Verne's eagerness to describe and classify belongs to the eighteenth century rather than to an advanced modernity. He reminds us of Linné and Linné's traveling apostles, not least in regard to questions of natural history. He admires the great variety of species but considers nevertheless that dangerous and unnecessary wild animals, i.e. those that are not of any use to human beings, should be exterminated. They only serve to make classifications, asserts the geographer Paganel in *Captain Grant's Children*. In Verne's novels there are plenty of fanatical hunters who go around killing in and out of season, quite often just for pleasure. The idea of exchange and balance in nature seems not to have a place in his philosophy; on this point he was not as farsighted as Linné, a precursor of ecological thinking. His understanding of medicine is not notably advanced, either, which is astonishing given the crucial advances made during his lifetime, not least in the field of infectious diseases, with Pasteur leading the way.

Nor does he seem to have had much respect for mathematics as a theoretical science. Like Alfred Nobel, he recognizes only its instrumental value, and he likes to lampoon mathematicians in portraits showing them as lost in the abstract, like Palmyrin Rosette (the name alone so ridiculous!) in *Hector Servadac, or the Career of a Comet*, an angry, egoistic oddball of a clearly asocial nature. True, there's a lot of calculating done in Verne's novels: engineers have to make the most detailed computations, as too do seafarers and pilots, but mathematics is reduced to a simple tool for immediate use, which seems quite short-sighted at a time when a great deal was being done on this front by Lobachevski, Cantor and other bold regenerators. Even if one stays on a more terrestrial plane, one will observe Verne's peculiar predilection for old-fashioned units of measurement, ones that are awkward to use, such as foot, inch, fathom, ounce, pound, etc., which hardly facilitate calculating; only exceptionally does he talk about kilograms and meters. This underrating of the metric system, an important inheritance from the French Revolution, seemed almost reactionary. Can it be explained by Verne's evident anglomania, which caused him to prefer words such as *railway, steamer,* or *bank-note* rather than *chemin de fer, vapeur* or *billet de banque,* even in a French context? Progress was something he liked to associate with England and the United States, and it is a well-known fact that it took a very long time for the Anglo-Saxon world to see the advantages of the metric system and embrace it!

Against the background of Verne's didactic efforts, one may well wonder in the end why he doesn't devote more space to the great scientific debates and controversies of his time. They can be glimpsed here and there, without being objects of any serious treatment, as if Verne was only interested in recording successful results, and had no real passion for their prehistory, that is, the competing theories that lay behind them. Nevertheless, there was a lot to come to grips with, e.g. discussions about mankind's origin, competing theories of evolution, conflicting ideas about the stages of development, the controversy between Pouchet and Pasteur about spontaneous generation extending from 1858 up to the latter's decisive victory in 1870. Jacques Boucher de Perthes (1788–1868), one of the founders of paleontology and a source of inspiration for *Journey to the Centre of the Earth*, Lamarck, Darwin and others appear between the lines, e.g. in *The Mysterious Island* and *The Village in the Treetops*, and also Pasteur, whom he certainly drew on for a great deal of sound advice about hygiene, but no more than that. It can naturally be explained by the requirements of the genre: to do too much lecturing can kill the suspense.

"From a perspective of the history of ideas, Jules Verne's profound importance lies in the fact that he embodies the great dream of industrialism: control of nature, both in a technical and in a geographical sense," affirms Tore Frängsmyr. And one may add that it wasn't until during Verne's time that the Promethean dream could be realized. For this reason, his interpreters have concentrated above all on the representation of technology in his work. Jules Verne came to symbolize our modernday's ambitious project to systematically develop nature's resources for the benefit of humankind. The preface quoted by Tore Frängsmyr which Nordenskiöld wrote for the 1892 edition of *Journey to the North Pole* (Volume 1 of *Voyages and Adventures of Captain Hatteras*) is typical in this respect. The famous explorer sees in Verne a mouthpiece for the longing "to make all the forces of nature servants of the human will," for scientists' efforts "to fathom out new means of controlling the forces of nature and harnessing them at will for the uses of humanity"—a variation on a well-known theme, Descartes' exhortation to become "Nature's lord and master." Verne understood the revolution that was taking place, realizing that energy, coal and electricity constituted the soul of the industrial world. The question is whether he approved of this development. He worries about the excessive exploitation of coal, which in his opinion will be exhausted within three centuries—this point of view is taken up in *Journey to the Centre of the Earth* and constitutes the main theme in *The Black Indies*. And he definitely disapproves of factories. He deplores the pollution they are guilty of causing without imagining that technology would also be able to find a cure for it in the long term. For him, industry remains an alien world. He displays remarkably little interest in the means of production and ignores the continuity of the process. His industrialists still seem to be mill owners. Yet it is during his lifetime that machines start to replace human beings, with the mechanization and, in due course, automatization of human work.

Verne has touched on and described most branches of technology: metallurgy, industrial processes, mining, navigation and shipbuilding (the sea was his great love, along with music and liberty), and of course all kinds of transportation, in accord with his interest in geography and voyages of discovery. His technological interests are focused very strongly on methods of transport, which during his lifetime became ever faster. He is particularly gripped by the much-debated question as to how objects heavier than air might fly. But it isn't always the most current concerns that capture his attention. He has a definite predilection for balloons (the first montgolfier went up in 1783), and he also praises traditional, not to say ancient, means of transport such as elephant-back, wind-propelled carriages, sledges, and so on (without such primitive means Phileas Fogg, despite everything, wouldn't have been able to travel round the world in 80 days!). The settlers in *The Mysterious Island* do indeed construct an electric telegraph link between their house and the place where they raise cattle, but at the fateful hour it is nevertheless the dog with a message around its neck that is the most reliable messenger. The prophet of modernity clearly doesn't recoil from the archaic.

Verne's technological fantasies sometimes have their roots in precursors' dreams. The submarine in *Twenty Thousand Leagues under the Sea* recalls the prototypes people tried to draw during the Renaissance, like the mystical fire-propelled flying ship in *Robur the Conqueror*. Leonardo da Vinci and Swedenborg had already had similar visions. As a rule, however, Verne built on whatever already existed, though with a rare ability to magnify, extend and to a certain extent project what existed into the future. One example: a fire-driven car establishes a record in May 1899 with a speed of 105.850 kph; five years later the record speed is 240 kph that Verne ascribes to *L'Épouvante*, Robur's universal vehicle which serves simultaneously as car, boat, submarine and flying-machine (*The Master of the World*, 1904)! He also goes on spinning tales on what would doubtless become reality in the foreseeable future, e.g. in the field of image and sound, an area in which many discoveries were made during the second half of the nineteenth century: electromagnetic waves (Hertz, 1866); microphone (Edison, 1877); photoelectric cell (Elster and Goitel, 1893). *Carpathian Castle* testifies to the imminent birth of cinematography (three years later) and Edison's inventions. Orfanik, a scientist, invents new devices along the lines of modern audiovisual equipment. The reproduction of images foreshadows film and television, but Verne doesn't seem aware of the potentialities of such an invention; Orfanik's only ambition is to give the illusion that a dead woman is alive.

There are plenty of wonderful machines in his work, but no information on the modus operandi: it is only the magic of the text that gets them to function. Nor is there anything about how they have come to be—it is only in the case of voyages of discovery that he gives his readers some historical background. The results are amazing and the principles fairly well set out, but there are no details about these machines, which never seem to need servicing, as if wear and tear was an unknown concept. As for the supply of energy, electricity is the miracle solution. But where does it come from, and how is it produced? It resembles a fairy with mystical properties, and Verne doesn't seem to realize that the production of electricity requires primary energy sources, for example, coal, which in his view will be exhausted in a few hundred years as a result of over-exploitation. Despite this alarming diagnosis, he fails to mention any possible alternative sources of energy, either. Nothing on wind power (Paris in 1960, as he imagines it, possesses 1,830 windmills, but of the conventional type), and not much more on water power and renewable energies. On this point Strindberg, who pleaded for a greater use of what nature offered for the taking—wind, water, and sun—as a counterbalance to the plundering of the earth's fossil resources, revealed much greater foresight.

There has been so much talk about Verne's fantastic predictions that it is time to mention also what he missed. He does not arrive at all the consequences of the communications society, which are constantly being developed during his lifetime. In his world there is a lot of traveling but astonishingly little communicating being done, despite so many revolutionary technological innovations: the telegraph (1840), the first cables under the sea (1851), the telephone (1878). To attest his arrival in various places, Phileas Fogg collects signatures: he doesn't send telegrams. Horse, dog, pigeon, bottles in the sea are as likely to occur as advanced technological devices. In his book *In the Twenty-Ninth Century: The Day of an American Journalist in 2889,* he predicts that journalism has become telephonic and that television news is sent all over the world, but we have already reached that stage and even gone a step further. He did not foresee the arrival of a post-industrial information society, such as started to develop in the 1970s. He imagines that the machine can, to a large extent, take over, especially in seafaring, but even if a large ship like *The Great Eastern* can easily be maneuvered with a lever, there is no question of an autopilot: it is always a human being dealing with information. And he hasn't always seen what social changes the march of technology would bring about. His sociological visions are actually more modest than the purely technological. In his eyes, the car is nothing other than a toy for the well-to-do. He doesn't suspect that it will be democratized and mass-produced, for good or ill. Just as he predicted little about motoring as a phenomenon of civilization, so he predicted little about air travel becoming widespread, even if he was well aware of its potential. Not even when he portrays Paris in the year 1960 or an American journalist's working day in the twenty-ninth century can he visualize the consumer society that arises as a consequence of mass-production, a throw-away culture and with it the growing problem of waste disposal. But life in his day is such that far less is thrown away: all one's rubbish is stored in the attic!

Tore Frängsmyr writes: "That technology was first and foremost a game played in the writer's imagination seems to me to be a point of view that has escaped many of Jules Verne's interpreters. It also seems to be the case that this possibility was overlooked by enthusiastic contemporary readers." One can only agree. All in all, Verne's predictions were perhaps not so remarkable. Many of them could be deduced by rational means. Presented as a prophet of genius, he yet makes mathematical-physical calculations that seem amateurish, and sometimes downright parodic. The bullet that flies between the earth and the moon resembles a stage device. And what about astronauts who open the porthole to obtain fresh air or to dispose of some rubbish! His age is marked by the Industrial Revolution, by railways, coal and steel—which continued to form the pillars on which Europe would be rebuilt after the Second World War. And it could be said that his predictions extend to the 1950s, but hardly beyond. To relativize the prevalent picture of Jules Verne is not to belittle him, insofar as his unique genius can be shown to lie on another plane!

In Tore Frängsmyr's opinion, Verne was above all a moralist, and a pessimistic one at that. From this viewpoint it may be interesting to take a closer look at the people in his novels who embody the heroes of his age, scientists and engineers. Do they do any work? Any research? The few times one sees a scientist at work he is busy doing calculations (immediate use) or classifying (descriptive science, of a literary nature). No laboratories, no scientific environments, hardly any experimenting. The application of methodical questioning does not appear anywhere, nor teamwork, nor any fumbling with trial and error. The inventors immediately achieve perfection—and hardly tolerate counterargument even if they meet any. Quite simply, they *know*. A Nemo, a Robur—such men have produced perfect machines, ones that cannot possibly be regarded as prototypes, for their originators don't care about development or follow-up: they have created something unique which they are not going to hand down. Nemo seals his fate by sinking himself and his submarine in the ocean depths. They are more like self-conscious artists than modern entrepreneurs, who concentrate their resources on reproduction and mass-manufacture to build up their empires.

There are plenty of examples of parody in Verne (he was of course a contemporary of Offenbach's and had himself tried his hand at genres such as operetta and boulevard theater), not least in the form of various caricatures of scientists. He also likes to show his disapproval of literal-minded, boorish individuals, who have no feeling for poetry (c.f. *The Green Ray*, for example). But his dominant characters have a completely different stature. They possess above all an indomitable energy, often combined with an equally pronounced stubbornness. Progress only comes from exceptional individuals. Verne clearly approves of romantic characters, rebels with an anarchic streak. This aristocratic elitism also has a drawback, for the enormous expertise achieved by these geniuses may also be misdirected. His works are populated with many ambivalent Prometheus figures, who are only too willing to exceed the limits. There is a demonic streak among brilliant recluses like Captain Nemo or the engineer Robur, who live outside society and establish their own laws. Arrogance and megalomania are constant threats. The superman Nemo reigns supreme over his own kingdom and follows his own moral code, sometimes cruel, sometimes generous, while Robur, an eminently successful engineer, manages to become master of space (*Robur the Conqueror*, 1886) but is gradually overcome by hubris, challenges God and misuses his colossal technical power, which in the end dispenses its own punishment (*The Master of the World*, 1904). It is perhaps Professor Schultze in *The 500 Millions of the Begum*, a Nazi before Nazism, who best exemplifies what can happen when advanced technology falls into the hands of criminals, when "the industrial-military complex" is allowed to run riot in the name of social Darwinism, with the justification that the struggle for existence is as indisputable and unquestionable as the law of gravity. In this case, for this specialist on racial biology from Jena, it is a question of subordinating all peoples to the Germanic nation, *das grosse Vaterland*. Of course, there is a polemical barb in this novel: the patriot Verne didn't like the Germans very much after the French were defeated in 1871 in the war against Bismarck's Prussia; against the background of this and most of the other novels he wrote, at least he can't be accused of romanticizing technology. Technology can be destructive: everything depends on how it is handled.

The whole of Verne's literary output, which extends from the 1860s to 1905, falls into a period of dramatic change. An enormous amount is happening, on all fronts. The emergence of the industrial culture gives rise to both hopes and fears. Threats are actually built into technical progress, which in itself provides no guarantee of prosperity or freedom. Verne has registered all these tensions and conflicts with seismographic sensitivity. In spite of scientific achievements, there is a constant threat of catastrophes, principally on account of mankind's inordinate longing to go ever further and further. The machine seems to upset the irrational, while yet bearing its seed. Already in his debut novel, *Five Weeks in a Balloon*, we read that people go on inventing machines until they are swallowed up by them, and Verne imagines the end of the world in the shape of a giant boiler that explodes, blowing the whole earth into the sky.

The pessimistic undertone in *Five Weeks in a Balloon*, as mentioned above, didn't escape Tore Frängsmyr. When he published *Framsteg eller förfall* [Progress or Decline] he could not have known about *Paris in the Twentieth Century*, written about the same time as *Five Weeks in a Balloon*, but not published until 1994 (*Paris au XXe siècle*; a Swedish translation appeared the following year). From this novel he had acquired additional evidence that the younger Verne was anything but blue-eyed. His portrayal of the future Paris is nothing less than an attack on our materialistic civilization. Social development proceeds towards mechanization and automatization, with the quantitative suffocating the qualitative along the way. Only material values associated with industry and finance are seen as legitimate. In this world there is no room for "parasites" or "dissidents," in other words, artists, intellectuals, classicists. Art has been outlawed. Latin and Greek are replaced by modern languages regarded as useful, and French itself becomes ever more leavened with foreign elements, mainly Anglo-Saxon loanwords. This jet-black pessimism didn't appeal to the publisher; Hetzel's rejection of the manuscript was probably not based on literary criteria (even if the novel in question definitely didn't come up to the standard of Verne's most enjoyable works). It could rather be seen as a type of censure, even if well-meaning. The publisher was frightened by this vision of a future dehumanization, which forebodes Kafka (*The Trial*, 1914), Huxley (*Brave New World*, 1932), or Orwell (*1984*, 1949). Verne was meant to be an optimist to order: he was expected to portray, in epic terms, humanity's march forwards and upwards. Talk about a chasm between the official façade and the author's deeply held philosophy!

That this early pessimism would get worse with the passing years is quite obvious. The author's dark side has been associated with aging, especially after 1886, when he was caught up in an obscure family drama. It seems as if he was losing his faith in humanity and progress more and more and was anxiously wondering where historical development would lead. Biographical circumstances surely played a role, but new currents of ideas were also making their presence felt in Europe. In a spirit of positivism, people had become drunk with their own power, but had also paid a high price. The reigning relativism of knowledge had modern-age angst as its companion. The Promethean spirit of conquest and exile from the absolute went hand in hand. An unavoidable reaction occurred during the last decades of the nineteenth century, when positivism gradually froze into an extreme mechanistic-deterministic way of looking at things. Comte was no longer the leading intellectual, as other prophets came to the fore—von Hartmann, Nietzsche, Freud, Bergson . . . . Scientific rationality was seen in a different light. Even natural scientists were divided, many of them allowing themselves to be fascinated by magic and the supernatural, even if the victory of science over superstition was still the declared goal. This fin-de-siècle mood is found again in the aging Verne, who, among other things, shares some common features with the sixteen-years-younger Nietzsche.

All in all, Verne's pessimism can be detected on different levels. The moral dimension is obvious: technology without ethics proves to be downright dangerous; insanity always threatens, as does dissension among people; everything depends ultimately on the agents themselves, on their morality, but not even wisdom and harmony can guarantee progress. His work, however, also gives expression to a cosmic pessimism built on a circular conception of history. For the romantic idealist, the ultimate aim of the scientific revolution was development from the material to the spiritual, and this was also Verne's expressed message, but in a number of stories he adopts a different model, a cyclical one. Every cycle ends with a catastrophe, a fire or flood, after which a few survivors have to begin again from the beginning. Humans wander around, and when they have crawled upwards, they are dragged back down again, at regular intervals. The most drastic illustration of this is to be found in the posthumously published story "The Eternal Adam" (1910), as Tore Frängsmyr has pointed out, and which he has analyzed. The action takes place between the years 2000 and 3000. Decline is in terms of both geology and civilization. The sea expands at the expense of the earth, while the last human beings, wandering over sunken shores, lands and mountains, gradually lose their human characteristics. Agriculture, engineering, clothing and even language disappear in turn, and bestiality rules once again, until earth and soil emerge anew. Even *The Survivors of the Chancellor* (1875), written 30 years before "The Eternal Adam," deals with decline, with fall, with loss. For four months a small group of people suffer from every imaginable misfortune one can at sea: fire, navigational errors, grounding, shipwreck. The survivors, who gather on a raft, suffer from thirst and starvation to such an extent that cannibalism becomes a temptation. How much suffering can a human being endure? And how uncertain is mankind's constant battle against nature—and against his own nature—anyway!

Even *The Mysterious Island*, which from his own position Tore Frängsmyr finds "the only really bright view of the future" in Verne's works, shows the superiority of the forces of nature. The island has arisen through a volcanic eruption, and it is ultimately destroyed by a volcanic eruption. True, the courageous settlers are rescued against all odds, but that is one of the conventions of the genre.

This many-layered novel gives an insight into the author's strategy. In the principal story the whole history of mankind is rewritten, since "the shipwrecked of the air" are totally destitute when they crash-land on the island and have to begin again from scratch, in a material sense, at least. The group's natural leader, Cyrus Smith, is a 45-year-old engineer, fully conversant with the science and technology of the time. We learn how these people, frozen and starving, gradually get on top of their situation. The first thing Smith manages to do is to make fire by creating a burning-glass out of two watchglasses. Little by little it is no longer a question of survival only; the harmonious little group transforms the desert island into a flourishing colony. Verne pays special attention to the mighty possibilities of chemistry but introduces other useful sciences as well. Everything seems completely rational; the logical sequence is inexorably followed from primitive man's condition to that of the nineteenth century's civilized Westerner. Metaphorically, it is a grandiose illustration of the Promethean industrial culture, just what one might expect from Verne.

But another type of story creates a counterpoint to this one. From the very beginning things occur which the settlers cannot explain. One fine day Top, the dog, growling and clearly nervous, discovers a well in the middle of the cave, which has been converted into a comfortable home. It seems as if an unknown providence, an invisible benefactor, is hidden in the abyss, watching the group via the well and protecting it when danger threatens. But this secret power doesn't get involved until the members of the group have shown their capacity to survive various trials. The mystery deepens all the time, puzzling signs accumulate, new questions arise continuously. Jules Verne has indeed wanted to spice his novel with exciting ingredients as a counterweight to the main plot's serious messages, one may think. But not only that. The settlers, who have no control over the situation, decide in the end to try to contact the anonymous, supernatural power. At long last they are summoned to "God," to the sacred place, in an ever-thickening religious atmosphere: "They couldn't control themselves any longer, and an irresistible power of attraction drew them to this mystical point, just as a magnet attracts iron." All this resembles a walk in the kingdom of death, in Hades, with all its stage props: a thunderstorm, a thread to follow (Ariadne), a hazardous descent, a boat trip across a dark bay (the Styx). Brutally abducted from the profane world, they finally meet their benefactor, Captain Nemo no less, whose ship *Nautilus* had disappeared in the maelstrom at the end of *Twenty Thousand Leagues Under the Sea*. He symbolizes the initiate, one who possesses the highest knowledge, and the whole story follows the classical initiation sequence, entailing a three-stage process: preparatory phase, peregrination in Hades, rebirth.

But there is also colossal ambiguity in the fact that it is Nemo, an outlawed misanthrope, a black angel, more Lucifer than Messiah, who has to stand for the sacred. And that is not all. Nemo is dying, and he is filled with doubt when he looks back on his life. His last act is to warn the settlers of an approaching danger, which no one can overcome: the extinct volcano is becoming active again, and the island blows up. Once again Verne illustrates his conviction that our victories are only provisional and fragile, that the forces of nature will have the last word, come what may. This grandiose, gloomy vision can be seen as the novel's actual final vignette. The romantic Verne would really have preferred to have his novel finish on a note of doom and destruction, some kind of Ragnarök, with a titanic struggle between fire and water, but Hetzel, his publisher, required a happy ending, a miraculous rescue of the settlers, which was considered more suitable for the young readership.

The Mysterious Island gives a fair idea of how Verne gets to work as a literary strategist. Two parallel tales are woven together: the one lectures and explains, the other leads us back to the archaic, the primeval. The rational narrative is made plausible thanks to technical precision, scientific lexis, rigorous chronology, detailed maps, and historical background. All this creates a strong impression of reality, but perhaps that is only a cunning way of silencing our critical sense, of anaesthetizing our reason. And then we become susceptible to something else, to a poetic creation, which builds on age-old archetypes. These, however, cannot be used by modern novelists in whatever way they wish; they stand in stark contrast to modern science and are difficult to bring into accord with the nineteenth-century mind, which is rational, discursive, and historically aware. Verne poses questions without providing answers: his recipe consists in playing exactly on this ambivalence. He gives expression to the age-old longing to triumph over time and death by "mystical births" (in the words of Mircea Eliade), but only symbolically, on the aesthetic-literary plane.

A novel such as *The Black Indies*, which is lighter but stirs the imagination, can be subjected to a similar analysis. The reader is invited to take part in a multidimensional journey. Superficially it is about a tourist trip to Scotland; then it is an encyclopedic journey, a scientific orientation in geology and economic geography; and finally the journey changes direction completely, being given a strong metaphysical stamp: the theme is the soul's journey from darkness to light. Three different levels can also be discerned in the vertical dimension: the subterranean (the dark pit of the mine), the superterranean (the electrically lit Coal-City, a gigantic cavern where the miners work), and the superterrestrial (the sun).

A girl, Nell, has had to grow up alone at the bottom of the mine in the custody of her grandfather, a devilish old man. This Silfax (= Lucifer) could represent a negative theology; his constant companion, a snowy owl, is the exact opposite of the biblical dove. A young man, Harry, discovers the girl in the mine, and they fall in love. Nell, who has always lived in darkness, has to be led towards the light. First, she comes upon the artificial lighting in Coal-City (an "electric sun"), and finally, in a chapter permeated by a strong religious atmosphere, the real, triumphant sun. Sunrise, "a sublime spectacle," induces ecstasy and rapture.

The whole fable, which describes successive rebirths, vividly brings to mind *The Magic Flute* (and with it Freemasonry's initiation and purification rites). The cast of characters is conspicuously similar in the two works, the novel and Mozart's opera— this Mozart whom Verne regarded as "the master of masters." Nell and Harry correspond to the couple Tamino-Pamino, and Silfax the Queen of the Night (and his snowy owl Monostatos), while a happy, naïve troubadour shoulders the role of Papageno, and the engineer responsible for the mining operations appears as a sound Sarastro (the fact that his name is Starr only reinforces the analogy!). One can also refer to Plato's allegorical epistemology (*Republic*, VII), in which different levels can be discerned, from sensual perception, which only provides silhouettes of true reality, to pure ideas, which are the prototypes for everything. The same scale is found in Verne's novel: the subterranean (imagination, illusion, with the mine's entrails as symbol); the superterranean (where the electric light stands as metaphor for rational, yet limited, knowledge); the superterrestrial (the revelation of the sun— contact with the Absolute).

Reading Verne in this way, one can come to the conclusion that his visions of the future may be nothing but the tree that conceals the forest. Or more accurately: anticipation is one side, repetition of the past or return to the source is another. Here, *Journey to the Centre of the Earth* speaks for itself. In the bowels of the earth, in the middle of a primeval forest, the fantasy travelers are confronted by tellurian forces; they meet even the primitive Adam, a giant with a buffalo's face and a lion's mane, who guards antediluvian monsters (the Minotaur!). In this passage, which has a strong romantic character, the atmosphere is both dreamy and anxiety-ridden.

Verne a Romantic? Yes, but not in any old way. He doesn't renounce past dreams but refashions them in a revolutionary way, which is evident, for example, in his reinterpretation of Edgar Allan Poe, one of his household gods. In 1897, he published *Le Sphinx des glaces* (of which to the best of my knowledge there is no Swedish translation; English translation, 1898: *An Antarctic Mystery*), which is quite simply a continuation of Poe's *The Narrative of Arthur Gordon Pym*, an unfinished story from the year 1837. Verne does all he can to make readers believe that Poe's novel was not a product of the imagination but a description of real events. In Poe's fiction, the ship *Jane* had disappeared in the Antarctic. Verne pretends to organize the expedition, which will attempt to locate her. From one perspective, he is true to the original (the fifth chapter is quite simply a summary of Poe's novel); seen from another, everything is turned upside down. Both authors were anchored in the romantic world, but Poe still strove to find a way beyond the world of the senses, to let the soul melt into an undifferentiated, cosmic oneness. Fifty years later this had become a beautiful but bygone dream. Verne had in no way cut off his moorings to the romantic world, but there was a crying need for modernization against the background of industrialization and the geographical discoveries. The myth of the boundless, the absolute, was hardly current any longer. Thus Verne constructs his own tale on methodical observations, such as scientific rationality requires. He systematically deconstructs his predecessor's romantic myth to come up with his own, also romantic, but with modern, scientific key signatures. The mountain range that looks like a sphinx is in fact a gigantic magnet. The whole globe appears as an enormous induction magnet, which continuously gets its energy from trade winds in the atmosphere. It is the triumph of magnetism, which is being staged now. Pym's skeleton is finally discovered, stuck fast to the magnetic mountain—he had been drawn to it on account of the gun he carried in a strap over his shoulder. If one reads Poe and Verne alongside each other, one is confronted with the two faces of Romanticism, corresponding to two phases in its development. The dream of a mystical association arises in both cases, but for Verne electricity now represents the very soul of the world. This can possibly lead one's thoughts to Comte's "positive stage," but it just as readily admits of metaphysical, perhaps even theological, elements.

If one scrapes the surface, the whole of Verne's published work shows similar gradations. Though appearing as a perfect mouthpiece for positivism, he is constantly searching for ways to escape from it. According to his own evidence, he wishes to portray "known and unknown worlds." Unknown worlds? He could, of course, be referring to the Polar Regions, the Amazonian rain forest, various deserts and even the moon, but he may also have had timeless mental structures in mind, just at the time when psychoanalysis was beginning to take shape.

Long before Verne, someone else had cherished the ambition of describing known and unknown worlds, and that was Homer, who had drawn up an inventory of contemporary codified knowledge—science and technology together with history with its tragic dimension and violent character—at the same time as he also recorded all kinds of legends and fables. In one way Jules Verne can be characterized as a modern-day Homer. Many of his heroes are placed between Scylla and Charybdis, like the much-tried Odysseus. The author dipped into the Ancient Greek heritage, with the Orpheus myth as the central element. Harry (*The Black Indies*) is an Orpheus searching for his Eurydice (his soul?), embodied in Nell. As for Nemo, the one who knows everything and is nobody, one thinks of Odysseus' reply to Polyphemus: "I have no name"; and the mysterious captain may be a variant of Saturn, who, according to Plutarch, spent time on an island surrounded by mists. Also present in Verne's works are many allusions to Telemachos (the search for a father under the leadership of a mentor, as in *Captain Grant's Children*). In almost all his stories, dramatic events occur that cause the journey to change its character, becoming transformed into a pilgrimage, with increasing difficulties and attendant trials. Very different stories such as *Journey to the Centre of the Earth* and *Michael Strogoff* have the initiation process as their basic structure. In the former—a kind of novel of courtly love in a scientific key—it is the young Axel, likewise an Orpheus of the Underworld, who will be forged through confrontation with the Earth's fire and other life-threatening entities, and wandering in Hades undergoing various purifying rites, in the subterranean sea, among other places. In *Michael Strogoff*, where the eponymous hero also has to suffer many trials, it is the Oedipus myth that is revived. It is not only the fable as a whole that recalls the Greek myth but also a mass of detail: the mother-son relationship is mentioned (or at least intimated); patricide (not the murder of the biological father, who is already dead, but of an infernal father who has to be removed) also comes into the picture. And Michael obtains higher clarity only when the Tartars blind him—what he did not see when he had his sight becomes clear to him when he is robbed of it. Then his initiation is complete: he knows himself and understands his fate. His guide, Nadia, is reminiscent of Antigone—or perhaps of Dante's Beatrice.

In actual fact it is the whole Western tradition regarding myths and esotericism that is hidden behind all the updated knowledge and exciting exoticism. Even the Bible is there. *The Adventures of Three Englishmen and Three Russians in South Africa* (1872) is a novel whose content appears to be strongly scientific (geodesics, trigonometry, astronomy). We get to follow an expedition whose task it is to measure a meridian curve in the Cape Colony in order to provide the metric system with scientific bases. Verne takes the trouble to give his readers as much information as possible about older degree measurements that were of importance in confirming Newton's theory about the shape of the earth, in the first place the two French expeditions in the 1730s to Peru and Sweden. These Russians and Englishmen undertake a scientific expedition, which recalls the one led by Maupertuis and Celsius in Tornedalen. One of the Russians is called Niklas Palander, a Swedish-sounding name which was surely not chosen by chance. A certain Gabriel Palander, a lecturer at Åbo University, had been Jöns Svanberg's colleague in Tornedalen from 1802 to 1804, when it had been their task to check the results of the previous expedition. And during Verne's time, Finland was part of the Russian Empire. Whatever the case, it is a thorough, instructive account of procedures in connection with base-line measuring, goniometry, and astronomical fixing. What drives the action is the tension between the universal search for truth and national rivalry (a war between Britain and Russia forms the background).

Behind the didactic content and the political message is concealed, however, another story. And this story is Exodus, about the Hebrews' migration across the desert and discovery of the Promised Land. Our travelers journey from the River Orange to the Zambezi via the Kalahari Desert—but these might just as well be called the Nile, the Jordan and Egypt. And they are exposed to torments such as blood rain, venomous snakes, locusts that devour everything, wandering in the wilderness, thirst and despair, before they discover the Promised Land. Parallels can be drawn to a quite considerable extent with Holy Scripture. Moses and Aaron have their correspondences in Verne's novel: the Ten Commandments are also there in burlesque form (the metric system as the French Revolution's Decalogue), and the same number mysticism is present in both texts (with the numbers 3, 7, and 10 recurring in various connections). In both cases it is a matter of a lost script which must be found (with Verne, the triangulation report which is stolen by a baboon), and where should it be found if not on a remote mountaintop? The basic structure is thoroughly biblical, while the content is modern and scientific.

And this, as far as I can judge, is Verne in a nutshell. Both positivist and mythological readings are reasonable—to a certain extent. But both fall short if they claim unilaterally to capture the phenomenon "Verne." For it is precisely the force-field between the modern and the archaic that gives the writer his unique profile. The author succeeded in constructing multilayered stories that accommodate opposites and create tensions, and it is in this that his genius lies. At the point of intersection between the most modern and the most primitive, he is able to anchor the nineteenth-century's dream of science and technology in a timeless anthropology. "If you go to the texts, you will find a far more interesting author, an artist and a philosopher, who had visions beyond the material conditions of crude everyday reality," writes Tore Frängsmyr. I would only add that his works are so very multidimensional that every generation, from their own horizon of expectations, can rediscover Verne with new eyes. Hence his enduring place among writers of all ages.

Toward a History of Reason

LORRAINE DASTON

Introduction: History and Philosophy of Science Estranged

In his 1874 harangue against historians and historicism, "On the Uses and Disadvantages of History for Life," Friedrich Nietzsche complained about what he called "the malady of history." According to Nietzsche, scholarly history, "objective" history, as it had come to be practiced by Leopold Ranke and his disciples, had forsaken the present and compromised the future in its preoccupation with the details of the past. These "historical virtuosi" had "set themselves the task of seeking 'pure, self-subsistent' knowledge or more clearly, truth that eventuates in nothing. There are very many truths that are a matter of complete indifference; [. . .] In this region of indifference and absence of danger a man may well succeed in becoming a cold demon of knowledge." The pallor of the archives on their cheeks, the dust of centuries in their souls, these demon-historians had, Nietzsche claimed, been born old, for "[h]istorical culture is indeed a kind of inborn grey-hairedness." The effect of historical culture on the young was paralyzing, for it fostered "doubts [about] all concepts and all customs." In their ant-like accumulation of trivial facts, in their cult of the footnote, in their quasi-religious prohibition of anachronism, in their worship of objectivity, the scholarly historians offended Nietzsche: "One goes so far, indeed, as to believe that he to whom a moment of the past means nothing at all is the proper man to describe it. This is frequently the relationship between classicists and the Greeks they study: they mean nothing to one another—a state of affairs called 'objectivity'!" Nietzsche's proposed therapy for the ennui and enfeeblement of "the malady of history" might be described as militant anachronism: "*If you are to venture to interpret the past you can do so only out of the fullest exertion of the vigour of the present.*"[1]

Although they make for strange bedfellows, I suspect that many philosophers of science may feel a certain sneaking sympathy with Nietzsche's sentiments when they survey the literature of the history of science of the past decade. They sigh over the ever-thicker monographs about ever more narrowly defined topics, they falter before the bottom-heavy apparatus of footnotes, they regret the territorialism of the standard response to counterexamples, viz. "*my* archives show . . .," they puzzle over the point of the avalanche of micro-studies that the history of science has largely become. Above all, they yawn. They cannot imagine that so many details, however thickly textured and artfully interwoven, can yield philosophically interesting theses, or even inductive generalizations about the nature of science. Under the micro-historical gaze, disciplines dissolve into traditions, traditions dissolve into local research groups, local research groups dissolve into highly particularized contexts. The mottoes of recent history of science—"local knowledge," "thick description," "science in context"—signal a level of craftsmanship and professionalism hitherto unknown in the discipline. Never have the narratives of historians of science been so ecumenical and ingenious in their use of sources, from the archival *Nachlass* to laboratory blueprints to the popular press; never have they been so intricately plotted, interleaving the many layers of context surrounding a scientific text; and never have they seemed less amenable to generalization, especially philosophical generalization.

The generality of this resistance to generalization is worth underscoring. It holds for sociological as well as for philosophical generalization. A few years ago, philosophers of science often explained their apathy towards the history of science as the result of an alleged sociological turn among historians. Certainly, historians read and discussed the work of Bruno Latour, Harry Collins, and Trevor Pinch; but I would wager that at least as many paid at least as much attention to the works of Nancy Cartwright, Ian Hacking, Philip Kitcher, and John Dupré. The telling point is how shallowly any of these stimulating accounts of science actually penetrated most of the narratives historians of science told. Certain phrases—"centers of calculation," "experimenter's regress," "scientific practices," "disunity of science"— turned up with some frequency in introductions and conclusions, but they did not provide the skeletons around which historians structured their stories. (Of course there are some notable exceptions to this picture of non-reception, but just because they are exceptions I beg your leave to put them aside for the moment.)

Nor can the practitioners of the history of science in context be accurately described as sociological in the more old-fashioned language of emphasizing external over internal factors. These historians are as passionately interested in the workings of a micrometer as they are in the emergent patterns of industrial funding of science, as attentive to the gendering of scientific labor as they are to graphing techniques. To a remarkable degree, their work has rendered the creaking, cumbersome language of the internal and external obsolete—*de facto*, not just *de jure*. Paradoxically, this achievement has made both the sociology and the philosophy of science appear equally irrelevant to the history of science.

From the perspective of twenty or thirty years ago, it is distinctly odd that historians of science should now be faulted for their failure to address big questions. They used to be notorious (at least among the larger tribe of historians) for their preoccupation with questions of paradigm shifts, confirmation and falsification, principles of symmetry, actor-network theory, and other themes imported primarily from the philosophy and sociology of science. For better or ill, the history of science was the most theoretically inflected branch of history, and its practitioners the least shy about generalizations. The effect of the historicist turn has been to undermine the bases of many of those old generalizations, often in fully justified ways. Historical research showing the "disunity of science" cast doubt upon generalizations about the nature of science *tout court*; fine-grained studies of science in context pitted the sharply etched specifics of any given episode against the blurry generality of attempts to amalgamate many such episodes into a larger, longer story. Emphasis upon the revelations of the local fostered indifference or incredulity toward the universal: was this any more than an illusion created by insufficiently detailed research? More subtly but no less effectively, the aesthetics of the discipline modulated in response to new methods of research, which privileged the archives, and of exposition, which featured the arresting vignette. "Rich" has replaced "incisive" as the term of highest praise. In short, historians of science have become more like other historians: guardians of the particular, skeptics about the universal.

The affinities of most history of science now lie with history, especially cultural history, and, at one remove, cultural anthropology. Historians of science, especially those trained in the past ten or fifteen years, have identified themselves closely with the methods (especially archival research), the values (especially the strict avoidance of anachronism), and the genres (especially the micro-history) of cultural historians such as Carlo Ginzburg and Natalie Zemon Davis.[2] Like the cultural historians, historians of science have turned to cultural anthropologists for inspiration. Although historians of science (and *a fortiori* philosophers of science) tend to use the words "social" and "cultural" interchangeably, social historians and cultural historians are keenly aware of their differences. In borrowing from cultural anthropology rather than from sociology or economics, historians of science have embraced the analytic categories of values and meanings rather than those of power and class, the symbolic coherence of cultures rather than the functional integration of societies, and the microscopic rather than the macroscopic view of historical events. Transplanting these perspectives into the history of science has certainly modified them in distinctive and interesting ways—"material culture" for example takes on a new halo of associations when applied to the gigantic, ultra-modern laboratories of post-World War II physics,[3] as does "local knowledge" when applied to the determination of the mechanical equivalent of heat.[4] But these modifications do not seem to have moved the history of science any closer to the philosophy of science, any more than historians have discovered new shared interests with philosophers.

If we exchange standpoints, and survey the philosophy of science with the eyes of a new-model historian of science, the prospects for a rapprochement do not improve. If the philosophers squint at the filigree of historical detail ("scholarly" is an equivocal compliment in their usage), the historians roll their eyes at the thicket of formalism enclosing many philosophical analyses of science. Even if they are initiated into Bayesianism and cognitive science algorithms, historians are skeptical that this kind of approach could ever yield results of historical interest. Because historians have come to believe deeply in the integrity and distinctiveness of various historical contexts, they automatically suspect formal approaches that homogenize differences and abstract from contexts. In the language of the Bayesians, historians tend to set the priors of these approaches very low indeed. But it is the rare historian who bothers to wade through a translation of the discovery of the Krebs cycle into information-processing algorithms, or a Bayesian evaluation of the evidentiary soundness of Ampère's reception of Ørsted's experiments. Again, although some historians of science do read the work of some philosophers of science avidly, in general the polite indifference of the philosophers is richly returned in kind by the historians.

I see no point in patching things up for old times' sake, when joint degree programs in the history and philosophy of science were many and flourishing, nor in bewailing a *fait accompli*. In the long history of classifications of knowledge, disciplines have ceaselessly drifted farther apart and closer together—recall that astronomy and harmonics were once neighbors, astronomy and chemistry once remote. There is nothing intrinsically regrettable about the estrangement of disciplines that once shared a border. But I nonetheless believe that the current state of mutual indifference that defines the relationships between the history of science and the philosophy of science is regrettable. I believe that it is based on a mistaken understanding on the part of both historians and philosophers of the implications of recent work in the history of science for the meaning and limits of rationality. More seriously, because more widespread, it has aided and abetted a spurious opposition of the cultural to the rational. And most seriously of all, it has blocked opportunities in both history and philosophy to pose, much less answer, new kinds of questions about science, culture, and rationality. In the short compass of this essay, I would like to address these misunderstandings (which are among the few things historians and philosophers still share), and offer some suggestions based on recent work as to how they might be overcome in a joint attempt to write a history of reason.[5]

The Historicist Turn

In the past twenty years, the history of science has been transformed by three movements: the embedding of science in its local context, the study of scientific practices, and the history of categories previously believed not to have a history. These three historiographic movements have been closely intertwined. Close attention to scientific practices such as collecting, experimenting, and reading depended upon and in turn enriched the understanding of science as firmly rooted in a particular historical and cultural context, be it the courts of Renaissance princes, the workshops of early modern European artisans, or the parlors of Victorian burghers.[6] A history of allegedly trans-historical entities such as evidence or objectivity would have been unimaginable had not these transcendent abstractions been translated into concrete practices that could be examined in specific contexts.[7]

Taken together, the impact of these movements has been radically historicist. This is most easily seen in the training and practices of historians of science themselves, who associate themselves ever more closely with historians (and ever less with sociologists and philosophers). But these developments have left still deeper traces in the work now published by historians of science, which strenuously avoids anachronism in terms and concepts (e.g. the use of the word "scientist" for the likes of Archimedes or Newton), tailors its narratives to the dimensions of micro-history, is defined less by discipline than by historical period and place (e.g. early modern Italy rather than physics), and, above all, vastly expands the domain of persons and topics deemed suitable subjects for dissertations and publications in the history of science. Studies of science in the family and in the pub,[8] of new scientific objects, both abstract and concrete,[9] of scientific seeing and note-taking,[10] of the botanical garden and the arsenal[11]—these are only a few examples of areas of research that would have been almost unthinkable two decades ago.

This enlargement of the possible has made the history of science, always blessed with porous disciplinary boundaries, still more open to scholars in other fields, especially those who also work historically and anthropologically. But it has also made history of science less interesting to other audiences, such as scientists (who miss studies of individuals and disciplines), philosophers (who find the history of concepts more congenial), and sociologists (who complain that social structures are obscured under the welter of historical detail). Even some historians of science who are otherwise well disposed toward the historicist turn yearn for the compelling grand narratives of yore and wonder how such local knowledge can be successfully universalized. Still more troubling, since the majority of researchers in the field are also teachers, is the absence of a general audience for the history of science (although the appetite for popular works on the subject seems to be voracious).[12] The situation is paradoxical: never have scholarly publications in the history of science been so richly textured, so vivid with local color and so animated with voices revived from the archives, so uncluttered with technical jargon and symbols, so inviting and accessible—and yet so apparently indigestible to students and the common reader.

All of these hesitations and criticisms of the historicist turn are linked to the role of contingency in new-style history of science. A chasm yawns between the earnest attempts of historians and philosophers of science some thirty years ago to discern patterns in the history of science and the almost utter indifference of their successors to such questions today. Whether the patterns were conceived as Kuhnian cycles of normal and revolutionary science or as the steady growth of knowledge through Popperian criticism or as the product of Lakatosian rational reconstructions, the shared assumption was that there were patterns to be found.[13] Plain old history might be just one damned thing after another, but the history of science, as the history of rationality, would itself display rational contours. This confidence was reinforced by claims about the unity of scientific method across disciplines and the universality of criteria of scientific evidence across epochs and locales. Contingency was in contrast associated with irrational social pressures, individual subjectivity, and local variability. Finally, just as the processes of science seemed to offer the best candidate for rationality, so the products of science seemed to be the most promising candidate for truth. Truth and rationality might be revealed over time, but in an orderly, almost Hegelian fashion: the slow but inevitable unfolding of the *Weltgeist*. To couple contingency with such transcendent concepts seemed oxymoronic—or heretical.

If the historicist turn had to be summarized in a single word, that word would be "contingency." Monograph after monograph hammered home the message: things could have turned out otherwise than they in fact did: "What is systematically left out of General Staff History is the contingency and the confusion of actual combat, the role of small groups of soldiers, the relationship between battle on the ground and the planning of generals. It would not be a flight of fancy to recognize in General Staff History a family resemblance to 'rational reconstructionist' tendencies in the history and philosophy of science. The 'von Moltkes' of the history of science have shown similar disinclinations to engage with actual scientific practice, preferring idealizations and simplifications to messy contingencies, speech of essences to the identification of conventions, references to unproblematic facts and transcendent criteria of scientific method to the historical work done by real scientific actors."[14] There was nothing inevitable about the ultimate triumph of truth over error—or even about the emergence of the criteria that distinguish truth from error: "A concept of evidence is a necessary condition for the stating of a problem of induction. A problem of induction does not occur in the earlier [prior to the seventeenth century] annals of philosophy because there was no concept of evidence available."[15] Everything, even the categories of fact and evidence, was historical, and everything that was historical was contingent.

These sorts of claims led many formerly avid readers of the history of science, especially the scientists and philosophers among them, to close their books with a bang. They understood contingency as incompatible not only with necessity but also transcendence. If truth and objectivity—not just the words or even the concepts, but the things themselves—were contingent, then, concluded these readers, they must be mere conventions. On this view, historicism was tantamount to relativism. To show that allegedly transcendent categories came into being in specific circumstances was to suggest that they might also pass away. To show that there had been, in other times and places, alternative epistemic ideals and practices was to imply that there was nothing privileged about the ones we currently embrace. Or so many scientists and philosophers interpreted the historicist turn—and therefore dismissed it as wild-eyed nihilism.

The Genetic Fallacy

Historicism has often been taken to imply relativism, albeit usually in the moral rather than the epistemological realm. But to infer from the existence of more than one alternative to either moral or epistemological questions to the conclusion that anything goes is a *non sequitur*. However, this false inference is nonetheless so often made that its persistence and prevalence demand an explanation, if not a justification. Although no one doubts that scientific achievements may continue to claim validity long after the specific contexts that generated them have ceased to seem plausible, the possibility of contextualized accounts of the scientific concepts and practices that are our best candidates for rationality is often regarded as an attempt to undermine them. For example, the fact that what came to be known as Kepler's Third Law was originally conceived in the context of a highly idiosyncratic cosmo-theology of celestial harmonies is rarely taken as grounds for dismissing the Third Law. Yet any talk of a history of rationality routinely provokes the suspicion that rationality is thereby under attack. A *mythology* of reason, very ancient, has apparently precluded the possibility of a *history* of reason. It is this mythology that underpins the genetic fallacy: the erroneous inference that a moral or epistemic value historicized is *ipso facto* a value debunked.

An abridged and allegorical form of this philosophical mythology might be told on hand from an actual myth. According to an ancient Greek myth, Athena, goddess of reason, springs full-grown and in complete armor from the head of her father Zeus. Although she is not an orphan—her parents are Zeus, hurler of thunderbolts and ruler of the gods, and Metis, goddess of practical wisdom—Athena has not been brought up; she is finished and self-sufficient from the first moment of her appearance, which she announces with a war cry. She has had no childhood, no phases of development, no education. She is not only immortal, like all the gods; she is also immutable. She did not always exist, but from the moment of her wondrous birth from the head of her father, she is complete: already an adult armed with all her attributes. She is also, at her own insistence, a virgin, in defiance of Zeus's offer of her in marriage to Hephaestus, god of fire and the forge. Like Artemis, virgin goddess of the hunt, she has foresworn not only the passions of the marriage bed, but also the subordination of wife to husband taken for granted in ancient Greek views of marriage. Hence there is no further *telos*, natural or social, towards which she might evolve: she was never a child; she will never be wife and mother.

With some allegorical liberties of interpretation, these peculiar features of Athena's myth still stamp our notions of reason and rationality. Rationality may have a date and place of birth—the candidates are numerous, from Babylonian astronomy to Indian logic to Greek philosophy—but it has no history, just as Athena has no biography. We imagine rationality to have been as complete and hence monolithic at its first appearance as Athena was at hers. That which is complete, "perfect" in the root sense of the word, contains no imperative to develop. Rationality has its own shield and spear, and its own battle cry. Forms of reasoning, such as the logical syllogism, the mathematical demonstration, or the philosophical dialectic, are the irresistible weapons of reason; a triumphant "Q.E.D.!" its shout of victory. Athena was also the goddess of war, and there is indeed something bellicose about the pursuit of rationality through arguments with an adversary, whether in Plato's Academy or in the pages of the latest issue of a scientific journal. Philosopher Robert Nozick once described the philosophical fantasy of a "knockdown argument," one so invincible that it would reverberate in the brain of a recalcitrant opponent until he fell down dead, impaled upon Athena's spear.[16] Rationality, like Athena, remains agonistic.

It also remains opposed to passion and insistent upon autonomy. There is a key moment in the *Odyssey* in which Athena appears to the travel-weary Odysseus in the form of a shepherd and assures him that he has at last landed in his yearned-for Ithaca. Although the goddess then reveals herself to Odysseus and assures him of her favor, he refuses to show the slightest emotion at her disclosure that he has finally reached home, lest it be a ruse or expose him to renewed risks. Far from being offended, Athena praises him for his self-restraint, so obedient to the dictates of reason.[17] The autonomy of rationality does not stop at checking the passions that cloud a clear mind; it further demands independence of time and place. Rationality is eternal and universal. What is rational for the here and now will also hold for the then and there, transcending the specificities of history and culture. Just as Athena knows no familial ties, rationality does not bind itself to context.

In contrast, the presupposition of a history of reason is that rationality is best studied in its specifics, in the techniques that weld aetherial principles to in-the-world practice. The two claims, historicity and specificity, are logically independent of one another: even if rationality were as unchanging as Athena, a strong case might still be made for a close study of its specific practices. But as it turns out, the historicity of rationality is most clearly revealed by concentrating on specifics. Of course, facts, evidence, and objectivity are themselves still rather lofty, Olympian notions. Part of the task undertaken by a history of reason is to transform these abstractions into concrete ways of thinking, feeling, and acting, which are also the achievements of history.

The word "achievement" is used advisedly here. To show that epistemic ideals and practices such as objectivity or precision or certainty have a history is not thereby to challenge them. There is a stubborn but somewhat incoherent association between historicism and subversion, which in turn depends on a conflation of inevitability and validity. To assert that a custom, belief, practice, or ideal has come into being at a certain time and place is indeed thereby to deny its inevitability. But the claim of inevitability does not bear in any necessary way upon its validity. Even if the assertion of historicity is made stronger, by further positing that this or that aspect of the rational requires particular cultural preconditions in order to emerge and be sustained, there is no presumption of invalidity. Mathematical demonstration, controlled experiments, and statistical sampling were all developed in highly specific—in some cases, uniquely specific—historical contexts; moreover, it has been plausibly argued that the connection between context and rational technique was not accidental. The peculiarities of ancient Greek society, for example, may have mattered to the crafting of mighty weapons of proof such as geometric demonstration, whereas other cultures—no less geometrically skillful—were content to rely on the testimony of vision and intuition.[18] Moreover, the cultural preconditions for the successful prosecution of controlled experiments or double-blind statistical trials may be both precarious and rarely met.[19] But these considerations, however true, do not in themselves undermine the reliability of these ways of knowing.

Why, then, is it so often concluded that they do? Why is historicism regarded, especially in scientific and philosophical circles, with so baleful an eye, as a corrosive acid that eats away at revered values and principles? I would suggest that the root of the misunderstanding lies in the venerable view that truth and reason are, like the Athena of myth, timeless and autonomous—and hence that rationality, as the due exercise of reason in precept and practice, is as well. Both traits suggest that, in order to be worthy of the name, rationality should be independent of context, of the entangling particulars of this or that place or time. The venerable tradition of disdain for context within philosophy has been strongly reinforced by modern developments in logic and mathematics that have linked rationality ever more tightly to formal methods, such as Bayesian statistics and decision theory, which by definition abstract from the specific case at hand. Against this view of rationality stand three arguments: first, that the inference from historicity to invalidity is false; second, that in point of historical fact, forms of rationality undeniably did emerge in specific historical and cultural settings; and third, that these forms continued to evolve and interact over time.

First, the historical and cultural specificity of the origins of an idea or a technique—or the very fact that ideas and techniques have origins—*in se* implies nothing restrictive about their validity in time and space. Mathematical demonstration seems to have arisen only once in human history, in ancient Greece, although sophisticated mathematics was a far more widely distributed cultural attainment. Yet mathematical demonstration was successfully transmitted, most notably in countless editions of Euclid's *Elements*, to intellectual traditions as different as ninth-century Islam and seventeenth-century China. Hence, historicity per se should hold no terrors for those who defend the validity of various forms of rationality. Second, it is simply the case that forms of rationality emerge in different places and different times. Moreover, the history of these forms, pursued in its details, indicates that when and where they emerged was not an accident. Some historical conditions seem more conducive than others for the creation and cultivation of rationality. Here the plurality of rationality—that it has several forms—becomes significant. The milieu that promoted the syllogism may not be—in fact, was not—the one that originally encouraged the controlled experiment. Third, and following directly from this last observation, the history of rationality is ongoing. Techniques of inquiry and proof do not simply emerge and persist; they develop, ramify, and intertwine with one another. From the standpoint of current precept and practice, rationality embraces elements that were at one time both historically and conceptually remote from one another. It is not obvious how to combine experiment with mathematical demonstration, or statistical inference with causal analysis; there is thus a further story to be told about the integration of rationality—the forging and welding together of the various parts of Athena's armor into a close fit. Even if we focus on just one form—experiment or taxonomy or demonstration—careful historical study reveals that it, too, continues to develop. This development often goes by the name of progress. Providing Athena with a biography does not threaten her credentials as goddess of reason.

These three points have been amply documented during the past two decades by the work of historians of science, who have investigated in impressive detail the historical careers of proof, experience, and evidence in various disciplines. There are, to be sure, still large gaps in this program of investigation. The literature on the local origins of forms of rationality still outweighs that on the diffusion, development, and interaction of these forms. There is considerable speculation and debate about how ideals and practices deeply rooted in their cultures of origin travel to and settle in strikingly different cultures. How and why, for example, did Chinese scholars understand and make use of the translations of Aristotle and Euclid prepared for them by Jesuit missionaries?[20] For that matter, how and why did the Jesuits themselves read these texts from lost, pagan milieux?[21] Moreover, there is a lingering tendency even among historians to assume that once a form of rationality, e.g. experiment, emerges, its key features are then frozen once and for all. Yet the history of experiment did not end in the seventeenth century, no more than that of mathematical demonstration ended with Euclid. These forms continue to mutate in surprising and productive ways.[22] Finally, the intercalation of forms of rationality, especially at the level of specific practices, remains largely terra incognita. Among historians and philosophers concerned to identify and track the development of these forms, the metaphors have been of mapping successive strata or classifying distinct styles, rather than of confrontation and combination.[23]

There is an undeniable splintering tendency in these historical investigations of rationality. Rationality branches into forms; each form follows its own sinuous course of development; these meandering lines of development crisscross and merge into still more complex patterns. It begins to look as if Athena's biography will conclude with a diagnosis of multiple personality disorder. Yet history need not destroy the integrity of rationality, although it queries homogeneity and immutability. The forms of rationality are the species of a genus, distinguishable but recognizably related. What they share is a normative appeal to reasons, and the force of that normative appeal depends on a wide range of evidentiary support—intuitive, discursive, authoritative, observational, experimental, technological. Rational theology and rational meteorology lie poles apart, yet both legitimately lay claim to participation in the genus rationality. Rationality need not yield truth. Although both the axioms that guided the arguments of Thomas Aquinas and the assumptions built into the latest statistical model for weather prediction may be false, the ways in which premises imply conclusions, conclusions are checked against evidence, and the whole is patrolled for consistency qualify both as exercises in rationality. A purist might object to this ecumenical usage: the honorific "rationality" should be reserved exclusively for the perfect exercise of reason, which leads infallibly to truth. But this standard seems to me theological in the worst sense, an unattainable fantasy about a divinity rather than an actuality or even potentiality of human reason. The subject of a history of reason is rationality as it is and has been practiced.

Epistemic Virtues[24]

What would such a history look like, and what would be its relation to the history of science? First, the history of reason and the history of science are not identical. There are many exercises of rationality—for example, in ethical deliberation or economic transactions—not embraced by the history of science nor even by the more capacious *Wissenschaftsgeschichte*. And there is much within the history of science that does not bear directly on the ideals and practices of reason. Nonetheless, the fact that the modern understanding of what reason is has been more deeply shaped by science than by any other single factor does mandate a central if not exclusive role for the history of science. Second, the history of reason, building on the historicist turn in recent history of science, should be narrated primarily as a history of practices rather than as a history of concepts about reason, although these latter are also important witnesses to the consolidation and spread of new forms of reason as normative. An echoing abstraction like Reason writ large defies history by its very blankness, but concrete practices like the method of least squares applied to a scatter of observational data or the use of computers to prove mathematical theorems by iteration of cases positively invites it. Moreover, concepts and norms usually crystallize out of established practices, so telling the story in this order keeps chronology straight. Third, the history of reason must ultimately be integrative: it must show how and why quite disparate practices that have emerged from diverse contexts have fused into a unified if not monolithic practice and ideal of reason.

The challenges of this third desideratum should not be underestimated. The practices of reason, even when eventually conceptualized in terms of epistemic virtues like objectivity, certainty, communicability, or precision, do not always pull in the same direction. Once again, this is easier to see in concrete practices than in abstract concepts. Take the case of communicability and precision. Since the late seventeenth century, strong arguments have been advanced against secrecy and inscrutability and in favor of open publication in science. Although there have been and continue to be violations of the ideal of communicability, these are controversial and often used as exceptions that prove the rule.[25] Ideally, scientific results should be publicly available and intelligible, and in principle replicable. At latest by the mid-nineteenth century, precision measurement had also become a widely acknowledged epistemic virtue in science. Yet the more precise the measurement, the greater the skill involved to make it and the less communicable and replicable the measurement.[26] Or the case of objectivity and accuracy: many nineteenth-century microscopists were prepared to countenance what they knew to be artifacts in their photographs of slide preparations because the photograph was deemed less subjective than drawings, which, in the words of the bacteriologist Robert Koch, "were always more beautiful than the original, with sharper lines or a darker shadow in the appropriate place in order to give the picture a completely different meaning."[27] Or, finally, truth and certainty: to discard all of those beliefs that are not absolutely certain but merely probable, as Descartes recommended, is to abandon many that are true.[28] There is no guarantee that the various practices of reason will always harmonize with one another. One of the advantages of focusing on the practices and specific decisions as to when and when not to apply them is to reveal such fissures in the smooth visage of reason viewed solely as a concept.

These fissures in that cool marble visage may rekindle the alarm of philosophers. It should be said straight away that the tensions among epistemic virtues are occasional, not constant. There are cases, perhaps even the majority, in which they can be successfully orchestrated. But this orchestration is itself an achievement, not a given, that begs for historical and philosophical scrutiny. And even when orchestration fails, the situation is not as catastrophic as visions of monolithic reason may lead one to conclude.

The predicament is analogous to that in which we find ourselves with respect to moral virtues. Our society may count both justice and benevolence to be moral virtues. It is, however, quite plausible that these virtues emerged as such in quite different historical circumstances, just as it is easy to conceive of cases in which they might clash. Just as we might ordinarily strive for both justice and benevolence but nonetheless be obliged by circumstances to sacrifice one for the other, so scientists may ordinarily strive for both accuracy (or certainty, or predictive reliability) and objectivity, but sometimes be forced to choose between them, and to choose in ways that involve quite concrete consequences for their inquiry: which research topic to pursue, which instrument to use, how to analyze the data. And just as we can imagine societies in which justice is an ancient social virtue, but in which benevolence attains this status only later, and within a specific cultural context, so we can at least imagine sciences in which truth is an ancient epistemological virtue, and objectivity a relative latecomer. Several philosophers, well aware of the historical diversity of visions of the good even within the Western cultural tradition, have argued that it is illusory to imagine that we will ever find a way either of reconciling these various visions or crowning one the victor over the others by proof and persuasion.[29] But this is not meant as a counsel of despair, much less as an invitation to mayhem. Rather, it is an invitation to reasoned argument, both to individuals and to whole societies, in order to judge the sometimes (but not always) competing claims of moral virtues in light of the shifting circumstances and aspirations of a society.

The message of a history of reason would be analogous: reason is historical and multifarious. The catechism of epistemic virtues has more than one entry. Even if all epistemic virtues ultimately serve truth, they do so in remarkably diverse ways—which in turn expand the meanings of truth, both abstract and concrete, into a spectrum. As historians and philosophers of science have repeatedly observed, explanatory coherence and predictive accuracy cannot always be simultaneously maximized. Pragmatic and correspondence criteria for the truth of a scientific theory may not always converge. But once again, this is not a counsel of despair but rather a call for reasoned argument among alternatives created by science, made visible by history, and pondered by philosophers.

Conclusion: Reason in Motion

One of the uses of a history of reason is to reveal such competing claims for what they are: conflicts among different epistemic virtues, each of which deserves to be called such, but with distinct histories and sometimes divergent implications for scientific practice—as opposed to a simple confrontation of epistemic virtue and vice *tout court*, objectivity squaring off against subjectivity, truth against error, precision against sloppiness, certainty against doubt. The scientist who does not publish all values of an experiment registered in her lab notebook may be guilty of fraud—but she may instead be exercising seasoned judgment concerning when the apparatus is working optimally and when not, siding with accuracy over objectivity.[30] Reinterpreted as collisions between objectivity and accuracy or certainty and truth, at least some of these episodes would become the occasion for discussion and deliberation rather than accusation and reproach. This would be a history of science that would hopefully re-engage the attention of both scientists and philosophers.

It would also be a history that could stimulate a fundamental rethinking about the nature of reason—or rather a rethinking of reason without foundations and perhaps without a fixed nature. If there is one lesson to be drawn from the history of modern science, it is that of Heraclitus: everything flows, no scientific truth is forever. As Ernst Mach wrote in 1872: "In fact, if one learned nothing more from history than the [mutability] of scientific views, then it would still be invaluable. . . . The attempts to hold fast to the beautiful moment through textbooks have always been futile. One gradually accustoms oneself [to the fact] that science is incomplete, mutable."[31] Although this may have been a melancholy realization for Mach and his contemporaries,[32] its obverse is the restless, inexhaustible inventiveness and dynamism of science. Since the sixteenth century, the history of science has been a history of novelty: new discoveries, new inventions, new theories, new techniques, new methods, new objects of inquiry. But history of science since circa 1980 has laid bare still deeper levels of innovation: experiment, demonstration, facts, objectivity, and other girders of scientific rationality have also been given their histories. Science invents and reinvents itself down to its core. This is a mobile, expansive, mutating kind of reason—perhaps the most powerful kind of reason ever articulated—that beggars metaphors of static foundations. If the history of science's contribution to the larger history of reason is taken at its full measure, a new concept of reason will be needed to do it justice: reason in motion.

Notes

1. Friedrich Nietzsche, "On the Uses and Disadvantages of History for Life," [1874] in *Untimely Mediations* [1873–76], trans. R. J. Hollingdale (Cambridge/New York: Cambridge University Press, 1983), pp. 57–124, on pp. 89, 101, 98, 93, 94 (emphasis in the original).
2. Influential examples of microhistory include: Carlo Ginzburg, *The Cheese and the Worms: The Cosmos of a Sixteenth-Century Miller* [1976], trans. John and Anne Tedeschi (Harmondsworth: Penguin, 1980); Natalie Zemon Davis, *The Return of Martin Guerre* (Cambridge, MA: Harvard University Press, 1983); and Giovanni Levi, *Inheriting Power: The Story of an Exorcist* [1985], trans. Lydia G. Cochrane (Chicago: University of Chicago Press, 1988).
3. Peter Galison, *Image and Logic: A Material Culture of Microphysics* (Chicago: University of Chicago Press, 1997), pp. 313–431.
4. H. Otto Sibum, "Reworking the Mechanical Value of Heat: Instruments of Precision and Gestures of Accuracy in Early Victorian England," *Studies in History and Philosophy of Science* 26 (1995), 73–106.
5. I shall use "reason" to refer to an intellectual faculty, "rationality" to refer to the expression of that faculty in thought, word, and deed.
6. Paula Findlen, *Possessing Nature: Museums, Collecting, and Scientific Culture in Early Modern Italy* (Berkeley: University of California Press, 1994); Pamela H. Smith, *The Body of the Artisan: Art and Experience in the Scientific Revolution* (Chicago: University of Chicago Press, 2004); and James A. Secord, *Victorian Sensation: The Extraordinary Publication, Reception, and Secret Authorship of The Vestiges of the Natural History of Creation* (Chicago: University of Chicago Press, 2000).
7. James Chandler, Arnold I. Davidson, and Harry Harootunian, eds., *Questions of Evidence: Proof, Practice, and Persuasion across the Disciplines* (Chicago: University of Chicago Press, 1994); Lorraine Daston and Peter Galison, *Objectivity* (New York: Zone Books, 2007).
8. Deborah R. Coen, *Vienna in the Age of Uncertainty: Science, Liberalism, and Private Life* (Chicago: University of Chicago Press, 2007); Anne Secord, "Science in the Pub: Artisan Botanists in Early Nineteenth-Century Lancashire," *History of Science* 32 (1994), pp. 269–315.
9. Lorraine Daston, ed., *Biographies of Scientific Objects* (Chicago: University of Chicago Press, 2000); Ursula Klein and Wolfgang Lefèvre, *Materials in Eighteenth-Century Science: A Historical Ontology* (Cambridge, MA: MIT Press, 2007).
10. Wolfgang Lefèvre, Jürgen Renn, and Urs Schoepflin, eds., *The Power of Images in Early Modern Science* (Basel: Birkhäuser, 2003); Ann Blair, *The Theater of Nature: Jean Bodin and Renaissance Science* (Princeton: Princeton University Press, 1997).

11. See for example the chapters in "Part II: Personae and Sites of Natural Knowledge," in Katharine Park and Lorraine Daston, eds., *The Cambridge History of Early Modern Science* (New York: Cambridge University Press, 2006), pp. 79–364.

12. Steven Shapin, "Hyperprofessionalism and the Crisis of Readership in the History of Science," *Isis* 96 (2005), pp. 238–243.

13. For the flavor of these debates, see Imré Lakatos and Alan Musgrave, eds., *Criticism and the Growth of Knowledge* (Cambridge: Cambridge University Press, 1970).

14. Steven Shapin and Simon Schaffer, *Leviathan and the Air Pump: Hobbes, Boyle, and the Experimental Life* (Princeton: Princeton University Press, 1985), pp. 16–17.

15. Ian Hacking, *The Emergence of Probability: A Philosophical Study of Early Ideas about Probability, Induction and Statistical Inference* (Cambridge: Cambridge University Press, 1975), p. 32.

16. Robert Nozick, *Philosophical Explanations* (Oxford: Clarendon Press, 1981), p. 4.

17. *Odyssey*, XIII. 415–419.

18. S. C. Humphreys, " 'Transcendence' and Intellectual Roles: The Ancient Greek Case," in her *Anthropology and the Greeks* (London: Routledge & Kegan Paul, 1978), pp. 209–241.

19. David Gooding, Trevor Pinch, and Simon Schaffer, eds., *The Uses of Experiment: Studies in the Natural Sciences* (Cambridge: Cambridge University Press, 1989); Harry Marks, *The Progress of Experiment: Science and Therapeutic Reform in the United States, 1900–1990* (Cambridge: Cambridge University Press, 1997).

20. Robert Wardy, *Aristotle in China: Language, Categories and Translation* (New York: Cambridge University Press, 2000); Peter Engelfriet, *Euclid in China: The Genesis of the First Chinese Translation of Euclid's Elements Books I–VI (1607) and its Reception up to 1723* (Leiden: Brill, 1998).

21. Peter Dear, *Discipline and Experience: The Mathematical Way in the Scientific Revolution* (Chicago/London: University of Chicago Press, 1995), pp. 32–62; Rivka Feldhay, "The Cultural Field of Jesuit Science," in John W. O'Malley, Gauvin Alexander Bailey, Steven J. Harris, and T. Frank Kennedy, eds., *The Jesuits: Cultures, Sciences, and the Arts, 1540–1773* (Toronto: University of Toronto Press, 1999), pp. 107–130.

22. For contemporary innovations, see (on experiment) Peter Galison, *Image and Logic: A Material Culture of Microphysics* (Chicago/London: University of Chicago Press, 1997); and (on mathematical demonstration) Donald MacKenzie, "Proof and the Computer: Some Issues Raised by the Formal Verification of Computer Systems," *Science and Public Policy* 23 (1996), 45–53; idem, *Knowing Machines: Essays on Technical Change* (Cambridge, MA/London: MIT Press, 1996), pp. 165–184.

23. Ian Hacking, " 'Style' for Historians and Philosophers," *Studies in the History and Philosophy of Science* 23 (1992), pp. 1–20; idem, "Historical Meta-Epistemology," in Wolfgang Carl and Lorraine Daston, eds., *Wahrheit und Geschichte: Ein Kolloquium zu Ehren des 60. Geburtstages von Lorenz Krüger* (Göttingen: Vandenhoeck & Ruprecht, 1999), pp. 53–77.

24. See Lorraine Daston and Peter Galison, *Objectivity* (New York: Zone Books, 2007), pp. 39–42 *et passim* for a more extended account of epistemic virtues.

25. See for example Myles W. Jackson, *Spectrum of Belief: Joseph von Fraunhofer and the Craft of Precision Optics* (Cambridge, MA: MIT Press, 2000).

26. M. Norton Wise, ed., *The Values of Precision* (Princeton: Princeton University Press, 1995).

27. Robert Koch, "Zur Untersuchung von pathogenen Organismen," *Mittheilungen aus dem Kaiserlichen Gesundheitsamte* 1 (1884), pp. 1–48, on pp. 10–11.

28. Bernard Williams, *Descartes: The Project of Pure Enquiry* (Hassocks, England: Harvester, 1978), pp. 69–70.

29. See for example Isaiah Berlin, *The Crooked Timber of Humanity: Chapters in the History of Ideas*, ed. Henry Hardy (London: John Murray, 1990); and Stuart Hampshire, *Justice is Conflict* (London: Duckworth, 1999).

30. See for example Gerald Holton, "Subelectrons, Presuppositions, and the Millikan-Ehrenhaft Dispute," *Historical Studies in the Physical Sciences* 9 (1978), pp. 161–224; and Allan D. Franklin,

"Millikan's Published and Unpublished Data on Oil Drops," *Historical Studies in the Physical Sciences* 11 (1981), pp. 185–201.

31. Ernst Mach, *Die Geschichte und die Wurzel des Satzes von der Erhaltung der Arbeit* [1872], 2nd unrev. ed. (Leipzig: Johann Ambrosius Barth, 1909), p. 3.

32. Lorraine Daston, "The Historicity of Science," in Glenn W. Most, ed., *Historicization—Historisierung* (Göttingen: Vandenhoek & Ruprecht, 2001), pp. 201–221.

Experience-Experiment: The Changing Experiential Basis of Physics[1]

H. OTTO SIBUM

Es ist ein entscheidender Charakterzug der Physik,
dass in ihr das Experiment die Beobachtung fast völlig verdrängt hat.[2]

*I*n 1923, one of the first German theoretical physicists, Felix Auerbach, told his readers that experimental physicists unlike botanists or geologists, do not merely observe nature anymore but rather artificially create physical phenomena in their laboratories. He made a rather bold claim, namely that x-rays for example were not discovered by Röntgen but were invented by him.

> X-rays are not a 'natural phenomenon,' until Röntgen there weren't such, they have been invented by him (this expression is more appropriate than the conventional 'discovered'); and in case it turns out that there will be such rays in nature, this does not change the issue essentially.[3]

This was certainly a contentious position, not in placing experiment at the heart of physics but in stressing the method of physics as artificial in character (what he termed its "technical" character). Auerbach's reflections on the changing experiential basis of physics are not just important expressions of his time but an integral part of a long historical process of settling the controversial positions about the epistemological status of experiment and experience. This debate goes back at least until the 17th century, but I would like to concentrate here on a time period in which the scientific persona of the experimentalist became fully established—the mid-18th century until the late 19th century. It was in this time span that the engineer came to be regarded as the "third man"—a novel actor capable of bridging the divide between theorists and practitioners, science and the arts. This coming-into-being of a new persona was contemporaneous with and coupled to an institutional and social process: the establishment of experimental physics as an academic discipline. Establishing experimental physics challenged the received scholarly tradition; it meant revising the sense of experience and experiment, of the relationship between the natural and the artificial—in short, a new sense of reality and how it could be studied.

Scholars and the Art of Experiment

Since the early modern period, scholarly opinions on "the art of experiment" have ranged from denying it had any epistemological value to the nineteenth-century conviction that this form of inquiry was the only way to make sense of natural causes. One of the underlying issues in these controversies about the meaning of experiment was that the physical manipulation of objects was seen as not belonging to the scholarly tradition, in which a clear distinction between doing and knowing still predominated. Even enlightened philosophers like Denis Diderot, who described the arts as a form of knowing, conceded that this knowledge operated beyond the enlightened discourse. His encyclopedic project was one answer to the dilemma of how to give the practitioners' knowledge a language, which could be understood by anyone. But together with many other literary approaches, the rising bourgeois culture reduced these complex forms of knowing to visual representations or verbal/textual descriptions of manual techniques, which finally maintained the boundary between epistemology and practice. In order to bridge theory and practice, since the mid-eighteenth-century the engineer had been seen as the ideal candidate—*the third man*. Christian Wolff articulated the need for such an intermediary figure and coined the term "the third man" in his introduction to the German translation of Belidor's *Architectura*:

> In such circumstances, a third man would be needed, who could in himself unite science and art, in order to correct the *theorists'* infirmities and to combat the prejudice of the lovers of the arts, as if they could be therein complete without the *theory*, and leave it [theory] to the idle heads good-for-nothing in the world [. . .]⁴

However, from the engineers' perspective, embodying this third man was an important but unsatisfactory position. Wolff described the difficulties involved by referring to the engineer who compared himself to a bat:

> Hence [...] he [the engineer Leupold] compared himself to a bat, tolerated among neither birds nor quadrupeds, and he complained that he was hated by the practitioners of art as well as despised by the *theorists*, for he wanted by his nature to be celebrated as a remarkable man by both, and to share fame in the learned world with the latter and happiness at court with the former.[5]

In establishing *physica experimentalis* within the *Gelehrtenrepublik* ("Republic of Letters"), experimentalists were experiencing the advantages and disadvantages of the third man's position. Like bats, experimentalists were difficult to classify. Did their studies of nature, practiced with head and hand—i.e. the art of experiment—lead to a specific form of *knowledge*, did it qualify as *science*? Answers to this question depended on the actors' stance towards the implicit distinction made in those days between experimental knowledge and science, or knowledge in general and scientific knowledge in particular. This distinction has a largely unwritten history of its own and is intimately linked with the social history of those who work with their hands and those who work with their heads. Furthermore, the dominant understanding of scientific knowledge as universal, autonomous, and permanent was intimately linked with the hegemony of the written text in the scholars' form of life. Hence even from the mid-eighteenth century onwards, several generations of experimental natural philosophers were required to free the art of experiment from its epistemological stigma and to position their knowledge within the Republic of Letters. The rising experimental research on electricity and magnetism played a key role in changing scholarly opinion about the epistemological status of the art of experiment. Hitherto unknown effects, created daily in these experiments, challenged the traditional and in the first half of the eighteenth century the widely accepted scholarly position about the twofold meaning of experience in physics:

> Experience gained in physics through the senses is of a twofold kind: one sort we take from God's creatures, from fire, air, water, earth, from the stars, flowers etc. the other we gain from artificial things, which are made by human hands [...] But we have no cause to make a great show of it, as if one could discover new and hitherto unknown physical truths through them [artificial things].[6]

In the course of development, this latter position became obsolete: "new and hitherto unknown physical truths" were in fact established by experiment and helped emancipate experimental physics from natural history. Already in 1755, the translation of an important contribution from the French experimentalist Abbé Nollet entered the German scene. In this inaugural lecture, Nollet had attempted to position *physique expérimentale* within the Republic of Letters. He insisted that this mode of investigation should be differentiated from the practices of natural history. The latter was a form of inquiry, which completed the "inventory of our wealth," that is, of existing knowledge, but did not investigate the "causes of what happens in the natural world." However, both were intimately linked with each other:

> For indeed, he who endeavors to investigate nature without understanding its history speaks at random and about things that he does not know in the least; but he who knows nothing else of nature than its history justly deserves a place among those natural philosophers who exercise their memory only. Accordingly, to practice experimental physics is nothing else than to investigate nature, not only with regards to its effects, but equally with the intention [of studying] the tools by which [nature's] effects are produced; in short it means to study what [nature] does, in order to be in a position to say how she does it.[7]

However, until the early nineteenth century, being a scholar at German universities first of all meant to be a writer who devoted much of his publishing activity to translations, textbooks, and compendia. And it is not a coincidence that the prime mover of experimental physics at Göttingen University, Georg Christoph Lichtenberg, was both man of letters and experimentalist. His editing of Johann Christian Polykarp Erxleben's *Anfangsgründe der Naturlehre*,[8] for example, shows quite clearly how the "new and hitherto unknown physical truths" established by means of experiment were challenging academic tradition from the bottom up—through appending lengthy footnotes to established texts, Lichtenberg demonstrated how rapidly the state of art of *physica experimentalis* was changing and how it affected the truth claims made in the script of nature. "Scientific invention" ("das Erfinden im Scientifischen") gradually replaced older encyclopedic mentalities.

The main challenge to traditional text-based scholarship was that experimentalists' investigation of nature's effects meant to develop and study instruments. From the engineers' point of view, instrumental intervention was unproblematic. But that was not the commonly accepted position amongst academicians. Especially the new field of inquiry, electricity and magnetism, was challenging because nearly every phenomena became observable only with the assistance of instruments or apparatus. A related problem involved the possibility of inference from small-scale models to real-world phenomena—something experimentalists would be bound to advocate.[9] In Germany, for example, the Rector of the University of Leipzig, Johann Heinrich Winkler, inferred from his artificially created illuminations in a vacuum tube that the Aurora Borealis was electro-magnetical. But not only macro phenomena became modeled, in the late eighteenth century the Italian physicist Alessandro Volta even succeeded in modeling micro-physical phenomena. He constructed a model of the electric fish, today known as the first electric battery, which for the first time demonstrated the existence of an electric current.[10] But the success of such experiments did not settle the theoretical controversies over their meaning.

Handwerksgelehrte and the Collective Refinement of Experience

Despite the immense practical achievements in creating "new physical truth," the status of experimental knowledge in 19th century scholarly form of life was still rather controversial. Very diverse knowledge traditions—those of artisans, instrument makers, natural philosophers and engineers—were in conflict about the true meaning and scope of their modeling practices. In the course of this process, the physical sciences based on experimentation were reshaped. However, each country integrated the experimentalist tradition into the elite academic culture in very different ways. As Thomas Kuhn has pointed out, in France the Abbé Nollet had been "a member of the somewhat motley section officially reserved for practitioners of *arts méchaniques*. There, but only after his election to the Royal Society of London, Nollet rose through the ranks, succeeding among others, both the Comte de Buffon and Ferchauld de Réaumur." In late-eighteenth-century Britain, a few outstanding instrument makers had become members of the Royal Society. In Germany, a few researchers succeeded quite early in establishing "Experimentalvorlesungen" at the universities: Georg Christoph Lichtenberg and in the 19th century Wilhelm Weber in Göttingen, Justus Liebig in Giessen, Robert Bunsen at Marburg, Gustav Magnus in Berlin, and so on. And as the cases of the Bavarian optician Joseph Fraunhofer, the brewer James Joule in Manchester, or the civil engineer Hermann Moritz Jacobi show, practical knowledge traditions played an important part in the formation of the experimental sciences. The founding of the German Physical Society in 1845, for example, was a major achievement in shaping the identity of *physicists*—a term to designate German students of nature.[11]

In the second half of the century, even a new term, *Handwerksgelehrte*, was coined which captures the amalgamation of the experimentalists' movement with the traditional academic elite. What had previously been regarded as quite distinct knowledge traditions, i.e. the experimentalists and the bookish scholars, now merged into a distinct community of experimental scientists in which ways of acting and ways of knowing were not seen anymore as epistemological antagonists. This process of amalgamation lasted the whole second half of the nineteenth century. Laboratories within most of the universities in Europe and North America were established. Practical physics as a new way of teaching ran parallel to this process. Chairs for experimental physics were set up and even methodology reflected the emancipatory process of the experimentalists. For example, Hermann von Helmholtz as well as James Clerk Maxwell, both chair holders of experimental physics, promoted an understanding of induction, which stressed the similarities between the intellectual work of the experimental physicist and that of the artist. Moreover, Maxwell even tried to investigate the commonalities between the intellectual work of theoreticians and that of the experimenters. The form of life common experienced by Cambridge Wranglers—with their extensive focus on mathematical practices—did not necessarily qualify them for doing experiments.

Speed and accuracy in performing knowledge, exhibiting mastery of abstract knowledge—the emblem of the Wrangler's world of mathematics—was regarded by them as the counterpart to the experimenters' accuracy of knowledge in manipulating objects. Furthermore, doing an experiment was associated with a loss of status within their colleges. Therefore, being the first professor of experimental physics in Cambridge University, the new intellectual work of the experimenter, involving head and hand, was hard to harmonize with the mathematicians' "tact." In the beginning, it was even difficult for Maxwell to attract a reasonable number of students for his classes. And later on in experiments of research, Maxwell tried very hard to make sense of the experimenters' knowledge in the traditional context of Cambridge Wrangler culture. Tellingly, he even tried to translate the often-unarticulated practices or phenomenological descriptions of the experimenters into formalized techniques of Wrangler mathematics.[12]

Finally, teaching practical physics became the strategy to change well-founded attitudes and to prepare the grounds for this new kind of research in Cambridge. Visitors to the laboratory were often astonished to see Maxwell and his students engaged in historical replications of experiments. But from his point of view, that was the secure way of leading students and himself into the experimenters' world. Following his general conviction "that the facts are things which must be felt, they cannot be learned from any description of them," he regarded it as an educational value to bring to consciousness the scholars' own tactile powers. Furthermore, reflections about the troubles in getting experiments to work made explicit the fact that it is only by the aid of their own senses that knowledge may be acquired. Or, as Maxwell put it by quoting Harvey:

> All this has been said more than two hundred years ago by one of our own prophets—William Harvey, of Gonville and Caius College. 'For whosoever they be that read authors, and do not by the aid of their own senses, abstract true representations of the things themselves (comprehended in the author's expressions) they do not represent true ideas, but deceitful idols and phantasms, by which they frame to themselves certain shadows and chimaeras, and all their theory and contemplation (which they call science) represents nothing but waking men's dreams and sick men's phrensies.'[13]

From the educational perspective Maxwell, sought to avoid drawing a line between the practices of Wrangler mathematicians on the one hand and experimental physicists on the other. He regarded exercises in the practical physics courses as being based on the same principle of tuition as those used in the coaching of Cambridge's mathematical elite. Similarly, Helmholtz explained to his audience at the Naturforscherversammlung in Innsbruck in 1869 the peculiar kind of work experimental scientists were performing:

> Besides the kind of knowledge that books and lectures provide, the researcher in the natural sciences needs the kind of personal acquaintance that only rich,

attentive sensory experience can give him. His senses must be sharpened [...]
His hand must be exercised that it can easily perform the work of a blacksmith,
locksmith, joiner, draftsman, or violinist.[14]

The new physical scientists have to be collectively trained in the refinement of
their sensuous experience. Only then they will succeed in discovering natural laws
through experiment.

> A law of nature, however, is not a mere logical conception that we have adopted
> as a kind of '*memoria technica*' to enable us to more readily remember facts. We
> of the present day have already sufficient insight to know that the laws of nature
> are not things, which we can evolve by any speculative method. On the con-
> trary, we have to *discover* them in the facts; we have to test them by repeated
> observation or experiment, in constantly new cases, under ever-varying cir-
> cumstances; and in proportion only as they hold good under a constantly in-
> creasing change of conditions, in a constantly increasing number of cases and
> with greater delicacy in the means of observation, does our confidence in their
> trustworthiness rise. Thus the laws of nature occupy the position of a power
> with which we are not familiar, not to be arbitrarily selected and determined in
> our minds, as one might devise various systems of animals and plants one after
> another, so long as the object is only one of classification.[15]

Helmholtz' plea for the collective refinement of experience marks an important change in the epistemic status of sensuous experience in science. Together
with Maxwell and others, he set sensuous experience center stage in the process of
generating scientific knowledge and of bridging the divide between theorists and
practitioners.

Experimenting Theory

And yet, despite these efforts and the practical achievements of the "third man" in
the age of natural science, reflections about the epistemological status of experimental physics in general and sensuous experience in particular continued. Not
only the new Handwerksgelehrte but also laypersons forcefully argued for media-
tion between knowing and doing, theory and experience. The German tanner
J. Dietzgen, for example, while engaged with philosophical problems in generating scientific knowledge, announced in 1869 the third man's problem as having
been resolved practically only. In his book *Das Wesen der menschlichen Kopfarbeit
dargestellt von einem Handarbeiter*, he wrote:

> The Christian opposition of spirit and flesh is in the age of natural science
> *practically* resolved. What's missing, in order to free the material interests from
> their evil reputation, is the theoretical solution, the mediation, the evidence
> that the spiritual is sensuous and the sensuous is spiritual.[16]

To him, the tension resulted from a conflict between two philosophical traditions about the sources of knowledge. The idealist regards the source of knowledge in reason only, the materialist in the sensually perceived world. But he saw a way out of this contradiction:

> The mediation of this contradiction requires the insight that both sources of knowledge are intimately connected with each other [. . .] Therefore even the lowest art of experiment which acts on the basis of experienced rules, is only gradually different from that scientific practice which is based on mere theoretical principles.[17]

Around 1900, the experiential basis of physics was changing around 1900 and evoked various reflections about these sources of knowledge. One important and widely shared understanding of experimental physics emphasized its technical—artificial—character. And it was the physicist Auerbach amongst others who spelled out most clearly this point:

> Experimental physics does not—as the term already suggests—practice observation of nature like other natural sciences, it deploys artificial experiments which are performed just for a specific purpose. Strictly speaking physics with regard to its method is not a natural science like astronomy, geology, botany etc; it does not deal with natural phenomena but phenomena produced artificially and through arbitrary acts of the researcher; in this sense we can speak of physics as a technical science.[18]

At the turn of the century, most of the physicists practiced precisely this technical science and regarded it as contributing to the higher goal of science—to achieve pure cognition. But the physics community was beginning not to speak with one voice any more and we could list here several different stances about the epistemological status of experiment and sensuous experience in generating knowledge. Experimentalists were just as concerned as theoretical physicists about the question: What was the source of physical knowledge? What role did sensuous experience play? Especially, the increasing number of techniques used to investigate microphysical objects like x-rays, electrons and so on around 1900 opened novel experiential spaces for the physicists and induced this increasing self-reflexivity about their tools and methods. They even were putting new demands on the quest for the unity of nature.[19]

For Auerbach, the source of scientific knowledge was always experience. His reflections on physics indicate that the third man's approach of bridging theory and practice even affected the culture of theoretical physics. This claim—he argued—that theoretical physics takes the material for constructing its general fundament from experience might appear as if physicists are arguing in a circle. How could one derive the facts of experience from a general schema and at the same time gain this schema by orientating one's self towards experience? In order to persuade his audience, he refers to the most striking invention of 19th-century electrical engineering: the dynamo—an invention about which he had done extensive research himself. In the 19th century, to many it might have seemed impossible to build a machine that could produce electrical energy out of mechanical work and at the same time feed this same machine with electric currents. But such a machine had been built; it was the Siemens dynamo electrical machine. In a way, such a machine starts automatically because of a trace of magnetism inherent in every piece of iron. A trace that suffices to produce weak electric currents, which take care of all the rest. In a similar way, we as theoretical physicists just need a minimum of experience. We do not want to address every question directly to nature; we would like to gain as much as possible knowledge about nature from a minimum of facts of experience (Erfahrungstatsachen). Of course, with some practice, we could then build theoretical physics directly out of our head with the foresight that a retrospective check against experience does not contradict the theoretical claim; but if this happens, we would have to restructure our building or eventually replace it through another one. [20] But it was important to note that this practice of theorizing has to be distinguished from another kind of theoretical physics whose promoters believed the general comes from a mere speculative inside of the researcher.

> [The speculative inside] constructs an ideal world, declares its satisfaction, if the real world matches the ideal. But in case of contradictions these theorists would go that far and declare the real world as false because it does not match with the ideal.[21]

When writing this, Auerbach clearly commented on an intellectual tendency among physicists, who held that experience and reason would remain divided into two separate domains. In contrast to that position, he was convinced that theoretical physics should take a middle position—to be the third man—between the mere speculative and the mere empirical. Although, methodologically speaking, he regarded physics more and more as a technical science; but as he reminds his readers:

> Understand me correctly: with regard to its method, for with regard to its goal physics is and remains a pure natural science, insofar as it does not aim (or at least not primarily) at technical applications, but serves pure cognition.[22]

Physics—and here he means his kind of theoretical physics—accepts both experience (Erfahrung) and reasoning (Denkarbeit). But although the changing experiential basis of science induced an increasing self-reflexivity in physics, he felt the needed to warn his students not to follow this other kind of theoretical physics that believes that mere cognition will grasp reality:

> Experience remains, of course, the sole criterion of physical utility of a mathematical construction. But the creative principle resides in mathematics. In a certain sense, therefore, I hold it true that pure thought can grasp reality, as the ancients dreamed.[23]

In Auerbach's days, the debate over the technical character of science and the related issues of experience, experiment and theory could not be settled satisfactorily and clearly deserves further investigation.[24]

Notes

1. This essay is a revised version of the "Experience—Experiment: The Changing Experiential Basis of Physics," in *Research Report 2002–2003: Essays* (Berlin: Max-Planck-Institut für Wissenschaftsgeschichte, 2004), pp. 89–97.

2. Felix Auerbach, *Die Methoden der theoretischen Physik* (Leipzig: Akademische Verlagsgesellschaft m.b.h., 1925), p. 3.

3. Felix Auerbach, *Entwicklungsgeschichte der modernen Physik: Zugleich eine Übersicht ihrer Tatsachen. Gesetze und Theorien* (Berlin: Julius Springer, 1923), p. 5.

4. Chr. Wolff, "Vorrede" in Bernard Forest de Belidor, *Architectura Hydraulica, Oder die Kunst, das Gewässer zu den verschiedentlichen Nothwendigkeiten des menschlichen Lebens zu leiten, in die Höhe zu bringen, und vortheilhaftig anzuwenden* (Augsburg, 1764), p. 2.

5. Ibid., p. 2.

6. "Die Erfahrung aber in der Physic durch die Sinnen ist hauptsächlich zweyerley: eine wird von den geschöpfen Gottes genommen, als vom Feuer, von der Luft, von dem Wasser, der Erde, von den Sternen, Blumen etc., die andere hingegen von künstlichen Dingen, die mit Menschenhänden gemacht sind [. . .] aber so großen Staat davon zu machen, als könnten dadurch neue und bisher unerkannte physische Wahrheiten erkannt werden, hat man nicht Ursach." Johann Georg Walch, *Philosophisches Lexicon, I*, 4. Auflage (Leipzig, 1775), reprint (Hildesheim: Georg Olms Verlagsbuchhandlung, 1968), pp. 1176–1177. See also Horst Schimank, "Zur Geschichte der Physik an der Universität Göttingen vor Wilhelm Weber (1734–1830)," in *RETE*, 2 (1974), pp. 207–252, 212–214.

7. For the relevant passages in French and German language, see J.A. Nollet, *Discours sur les dispositions et sur les qualities qu'il faut avoir pour faire du progress dans l'etude de la Physique Experimentale in Oratio Habita a Joanne-Antonio Nollet, Licentiato Theologo, Regiae Scientiarum Academiae Sopcio, Cum primum Physicae Experimentalis Cursum Profesor à Rege institutus auspicaretur. In Regia Navarra, die Martis decima-quinta mensis Maii, anno Domini 1753* (Parisiis: Universitatis jussu edita, MDCCLIII), p. 9 and J.A. Nollet, *Rede von der nöthigen Geschicklichkeit zur Erforschung der Natur, Welche er den 15. May 1753 bey dem Antritte seines öffentlichen Lehramtes, in dem Navarrischen Collegio. Gehalten. Aus dem Französischen übersetzt* (Erfurt, 1755), p. 12.

8. Johann Christian Polykarp Erxleben, *Anfangsgründe der Naturlehre* (Göttingen: Dietrich, 1772); see particularly 6th edition, with corrections and many additions by G.C. Lichtenberg (Göttingen: Dietrich, 1794).

9. On this topic see for example Simon Schaffer, "Die Reichweite experimenteller Wissenschaften: Modelle, Mikrogeschichten, Mikrokosmen," *Historische Anthropologie* 13, no. 3 (2005), pp. 343–366.

10. On the electrical battery as "artifical electrical organ," see Jörg Meya and H. Otto Sibum, *Das fünfte Element: Wirkungen und Deutungen der Elektrizität* (Reinbek bei Hambrurg: Rowohlt, 1987), pp. 125–141; Guiliano Pancaldi, *Volta: Science and Culture in the Age of Enlightenment* (Princeton: Princeton University Press, 2003).

11. Historians of science have dealt with this issue at length, but most convincingly for the early modern period, see here for example Steven Shapin, Simon Schaffer, *Leviathan and the Airpump: Hobbes, Boyle and the Experimental Life* (Princeton: Princeton University Press, 1985);

Peter Dear, *Discipline and Experience: The Mathematical Way in the Scientific Revolution* (Chicago: Chicago University Press, 1995). For the period under discussion here we still need further detailed studies than those pioneering ones of T.S. Kuhn, "Mathematical versus Experimental Traditions in the Development of the Physical Sciences," in *The Essential Tension: Selected Studies in Scientific Tradition and Change* (Chicago: Chicago University Press, 1977), pp. 31–65; or T. Frängsmyr, J.L. Heilbron and R.E. Rider, eds., *The Quantifying Spirit in the Eighteenth Century* (Berkeley: University of California Press, 1990). For the most recent studies, see Myles Jackson, *Spectrum of Belief: Joseph von Fraunhofer and the Craft of Precision Optics* (Cambridge, MA: MIT Press, 2000) and idem, *Harmonious Triads: Physicists, Musicians and Instrument Makers in Nineteenth-Century Germany* (Cambridge, MA: MIT Press, 2006); H. Otto Sibum, "Les Gestes de la Mesure: Joule, les pratiques de la brasserie at la science," *Annales: Historie, Sciences Sociales,* no. 4–5 (1998), pp. 745–774 and idem, "Experimentalists in the Republic of Letters," *Science in Context* 16, no. 1/2 (2003), pp. 89–120; M. Norton Wise, *Bourgeois Berlin and Laboratory Science,* forthcoming, and Simon Werrett, *Philosophical Fireworks: Science, Art, and Pyrotechnics in Early Modern Europe,* forthcoming with Chicago University Press.

12. For more details on this issue, see H. Otto Sibum, "Working Experiments: A History of Gestural Knowledge," *The Cambridge Review,* 116, no. 2325 (1995), pp. 25–37, 33–36. On the Cambridge Wranglers, see in particular Andrew Warwick, *Masters of Theory: Cambridge and the Rise of Mathematical Physics* (Chicago: Chicago University Press, 2003).

13. J. C. Maxwell, "The Telephone," The Rede Lecture, Delivered at the Senate House of the University of Cambridge, 24 May 1878, reprinted in R. Staley, ed., *The Physics of Empire* (Cambridge: Whipple Museum, 1994), p. 17.

14. Herman Helmholtz, "Über das Ziel und die Fortschritte der Naturwissenschaft: Eröffnungsrede für die Naturforscherversammlung zu Innsbruck, 1869," in *Populäre wissenschaftliche Vorträge von H. Helmholtz* (Braunschweig: Vieweg und Sohn, 1871), pp. 183–211, 183–185.

15. Ibid., p. 189.

16. Joseph Dietzgen, *Das Wesen der menschlichen Kopfarbeit: Dargestellt von einem Handarbeiter. Eine abermalige Kritik der reinen und praktischen Vernunft* (Hamburg: Meissner, 1869).

17. Ibid., p. 76.

18. Felix Auerbach, 1923, p. 4.

19. Detailed studies on the role this change of experiential space played in the process of the differentiation of scientific work like experimental and theoretical physics as well as in the strikingly self-reflexive turn in the sciences as formulated so clearly in the early writings of Ludwik Fleck, Michael Polányi, and Gaston Bachelard has been undertaken at the Independent Research Group "Experimental History of Science" in Max Planck Institute for History of Science, Berlin. See the forthcoming publication: H. Otto Sibum, ed., *Science and the Changing Senses of Reality circa 1900.* Thematic double volume of *Studies in History and Philosophy of Science,* 2008.

20. For the relevant source in the original language, see Felix Auerbach, 1925, p. 2.

21. "Denn es muß doch sofort gefragt werden, woher denn diese Wissenschaft das Allgemeine, von dem sie ausgeht, hernehme; und da scheiden sich die Wege; die reine Spekulation nimmt es aus dem spekulativen Innern, sie konnstruiert eine ideale Welt, erklärt sich für befriedigt, wenn die wirkliche Welt mit jener übereinstimmt, und geht, wenn sie die äußerste Konsequenz zu ziehen entschlossen ist, so weit, in Fällen des Widerspruchs zu erklären: die wirkliche Welt ist falsch, denn sie stimmt mit der idealen nicht überein." Ibid.

22. Felix Auerbach, 1923, pp. 4–5.

23. A. Einstein, *Ideas and Opinions,* based on *Mein Weltbild,* Carl Seelig, ed. and other sources. New translations and revisions by Sonia Bargmann (New York: Bonanza Books, 1954), pp. 270–276.

24. For a most recent attempt, see Paul Forman, "The Primacy of Science in Modernity, of Technology in Postmodernity, and of Ideology in the History of Technology," *History and Technology,* 23, no. 1/2 (2007), pp. 1–152.

Intermediate Theoretical Physics[*]

KARL GRANDIN

This paper examines the Swedish theoretical physicist Ivar Waller (1898–1991) against the background of the emergence of a modern theoretical physics in Sweden in the 1920s. How was the making of a theoretical physicist accomplished at a time when there was no consensus of what theoretical physics was or should be in Sweden?[1] Focus will be on Waller's way of doing what will here be labeled *intermediate theoretical physics*. Waller was not an average physicist. He was professionally rather successful and became professor of mechanics and mathematical physics at Uppsala University, member of and, for a period, head of the Nobel Committee for Physics, member of the Swedish National Atomic Energy Committee, and held several other central positions. So he was not intermediate in the sense of being average. During the 1920s, however, he did not hold any permanent position, and he was more or less constantly occupied with maneuvering for the means to pursue his career. By following this career, we can illuminate different parts of the Swedish physicist community.

Waller followed, throughout the 1920s, the direction already set out in his dissertation work—the problem of X-ray diffraction and scattering—in a way that can be investigated through the concept of "theoretical technologies."[2] Here *technology* should be understood in its more narrow sense, that is, technical methods, means, and skills. A theoretical technology should thus indicate a broad concept or notion including theoretical methods, means, and skills. Theoretical methods or mathematical techniques are not something that everyone involved has instant access to, or can easily change overnight, analogous to the apparently inertial or material part of the experimental physics culture that does not easily change. The analogous part of the theoretical physics culture would be the acquired physical concepts, mathematical techniques and schooling.

I use the term inspired by Andrew Warwick, who suggests that theoretical technologies "describe pieces of theoretical work that are not constitutive of a general theory, but which are used to solve particular problems and which are taken for granted by members of a local community."[3] Warwick also discusses in terms of local cultural resources his search for the nature of theoretical practices. It is also often assumed that theoretical physics can be divided into different levels. High-level theorizing or speculating would be one, calculating another, and model building a third. These distinctions might be pursued among the theoretical physicists: one being a speculating theoretical physicist, the other a calculating theoretical physicist, and so on. These divisions are of course oversimplifications, but can illuminate the breadth of the theoretical physics culture. Different levels of the theoretical enterprise interact in both directions, rather than only lower levels being influenced by a core of high theory. Waller's career can be seen as operating, not at a highly theorizing level, but at a calculating level—working more closely with experimentalists, but also influencing high-level theorists like P. A. M. Dirac, as shall be seen.[4]

Waller's way of doing theoretical physics can be characterized as an intermediate theoretical physics: intermediate in the sense that he *mediated* experimental and theoretical physics by direct cooperation with experimentalists; intermediate also in the sense that he cooperated with other theoreticians; intermediate in the sense that he mediated different theoretical technologies; intermediate in the sense that he served as a mediator between the center of modern theoretical physics back home to the periphery; and finally intermediate because Waller's work and career formed an intermediate state between an older, more mathematically inclined theoretical physics, and a more modern theoretical physics.

Reciprocal Arguments and Experimental Evidence

Bis jetzt gab es ja in Schweden praktisch fast keine theoretische Physik [. . .] ein Mißverhältnis besonders auch zur Experimentalphysik, die ja durch Siegbahn und Hulthén so glänzend vertreten ist. Nun braucht man in Schweden einen mit der *modernen* theoretischen Physik vertrauten Mann.[5]

Even though Sweden had been neutral during the First World War, the blockades and the naval warfare had imposed restrictions, rationings, and a high degree of military preparedness that left deep marks. Ivar Waller did most of his undergraduate work during this time, interspersed with military service. Swedish physics was dominated by experimentalists who above all cherished exact measurements. At Uppsala University, the professor of (experimental) physics conducted investigations into the electric arc, whereas in Lund the Siegbahn group was slowly gaining a reputation by doing more modern physics. Brilliant experimental physics, as

Pauli would allude to ten years later, was, however, not yet to be found in Sweden; and any modern theoretical physics was still beyond the future horizon.

Ivar Waller started his licentiate studies at Uppsala University in the autumn term of 1919, very much inspired by the professor of mechanics and mathematical physics C. W. Oseen's lectures in the summer course for high school teachers the same year on "Atomic conceptions in today's physics: Time, space, and matter."[6] A few years later, in 1922, Waller would publish an article, on the possibilities of testing Einstein's general theory of relativity, in the second volume of the newly founded Swedish Physical Society's yearbook *Kosmos*. Waller went through the three testing possibilities for the general theory of relativity: the perihelion effect on the planet Mercury, the deflection of light in the vicinity of the sun, and the red shift. The article was concluded with the cautious remark that, even if all three tests only supplied uncertain evidence taken one by one, it could not be mere coincidence if all three supported the general theory.[7] The need to, in a popular way, present an experimental evaluation of the relativity theories was truly a touchstone test for Waller himself. The problem of testing these theories might be seen as a focal point for the difficulties of the aspiring theoretical physics community. Given a strong program in experimental physics in Sweden with a concentration on precision measurements, the theoretical physicist at least had to accommodate the defiant theories of relativity with the ethos of exactitude, even though the theoretical physicists did not take any clear epistemological stance yet.

Oseen suggested to Waller that he should investigate heat conductivity in solids for his licentiate work. In 1913 and 1914, Peter Debye had published interesting papers on this subject.[8] Waller started to study the literature on lattice dynamics, where the works by M. Born, P. Debye, T. von Kármán and C. G. Darwin were the most important. Waller's interest shifted, however, from heat conductivity to the interaction of X-rays and crystals. Waller was interested in the scattering of X-rays in crystals and used Born's method of finding the normal coordinates, but was not satisfied since by doing so he ended up with 6N normal coordinates, whereas there only were 3N degrees of freedom. Born's normal coordinates could therefore not be independent, and by a qualitative argument Waller reasoned to only take the normal vibrations on one side of a plane through the origin of coordinates in the reciprocal lattice into account; if this lattice then was transformed, the correct number of normal coordinates was acquired. Waller then adopted his modification of Born's normal coordinates to Debye's theory for the thermal influence on X-rays diffracted in a crystal. He arrived consequently at results that differed from Born's and Debye's.[9] This led to the change from a e^{-M}- to a e^{-2M}-behavior for the intensity of the diffracted radiation (where M is a function of the frequency of the normal oscillations, the temperature, the angle of diffraction and the atomic masses).[10]

A lot of this work was done in 1923 while Waller spent four months in Göttingen. There he worked out a paper on the effect of heat motion on the interference of X-rays, and he sent the proofs to Peter Debye, who replied that what Waller proposed: "sehr gut richtig sein kann."[11] The professor of (experimental) physics in

Göttingen—James Franck—asked Waller to investigate the dispersion theories of X-rays further in connection with some experiments performed by one of his students. Together with among others Enrico Fermi, Waller attended the seminars headed by Hilbert and Born on "Struktur der Materie," but he did not, as we have seen, find Born's Chapter 19 on the normal coordinates in his "Atomtheorie des festen Zustandes (Dynamik der Kristallgitter)" in *Encyklopädie der mathematischen Wissenschaften* satisfying.[12]

For many years there had been an interest in investigating the analogy between ordinary light and X-rays refracting in a glass prism. Waller made some calculations on this back in Sweden and in 1924 he suggested an experiment to Manne Siegbahn, now professor of physics in Uppsala. The experiment was performed and a small refraction was indeed detected.[13] When this work was completed, Waller had also finished his doctoral dissertation, which arrived from the printers in the spring of 1925.[14] Later in 1925, as Manne Siegbahn had been awarded the retained Nobel Prize of 1924, we are given an illustration of the demarcation line being established between the two emerging cultures within Swedish physics—experimental atomic physics and theoretical physics. Hilding Faxén, another of Oseen's pupils, had to show Siegbahn curves based on some of Waller's dispersion calculations, because Siegbahn was not at all interested in seeing any actual calculations from Waller. Siegbahn said he was "a pure experimentalist who [did not] concern himself with theory."[15] A curve was enough. This might explain why Waller, writing a dissertation on X-rays, did not choose to stay in Uppsala to co-operate more with Siegbahn—one of the world's leading experimentalists in X-ray spectroscopy—despite Waller's own interest in X-rays. Waller's dissertation dealt with crystals—how to understand the nature of the atomic heat motion in a crystal lattice. Siegbahn was more concerned with precision surveying of elements for their X-ray spectra.[16]

Waller's dissertation contained his modifications of the theories of Debye and Faxén, as well as the X-ray refraction of the glass prism. This dissertation immediately aroused great attention from the international community of X-ray specialists.[17] Waller was awarded the highest marks for his dissertation. According to Oseen, these high marks were due to Waller's initiation of the joint investigation together with Siegbahn and Larsson.[18] This is worth noting: Oseen, who himself never co-operated with any experimentalist in that way, highly regarded co-operation between theoreticians and experimentalists.[19]

In 1925, Max Born saw the possible application of matrix algebra to Werner Heisenberg's idea of non-commutative quantities in order to describe atomic phenomena. Matrices were already used in Born's theory for crystal dynamics, where the eigenvalues and the eigenvectors gave the frequencies and the amplitudes of the normal vibrations. Waller, who was trained in this theoretical technology, Born's crystal dynamics, was familiar with matrix calculation—crucial knowledge in 1925.[20]

After Waller had been promoted and appointed "docent" at the University, he went on a study tour to Zurich.[21] There he continued his calculations from his dissertation and tried to improve them by including terms of the third and fourth degree in the expression for the potential energy of the crystal. Thus Waller was able to predict small deviations from the Bragg law of refraction. Debye, himself also working on this problem, became quite interested in this work and encouraged Waller to publish his results, but Waller was reluctant to do so, since he had problems with some integrals in the three dimensional case.[22]

During the first week of September in 1925, Waller attended the informal conference in Holzhausen by the Ammersee in Bavaria (held in the cottage of P. P. Ewald's mother). Oseen had encouraged him to go there and underlined the importance of making international contacts.[23] Attending this conference were: Lawrence Bragg, L. Brillouin, C. G. Darwin, P. Debye, A. D. Fokker, K. Herzfeld, R. W. James, M. von Laue, H. Mark, H. Ott, P. Scherrer, I. Waller, R. W. G. Wyckoff and P. P. Ewald.[24] These "Auktoritäten über die Intensitätsfrage bei Röntgeninterferrenzen," as Ewald named them, exchanged their views. Waller later recalled it as of "great value [. . .] to attend [the meeting and that it] was a great inspiration for future work."[25] This conference became a meeting ground for experimentalists and theorists or might be seen as what Peter Galison calls a trading zone.[26] They discussed such fundamental aspects as how to describe crystals: perfect or mosaic? The chief object of the meeting was to achieve consensus on interpretation techniques. Of course this was not easily achieved. For example, the British scientists felt a bit misunderstood, since Ewald had critizised their method of measuring the intensities.[27] One outcome of the conference was an agreement that the technique best suited to determine the degree of perfection in crystals was based upon X-ray intensity calculations.[28] At this meeting, Waller teamed up with R. W. James, an experimentalist specializing in low temperature X-ray diffraction techniques and working at W. L. Bragg's laboratory in Manchester.[29] James' results fitted well with Waller's calculations, and their co-operation made W. L. Bragg "delighted."[30] One might argue that Waller's and James' theoretical and experimental techniques respectively fitted well together. Their trading should prove successful.

Waller then spent eight months of the academic year 1925/26 in Copenhagen at the "Institut for teoretisk fysik." The papers on matrix mechanics had been recently published. That spring, Erwin Schrödinger started to publish his four papers on wave mechanics. Waller saw the possibilities of the new wave mechanics and started to work on the second order term Stark effects. Wolfgang Pauli and Schrödinger himself were calculating the first order terms.[31] The second order terms are, however, more complicated to calculate.

Waller was, after returning to Uppsala, now concerned with future possibilities for earning his living. There was an opportunity to lecture in Uppsala as "docent," "albeit this would entail a most severe break in the scientific work," as he put it in a letter to Oseen. Waller also felt that he would need to study some mathematics in order to manage the problems of the new wave mechanics.[32] Even if the wave mechanics is generally described in the literature as more accessible to most physicists than the matrix algebra, there is evidence that this or any mathematical equations requires an understanding of whole set of mathematical techniques or theoretical technologies. Waller was eager to return to Copenhagen and corresponded with Bohr on the calculation of the hydrogen atom using perturbation theory with relativistic considerations.[33] Waller's latest paper made Bragg ask if Waller still took an "interest in the problems of x-ray diffraction."[34] Despite all the excitement, Waller was still very much interested in his "old" subject: the theory of X-ray scattering, but now interested in basing it upon the new quantum mechanics. Thus he applied for a scholarship from the Rockefeller Foundation to travel once again.[35]

Waller was interested in applying quantum mechanics to X-ray scattering on an arbitrary configuration of atoms with thermal vibrations. This dispersion problem had already been treated by Schrödinger and Klein, who had modified the Kramers–Heisenberg dispersion formula. The result of Waller's investigation was that the scattered radiation appeared in a coherent and an incoherent part, where the coherent part corresponded to the classical case, and the incoherent part showed inelastic behavior—the vibration energy changed during scattering. This was, however, such a very small deviation that it could not be experimentally verified until the late 1930s.[36] Others also took an interest in these questions.

Max von Laue in his turn proposed altering the Debye–Faxén–Waller results by introducing a Doppler effect.[37] Waller showed, however, that the Doppler effect, suggested by von Laue, had no influence, and he also showed the good agreement between his own theory and James' new experimental results.[38] Waller's direct access to James' data was a clear advantage in this respect. James reported to Waller that, from the experimental activities at The Physical Laboratory in Manchester "it seems quite certain that your e^{-2M} is *much* better than Debye's e^{-M}."[39] Particularly, the difference could be seen for low temperatures. These results proved that Waller's modification of Born's and consequently Debye's theories were correct.

This collaboration proved fruitful and in 1927 Waller went to spend three months at the laboratory of Lawrence Bragg in Manchester. There the theoretician Waller and the experimentalist James could co-operate directly. Waller also began collaborating with the Cambridge theoretician Douglas R. Hartree. James had made investigations on rock-salt, from room temperature up to 600°C, in 1925. During the following years, together with E. M. Firth, he extended the measurements down to liquid air temperature (−183°C). Since Waller had modified the Debye theory on the temperature effect, these testing possibilities were of obvious concern to him. The low-temperature facilities made it also possible to investigate the existence of zero-point energy. Douglas Hartree had been calculating atomic electron distributions, and by using his invention of the self-consistent field, he was able to produce calculations of the atomic form factors.[40] Hartree is almost an ideal type of the more technical theoretical physicist. Because of the method that he had developed for calculating the atomic fields and charge distributions, he was asked by various people "to calculate results" for their particular cases. Of course—being the one who had invented this technique, i.e. to calculate atomic scattering factors— Hartree thought himself to be the best suited for the job. To disseminate the technique at too early a stage might also be a risk—so he chose to perform the calculations himself. This tiresome numerical labor, however, made him confess: "One really wants a staff of [human] computers for numerical work in connection with atoms[,] like the astronomers have for doing their numerical work!"[41] Following a request from James, Hartree thought of teaching someone at Manchester "the numerical technique of my method of working out atomic charge distributions, for F-curves." The general idea for the collaboration was that Waller should formulate a quantum mechanical theory for the scattering of X-rays by atoms; that Hartree should calculate charge distributions; that James should measure the scattering; and that Waller should discuss the temperature factor from these results.[42]

In Brussels, in October 1927, the fifth Solvay conference was taking place. The conference was opened by W. Lawrence Bragg, who presented the works of James, Hartree and Waller, while Waller was in Uppsala delivering lectures on mechanics.[43]

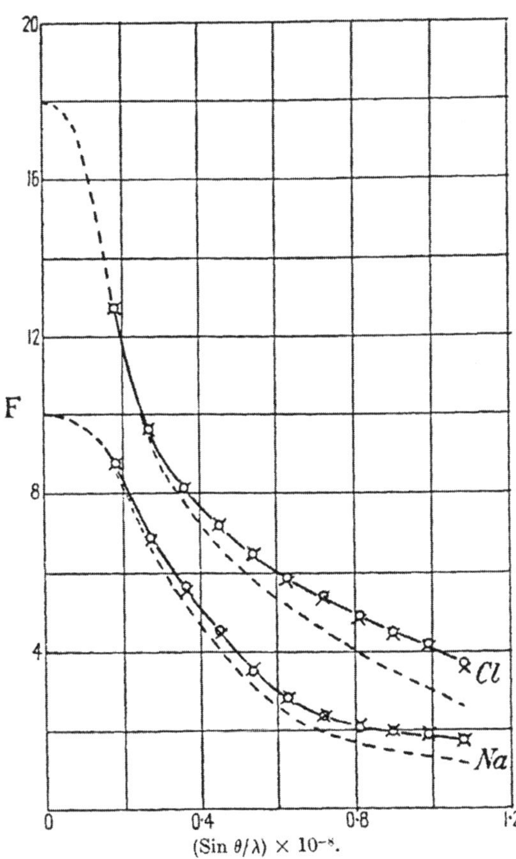

FIGURE 1.   The figure shows evidence of zero-point energy. Dotted line without zero-point energy, solid line with zero-point energy. ○ and × measured values.[47]

During the conference, James reported to Waller on Hartree's most recent work, which seemed to be in favor of the zero-point energy.[44] Bragg was excited by the agreement between theory and experiment and wrote: "It seems to me that the check between James' experimental results corrected by you for thermal vibration and zero-point energy, with Hartree's F curve calculated from the sodium and chlorine atoms, is quite striking, and personally I feel rather excited over the results."[45] In order to write their paper Bragg requested that Waller should state in a few words the argument of e^{-2M} instead of e^{-M}. Bragg thought "that if it could be stated in somewhat physical terms it would be worth including."[46]

The Waller-Hartree-James collaboration gives a fine example of co-operation on several levels. Waller and Hartree, with two different theoretical techniques, were able to intercalate their respective theories and put the combined result to the test in James' experiment. Waller supplied the theory for the lattice, Hartree the atomic form factors, and James the experimental technique for the wide temperature range. Waller's and Hartree's different theoretical technologies were both of a calculating sort, so here we have evidence of two different theoretical technologies of the same kind or level working well together, complementing each other. Waller and Hartree continued collaborating while Hartree was in Copenhagen a year later.[48] And, of course, we also have evidence of the fitting of theoretical and experimental technologies or a trading of different scientific practices.

Transformation Theory and Negative Energy

P. A. M. Dirac is considered the epitome of a theoretical physicist, or by a biographer described as being "indeed a mathematical physicist."[49] For Dirac: "empirical matters were subordinated to theoretical considerations, derived from principles."[50] Dirac, as almost every theoretical physicist of the 1920s, was greatly committed to relativity. The mathematical inclination of his way of doing physics was apparent in his transformation theory of 1927.[51] But it was not until 1928 that he made a serious attempt to deal with the problem of constructing a relativistic quantum theory. At this time, the connection between relativity and the concept of spin was not apparent. The attempts by Pauli to create operators that accounted for the spin, i.e. spin operators, led him to formulate the non-relative Hamiltonian termwise and to connect spin matrices with corresponding spin operators working on the two-component wave function.[52] The result of Pauli's work was the same as Heisenberg–Jordan had arrived at before him. This was still being more of an appending technique, rather than an incorporating technique, to solve the problem.[53] In Dirac's transformation theory, a formalism had been created that incorporated a number of theoretical technologies: matrix mechanics, wave mechanics, operator formalism, and Dirac's own q-number algebra. This theory needed, however, the equations of motion to be of the first order in the time derivative $\partial/\partial t$. This excluded the relativistic Klein–Gordon theory, for example. It has been suggested that the key motivation for Dirac's dismissal of the Klein–Gordon theory was not the non-positive particle density or the negative energies, but because of it not being of the first order in $\partial/\partial t$.[54]

It has been suggested that Dirac was standing well outside the spectroscopic camps of the Continent, which had introduced the notion of spin in order to explain doublet structures in the spectroscopic data. It has also been suggested that empirical problems were not the prime concern for Dirac's interest in spin.[55] And it is also characteristic of Dirac's high theoretical attitudes that he considered his transformation theory higher than the relativistic theory of the electron, which can be seen as a "mere" application of the more general transformation theory to the electron.[56] The equations were supposed to manage both the transformations according to the theory of relativity, and to quantum mechanics. Dirac made what has been described as an inspired mathematical guess, starting from a square root operator—the energy operator—but modifying the operator by extending the Pauli matrices from 2×2 to 4×4 matrices, and used a set of conditions for these matrices, to deduce their elements. Relativistic equations of the electron emerged from this work. These equations were invariant under Lorentz transformation, thus passing the relativity test; they were linear in $\partial/\partial t$ and $\partial/\partial x$ and they passed the quantum mechanical test as well. This was entirely a formal/mathematical performance, but Dirac put it to the test by introducing an electromagnetic field. By doing this, Dirac reached an equation where two additional terms appeared, i.e. additional compared to all previous theories. The correct magnetic and spin moments of the electron were deduced from this theory. Dirac's theory did not just explain this, but it was also a powerful conceptual contribution. The theory was immediately considered a masterpiece.[57]

One of the reasons for Dirac to discard the Klein–Gordon theory was that the equation did not exclude the apparently ambiguous negative energy solutions. Dirac realized this difficulty and proposed that: "One gets over the difficulty on the classical theory by arbitrarily excluding those solutions that have a negative W." Dirac continued by pointing out that:

> One cannot do this on the quantum theory, since in general a perturbation will cause transitions from states with W positive to states with W negative. Such a transition would appear experimentally as the electron suddenly changing its charge from $-e$ to e, a phenomenon which has not been observed. The true relativity wave equation should thus be such that its solutions split up into two non-combining sets referring respectively to the charge $-e$ and the charge e.[58]

The negative energy states were inherent in all relativistic theories. These negative energies, however, were considered insignificant and dismissed as "unphysical" until 1928/29—there had been other problems to attend to.[59]

Dirac was satisfied with his transformation theory and did not really bother to work out all the details of his relativistic theory for the electron. But with all the attention his theory aroused from the rest of the physicist community, there were several others who were willing and eager to do so. Oskar Klein, who in his own attempt to formulate a relativistic theory had dismissed the negative energy solutions,[60] drew attention to the situation with what was later to be named the "Klein paradox," in which electrons governed by the Dirac equation coming upon a potential barrier of sufficient height, more electrons were reflected than originally sent in.[61] This was much debated at the first Copenhagen conference during Easter 1929 where among others Bohr, Klein, Kronig, Holtsmark, Rosseland, Pauli, Jordan, Ehrenfest, Gamow, Darwin, Heitler, Waller, participated. Dirac was not there, however. Two weeks earlier, he had embarked on a trip to United States together with Heisenberg. There were many attempts to get rid of the negative energies, but none of these were successful. It turned out that the negative energy solutions were inherent in the Dirac theory.

Reinterpretation of Calculation

In a letter from Niels Bohr's institute in April 1928, Oskar Klein mentioned to Waller that Dirac would visit the institute for a week, and: "We are all very interested in his latest work. Nishina and myself have been giving the dispersion some thought according to his new equations, but we have not come anywhere yet. Perhaps you have been dealing with this problem?"[62] Waller and Dirac had been corresponding earlier and Waller was eager to meet Dirac. Waller went to Copenhagen, whence he wrote back to Sweden and reported to Oseen on the most recent developments within theoretical physics. This was probably in part to help Oseen survey possible candidates for the Nobel Prize.[63] Waller met Dirac there, but only for one day since Dirac continued on to Ehrenfest in Leyden. Dirac's efforts then, Waller told Oseen, seemed:

> to be to try to avoid the incompleteness still affecting the theory, i.e. the 'symmetry' of $+e$ and $-e$ and the consequential difficulty of having too many solutions. [Dirac's] view seems to be that at present this problem is of greater importance than the development of the theory to include several electrons.

But even if Dirac's equations were in vogue, Waller stated that: "Quantum theory now seems to have entered a more steady stage, after the more easily accessible problems have been dealt with. The statistical point of view seems to have become widely accepted."[64] Waller continued: "Dirac now seems to be taking a leading position within the field of quantum mechanics. This is extraordinary as one hears he is only 25 years old."[65] Waller reported back to Oseen that he would "come back, as soon as possible, with more complete notes concerning the guidelines according to which one now seems to work within the field of quantum mechanics." Since

Oseen was interested in having reprints of Dirac's papers, Waller told him that they would perhaps be reprinted and edited in the same manner as Schrödinger's. Waller assured Oseen that: "I did not find it suitable to mention the connection with the Nobel Prize and I am convinced this was of no consequence."[66] Waller answered Oseen's request for copies with: "I am sorry I cannot send you my copies of Dirac's work; I need them daily."[67]

In the summer of 1928, as we have already seen, Waller was in England cooperating with James and Hartree. In August, Waller went once again to Niels Bohr's institute. He wrote to Dirac telling him that he had almost finished his "calculations about the scattering of radiation according to your theories."[68] But it would take him some further effort before he had finished this work. In the beginning of 1929, he sent off a version to Dirac asking for comments and critique.[69] Dirac read the paper[70] that Waller had sent him. He thought it "quite correct" and made some technical remarks. He also pointed out that:

> You should mention that the scattering process 'zweiter Art,' those in which two light-quanta are emitted, occur for all frequencies v_r and v_s for which $hv_r + hv_s = H_0(J') - H_0(J''')$, not necessarily those for which v_r is the frequency of the incident radiation. When there is incident radiation it stimulates them in accordance with Einstein's law.
>
> [And on the section dealing with scattering from atoms with more than one electron Dirac underlined] that we look for *periodic* solutions of the wave equation. These periodic solutions are simpler than the non-periodic ones used in the earlier method. Since they are periodic, however, they represent stationary states of the system, *i.e.* states in which the physical conditions are not changing, so that we must have a continuous stream of incident particles (light quanta) and a continuous stream of scattered ones. The method is thus suitable only for studying scattering and dispersion, and not emission and absorption, since emission and absorption coefficients cannot even be defined without reference to a non-steady state of affairs.[71]

We have seen that at the first Copenhagen conference in the spring of 1929, Dirac's theory was much debated. Later the same year, Waller also received some opinions from Pauli on his recent paper on the scattering of short wave radiation from atoms according to Dirac's theory in the *Zeitschrift für Physik*.[72]

> Heute möchte ich Ihnen schreiben über die Frage der Herleitung den Klein–Nishinaschen Streuformel aus Diracs Spin- + Strahlungstheorie. Herr Rabi, der dieser Problem hätte behandeln sollen, hat nichts weiter von sich hören lassen und ich glaube nicht, daß er es zustande bringen wird; er ist übrigens wieder in Amerika. Infolgedessen würde ich er jetzt doch sehr wünschenswert fürden, wenn Sie die formel (42) Ihrer letzten Arbeit für den Fall freier Elektronen weiter diskutieren würden. Auch würde es mich interessieren, über das Resultat, wenn Sie soweit sind von Ihnen zu hören.[73]

The scattering formula (42) was the conclusive result of Waller's paper on the scattering of short wave radiation through atoms.

$$D\left(J'J''\right)=\frac{e^2c^2}{4\pi^2 vv'}\sum_{J'''}\left[\frac{\left(EA_{JJ'''}\right)A'_{J'''J''}}{E_{J'''}-E_{J'}-hv}+\frac{A'_{JJ'''}\left(EA_{J'''J''}\right)}{E_{J'''}-E_{J''}+hv}\right]e^{2\pi iv't} \qquad (42)$$

where

$$A_{JJ'''}=\int\tilde{\psi}_{J'}\,\rho_1\sigma e^{-ixx}\psi_{J'''}dv \qquad (42')$$

J' denotes the initial state, J'' the final state and J''' the intermediate state.

Waller noted that the intermediate states J''' in eq. (42) were associated with negative energies, even if he put these in quotation marks. Waller then, acting on Pauli's suggestion, extended his calculations of bound electrons to a use of Dirac's relativistic theory on the problem of scattering from free electrons.

The content of Waller's scattering formula for electrons bound in an atom implied that the atom existed in a certain initial state and then transformed into an intermediate state before transforming once again into the final state. The essential thing was the appearance of some intermediate states, which could be any state of the atom. This result was connected to the question of negative energies, since the essential part of the radiation, when there was an impinging light quantum of low frequency, originated from processes where the intermediate state was one of negative energy.[74]

Waller was not satisfied with the elusive interpretation of the negative energies, so he wrote to Dirac asking "Have you perhaps found some way to avoid that difficulty about the negative energy?" He continued hoping for more precise guidance elaborating his problem:

The other difficulty refers to the scattering formula which follows from your relativistic dynamics of the electron. This scattering formula may be deduced by using the Schrödinger–Gordon–Klein density method or your radiation theory. Denoting the electric vector of the incident radiation by the real part of $Ee^{2\pi i\left(vt-\frac{nx}{c}\right)}$, x being the position vector from a fixed point, and the initial state of the atom by J', it is then found, that the components of the radiation scattered in the direction n' are given as the sum of dipol[e] moments (the real parts of these should be taken):

$$\sum_{J''}^{H_0(J'')<H_0(J')+hv} D\left(JJ''\right)+\sum_{J''}^{H_0(J'')<H_0(J')-hv} D\left(J''J'\right) \quad,\text{where}$$

$$D\left(J'J''\right)=\frac{e^2c^2}{4\pi^2\nu\nu'}\sum_{J'''}\left[\frac{(EA_{J'J'''})A'_{J'''J''}}{E_{J'''}-E_{J'}-h\nu}+\frac{A'_{J'J'''}(EA_{J'''J''})}{E_{J'''}-E_{J''}+h\nu}\right]e^{2\pi i\nu't}$$

here

$$A_{J'J''}=\int\tilde{\psi}_{J'}\rho_1\sigma e^{-\frac{2\pi i\nu}{c}xn}\psi_{J''}dv,\ \cap A'_{J'J''}=\int\tilde{\psi}_{J'}\rho_1\sigma e^{-\frac{2\pi i\nu'}{c}xn'}\psi_{J''}dv,\ \text{the integra-}$$
tions being performed over the whole space (volume element dv). $\psi_{J'}$ are wave functions with the four components $\psi_{J'}^{(i)}$ ($i = 1, 2, 3, 4$). $E_{J'}$ the energy value corresponding to the state J', and $\nu'=\frac{1}{h}\left(E_{J'}-E_{J''}+h\nu\right)$. Neglecting relativity and spin corrections (i. e. using the ordinary Schrödinger equation) I have found that

$$(II)\quad D\left(J'J''\right)=-\frac{e^2}{4\pi^2m\nu\nu'}\left\{E\left(J'\left|e^{i\frac{2\pi i}{c}(n'\nu'-n\nu)x}\right|J''\right)+\textit{other terms}\right\}$$

We must now be able to get (II) from (I) by neglecting relativity and spin corrections. It is then found that the first term in { } of (II) which is the essential term for high frequencies ν, comes essentially from that part of $\sum_{J'''}$ in (I) which refers to negative energies $E_{J'''}$, whereas the other part of $\sum_{J'''}$ in (I) (gives) corresponds to the other terms in (II). (I have made the calculations in detail but I shall not trouble you about it). Now I really find it difficult to see how far it is justified to use your radiation theory for deducing (I), because one has then to consider transitions $E_{J'}\rightarrow E_{J'''}$ to negative energy values $E_{J'''}$. By using the density method these states enter more formally into the calculation, so the difficulty is perhaps not so serious there.[75]

The difficulty was serious, but Dirac at first was not convinced that the negative energies had to have any significant meaning. He replied:

Are you sure you are right about the important part of the sum

$$(A)\quad\sum_{J'''}\left\{\frac{(EA_{J'J'''})\bar{A}_{J'''J''}}{E_{J'''}-E_{J'}-\hbar\nu}+...\right\}$$

for large ν, coming from those terms J''' referring to a negative energy? If one worked out the corresponding expression for scattering with neglect of relativity and spin, I think one would find that it corresponds term by term, *i.e.* each term in the non-relativity expression would approximately equal one of those terms in the relativity expression (A) belonging to a positive energy. It therefore seems to me that (A) would be approximately correct if one neglects the negative energy terms. In any case I think the negative energy terms are very small when one takes the proper retardation effects into account in calculating the matrix elements.[76]

Waller found "it difficult, however, to see how it would be possible to exclude in the dispersion formula derived from your dispersion theory and your spin theory those terms which correspond to negativ[e] energies for the intermediate states J'''."[77] The questions in Waller's first letter apparently made Dirac think the problem over once more, and while pondering over this he invented the concept of a uniform distribution of electrons with negative energies, thus giving an answer to Waller's question about the intermediate states of negative energies inherent in the scattering process. When Dirac had concluded this idea, he wrote an often-cited letter (from the same day as Waller replied to Dirac) to Bohr explaining his new ideas.[78] The day after Dirac had written to Bohr explaining his new ideas, he also wrote to Waller. This was only nine days after the letter in which Dirac had thought of "neglect[ing] the negative energy terms."[79]

I am afraid my answer to your second question was wrong and I am sorry I doubted the accuracy of your result. I have been looking into the question of the scattering of radiation by a free electron on the relativistic spin theory (the Klein-Nishina work) and I find that the same thing that you found for atomic scattering occurs there, namely;—the greater part of the scattering comes from 'double transitions' in which the intermediate state is one of negative energy for the electron.

I have recently obtained, I think, a solution of the negative energy paradox and can on this basis give an answer to your question. My idea is to assume that there are so many electrons in the world that all the states of negative energy are filled up, and that there are some electrons left over, which are obliged by the Pauli principle always to have positive energy. This will mean an infinite density of electrons with −ve energy, but since their distribution is quite uniform they will not be observable. If there is a state of negative energy that is not occupied, the resulting 'hole' will have a positive energy, and will also move in an external field as though it had a +ve charge. We can therefore interpret this hole as a proton. I hope, by taking the Austausch interaction into account, to get an explanation of the different masses of proton and electron, but have yet not done this.

If we apply these ideas to the problem of scattering, we see that double transitions in which for the intermediate state the electron has negative energy must be excluded by the Pauli principle. On the other hand we can have double transitions in which firstly one of the distribution of −ve electrons jumps to a state of +ve energy and secondly the original +ve electron jumps down and fills up the hole. These processes will just make up for those excluded above, (in which the electron jumps to a –ve energy for the intermediate state) and justify our including the latter in our scattering formula.[80]

Waller later gave the following account:

The above-mentioned paradox about the scattering of radiation by an atom was almost automatically solved by this theory. Exactly the same formula as before was obtained. Only now the intermediate states could not be just any states. Starting from a state with positive energy, a transition to a state with negative energy cannot take place since the states with negative energies are all occupied. This is an important objection. Instead you have to reinterpret this calculated result. First you have to imagine that a transition takes place from a state with negative energy, and it has to be to a state with positive energy. The hole thus formed can then be filled with an electron, which has a positive energy. You get the same transitions where negative energy plays a role as you had before when the negative energy states were empty, only in the reverse order, and the matrix elements are the same—the whole difficulty has disappeared.[81]

After having his calculations confirmed by Dirac, Waller published a paper on the scattering of radiation from free and bound electrons, according to the new relativistic theory of Dirac and allowing for the negative energies.[82] In this paper, Waller could show that it was necessary to take both the positive and negative intermediate energy states into account. The contribution from positive or negative energy states to the scattering process depended on the frequency of the incident radiation. Pauli wrote to Waller after reading the proofs of this article: "Ihr Resultat betreffend die Klein–Nishina'sche Streuformel ist ja sehr befriedigend. Ich hoffe sehr, daß Sie Ihre Rechnungen bald publisieren werden!"[83] So Waller did, the paper was sent in a couple of weeks later. In a letter to Dirac, he complained though that there did not "seem to be any good measurements of the intensity of scattered radiation for very short wavelength available at present." He had heard that Lise Meitner was doing work on this so there might be results for him to test his calculations eventually.[84] Waller was anxious to intermediate good measurements with satisfying calculations. This was more important to him than Dirac's conceptual breakthrough, one might conclude. In this paper, he addressed the problem of scattering of radiation both from free and bound electrons according to Dirac's new theory, and the meaning of the negative energy states was taken into special consideration. The paper consisted of two sections: the first on bound electrons and the second on free electrons. For bound electrons, the scattering process resulted from absorption and an emission process following each other. For this case, it was important to take into consideration the intermediate negative energy states of the electron. However, Waller meant that the new ideas of Dirac would only be a formal contribution to his work.[85] His own formulas and calculations had already been around for a while.

For the case of long wave radiation, the contribution from negative energy states was very small and possible to neglect, but for the case of X-ray radiation Waller was clear; according to his calculations all the negative energy states were necessary to describe the scattering of X-rays from atoms. "Am schärfsten ist dies für die kohärente Streustrahlung von R. W. James in Manchester und seinen Mitarbeitern bewiesen. Dies zeigt in zwingenden Weise die Notwendigheit, in der Streuformel (10) Zustände negativer Energie zu berücksichtigen."[86] Waller could not bring his calculations to an end, given the experimental data from James, without including the troublesome negative energies. He could not just "naturally" exclude these energy states by referring to conceptual or epistemological difficulties as one had done before. With the support of Dirac's new idea, Waller's calculation of the scattering from atoms guided by the results from the Manchester experiments now was put on firmer ground. To include the negative energies did not create any formal or mathematical problems; but some of the conceptual problems remained even after Dirac's innovation. How to interpret the actual physics involved in the scattering process?

Waller had also dealt with scattering from free electrons. Here the electron did not move in any potential field and its energy eigenvalues were just the positive *and* negative energies of a relativistic particle. In this section of the paper, Waller referred to an unpublished investigation of Heisenberg, where Heisenberg had shown that when $h\nu_S/mc^2 \to 1$ in the classical Thomson scattering formula, negative energies appeared.[87] This was probably communicated to Waller by Pauli, because Heisenberg had written to Pauli one month after the "Leipziger Universitätswoche" in 1928, where Dirac had spoken "Über die Quantentheorie des Elektrons,"[88] and following the Diracian dispersion theory Heisenberg had arrived at the classical Thomson formula. Heisenberg concluded in this letter: "Es ist doch sehr amüsant, daß diese Formel gerade erst auf Grund der verrückten Übergänge herauskommt."[89]

Now Waller arrived at similar results as he performed his calculations, arriving at the Klein–Nishina formula for the total scattered intensity. He concluded, once more, that it was necessary to take into account both the positive and the negative energies. The process depended upon the frequency of the incident radiation. Waller concluded that when $h\nu_S/mc^2 \to 1$ there would only be negative energy states involved in the scattering process; and the result would be the classical formula, i.e. the Thomson formula.

Epilogue

Waller went to Uppsala to study during the First World War. He left Gothenburg, the Swedish town that has always faced Great Britain in more than a geographical meaning, to enter the traditional university of Sweden. At this university, German in style, Waller developed an interest in mathematical physics because of Oseen's lectures and the exciting news of the confirmation of Einstein's theory of relativity in 1919. After Oseen assigned him the problem of describing the specific heat, Waller changed to X-ray diffraction, a choice that would have proven reasonable, since Manne Siegbahn was about to establish his research school of X-ray spectroscopy in Uppsala. Finishing his dissertation at a time when international scientific contacts were slowly recovering after the World War, Waller made close contacts with both experimentalists and theoreticians within a clearly defined research field. He also did more technical, theoretical work in connection with the new quantum physics. Keeping up his old line of work and circling around the more general problem, X-ray scattering from electrons, Waller was led to the work of Dirac. Waller pursued his career in an intermediate, rather modern, way, by interacting with Oseen—the old mentor at home—and by taking part in the international work within quantum physics both with experimentalists and other theoreticians. This interaction clearly influenced the making of Waller as a theoretical physicist. We have seen how Ivar Waller acquired profound knowledge and competence within the field of X-ray scattering by cooperating with experimentalists such as Manne Siegbahn and R. W. James, but also by cooperating closely with other theoreticians such as Douglas Hartree, by participating in conferences, and by corresponding with other physicists. Waller's work during the 1920s might seem one-sided since he was exclusively occupied with the theory of scattering of radiation and this was commented upon by Oseen, who wrote that: "This problem is so closely connected with the central problem of the new physics, the understanding of the structures of atoms, the last and at present most fascinating chapter of theoretical physics."[90] So Waller's narrow focus ought to be excused he concluded.

We have seen how Waller's temporary collaborations both with experimentalists and other theoreticians were fruitful, and that he from his calculating practice could intermediate and influence development in both directions, guiding experiments and (high) theory. The notion of intermediate theoretical physics thus helps us to get a better understanding of the practices within physics.

Notes

*This paper is based on Karl Grandin, *Ett slags modernism i vetenskapen: Teoretisk fysik i Sverige under 1920-talet*, diss. (Uppsala, 1999), esp. chapt. 4. Anders Bárány has made important suggestions for this paper. Valuable comments on a previous version of this text was given by Sven Widmalm and John L. Heilbron.

    1. Geoffrey Cantor, "The making of a British theoretical physicist—E.C. Stoner's early career," *The British Journal for the History of Science*, 27 (1994), pp. 277–290. Cantor's account is an effort to adopt an "existential approach" to the scientific biography of E. C. Stoner, although the financial pressures and career motives are also stressed (pp. 288 ff). Cantor's account bears on Stoner's diary entries, whereas this paper originates in scientific correspondence, thus perhaps

not forming an existentialistic scientific biography. However, the term "making of a theoretical physicist" captures one of the main themes in this paper.

2. Andrew Warwick, *Masters of theory: Cambridge and the rise of mathematical physics* (Chicago & London: Chicago University Press, 2003); idem, "Cambridge mathematics and Cavendish physics: Cunningham, Campbell and Einstein's relativity 1905–1911: Part I: The uses of theory," *Studies in History and Philosophy of Science*, 23:4 (1992), pp. 625–656; idem, "Cambridge mathematics and Cavendish physics: Cunningham, Campbell and Einstein's relativity 1905–1911: Part II: Comparing traditions in Cambridge physics," *Studies in History and Philosophy of Science*, 24:1 (1993), pp. 1–25; Peter Galison, *How experiments end* (Chicago: Chicago University Press, 1987), chapt. 1 and 5, esp. pp. 69f, 244f; idem, *Image & Logic: A material culture of microphysics* (London & Chicago: Chicago University Press, 1997), chapt. 1 and 9.

3. Warwick, "Cambridge mathematics and Cavendish physics, I," pp. 633ff.

4. The question might arise what Dirac would have thought of such a distinction. "*The only object of theoretical physics is to calculate results that can be compared with experiment.*" (emphasis in original) P. A. M. Dirac, *The principles of quantum mechanics* (Oxford, 1930), p. 7.

5. Wolfgang Pauli wrote this in 1930 in a letter to Oskar Klein, in which Pauli congratulates Klein on his professorship at Stockholm Högskola. Wolfgang Pauli to Oskar Klein, 12/12 1930, [261]. Armin Hermann and Karl von Meyenn, eds., *Wolfgang Pauli Wissenschaftlicher Briefwechsel mit Bohr, Einstein, Heisenberg u.a. Bd. 2: 1930–1939* (New York: Springer Verlag, 1985), p. 43.

6. The fifteen lectures were published later in 1919 as C. W. Oseen, *Atomistiska föreställningar i nutidens fysik: Tid, rum och materia* (Stockholm, 1919).

7. Ivar Waller, "Prövningsmöjligheterna av Einsteins allmäna relativitetsteori," *Kosmos*, 2 (1922), pp. 212–225, on p. 225.

8. Peter Debye, "Zur Theorie der anomalen Dispersion im Gebiete der langwelligen elektrischen Strahlung," *Berichte der Deutschen physikalischen Gesellschaft*, 15 (1913), pp. 777–793; idem, "Interferenz von Röntgenstrahlen und Wärmebewegung," *Annalen der Physik*, 43 (1914), pp. 49–95. In 1912, Max von Laue had published on the X-ray interference and the thesis by P. P. Ewald had appeared. Max Born and Theodor von Kármán also contributed with important works from this time on the specific heat. Cf. Ivar Waller, "Memories of my early work on lattice dynamics and X-ray diffraction," *Proceedings of the Royal Society of London*, A371 (1980), pp. 120–124.

9. Ivar Waller, "Early history of lattice dynamics," in M. Balkanski, ed., *Proceedings of the international conference on lattice dynamics: Paris, September 5–9, 1977* (Paris, 1978), pp. 3–5. Waller, "Memories of my early work." Also see Hilding Faxén, "Die bei Interferenz von Röntgenstrahlen durch die Wärmebewegung enstehende zerstreute Strahlung," *Annalen der Physik*, 54 (1918), pp. 615–620.

10. Ivar Waller, "Zur Frage der Einwirkung der Wärmebewegung auf die Interferenz von Röntgenstrahlen," *Zeitschrift für Physik*, 17 (1923), pp. 398–408. See also idem, "Memories of my early work."

11. Peter Debye to Ivar Waller, 6/7 1923, Ivar Waller Archives, Center for the History of Science, Royal Swedish Academy of Sciences, Stockholm—hereafter IWA. The paper was Ivar Waller, "Zur Frage der Einwirkung der Wärmebewegung auf die Interferenz von Röntgenstrahlen," *Zeitschrift für Physik*, 17 (1923), pp. 398–408.

12. Max Born, "Atomtheorie des festen Zustandes (Dynamik der Kristallgitter)," V: 25 of *Enzyklopädie der mathematischen Wissenschaften* (Leipzig, 1923). Ivar Waller to C. W. Oseen, 19/6 1923, Oseen Family Archives, Landsarkivet, Lund, hereafter OFA. On Born cf. Nancy Thorndike Greenspan, *The End of the Certain World: The Life and Science of Max Born* (New York: Basic Books, 2005), pp. 54ff.

13. A. Larsson, M. Siegbahn and I. Waller, "Der experimentelle Nachweis der Brechung von Röntgenstrahlen," *Die Naturwissenschaften*, 12 (1924), pp. 1212–1213.

14. Ivar Waller, *Theoretische Studien zur Interferenz- und Dispersionstheorie der Röntgenstrahlen* (Uppsala, 1925).

15. Hilding Faxén to Ivar Waller, 13/11 1925, IWA.

16. In early 1926, Waller was certain that the co-operation with W. L. Bragg would *not* lead to any awkward situations of the kind experienced before, when co-operating with an experimentalist, i.e. Siegbahn. Ivar Waller to C. W. Oseen, 23/1 1926, OFA.

17. Max Born wrote in 1942 that in Waller's "brilliant dissertation (Uppsala, 1925) which contains the complete formulæ for the most general types of lattices; nothing essential has been added to these formulæ by later theoretical research." Max Born, "Theoretical investigations on the relation between crystal dynamics and X-ray scattering," in *Reports on progress in physics*, IX (1942–43), pp. 294–333, on p. 295. At the first international conference on the interaction between phonons and X-rays and neutrons, Ewald suggested that the formula for the temperature dependence of the X-ray and neutron diffraction by crystals should be named the Debye–Waller formula. According to "Some references to papers by I. Waller," ms., IWA.

18. According to C. W. Oseen's expert advisory report on Waller as applicant for the professorship of Mathematical physics and mechanics at Stockholm Högskola 1930. *Professuren i Matematisk fysik och mekanik vid Stockholms högskola 1930*, O22.

19. Niels Bohr also pointed out Ivar Waller's cooperation with experimentalists. See Niels Bohr's expert advisory report on Waller as applicant for the professorship of Mathematical physics and mechanics at Stockholm Högskola 1930. *Professuren i Matematisk fysik och mekanik vid Stockholms högskola 1930*, B12. Waller's way of doing theoretical physics—in close contact with experimental physicists—depended of course on the area where he had his theoretical interests. Cf. also the co-operation between the Swedish experimentalist Manne Siegbahn and the German theoretical physicist Arnold Sommerfeld in Thomas Kaiserfeld, "When theory addresses experiment: The Siegbahn–Sommerfeld correspondence, 1917–1940," in Svante Lindqvist, ed., *Center on the periphery: Historical aspects of 20th-century Swedish physics* (Canton, MA: Science History Publications, 1993), pp. 306–324.

20. Ivar Waller, "Memories of my early work," p. 123.

21. Uppsala universitet, filosofiska fakulteten, Matematisk-naturvetenskapliga sektionen, protokoll 1925, §7 protokoll 6/5 1925. Interesting to note is that Waller had applied to be "docent" in (the more modern terminology) *theoretical physics* but the faculty appointed him associate professor in *mathematical physics*, which was the proper, official name.

22. Ivar Waller to C. W. Oseen, 2/8 1925, Carl Wilhelm Oseen Archives, Center for the History of Science, Royal Swedish Academy of Sciences, Stockholm, hereafter CWOA.

23. "It is by such relations you are secured from the hush(ing) up of your work! This is also of importance for the academic career." C. W. Oseen to Ivar Waller, 15/12 1925, IWA.

24. Ivar Waller, "Memories of my early work," p. 122.

25. Ibid., p. 123.

26. Galison, *Image & Logic*, pp. 46ff and chapt. 9.

27. C. G. Darwin to Ivar Waller, 10/9 1925, 26/2 1926; R. W. James to Ivar Waller, 2/10, 25/11 1925, IWA.

28. R. W. James, "Early work on crystal structure at Manchester," in P.P. Ewald, ed., *Fifty years of X-ray diffraction: Dedicated to the International Union of Crystallography, July 1962* (Utrecht, 1962), pp. 420–429, on pp. 423f. See also [R. W. James], "Early work on crystal structure at Manchester," dupl. ms. from P. P. Ewald in IWA.

29. Waller and James co-authored three papers: "On the temperature factors of x-ray reflection for sodium and chlorine in the rock salt crystal," *Proceedings of the Royal Society of London*, A117 (1927–28), pp. 214–223; "Is crystal reflection of x-rays entirely a classical phenomenon?" *Nature*, 122 (1928), pp. 132–133; and "An investigation into the existence of zero-point energy in the rock-salt lattice by an x-ray diffraction method," *Proceedings of the Royal Society of London*, A118 (1928), pp. 334–350, which was co-authored with D. R. Hartree.

30. W. L. Bragg to Ivar Waller, 13/11 1925. Cf. also W. L. Bragg to Ivar Waller, 25/11 1925; Note for scholarship application in W. L. Bragg to Ivar Waller, 18/1 1926. Bragg was interested in Waller's

work but later confessed: "I wish I knew more mathematics and could follow all your work better." W. L. Bragg to Ivar Waller, 22/3 1926, IWA.

31. Wofgang Pauli to Ivar Waller, 7/6, 17/6 and 18/6 1926, IWA. In this correspondence, Pauli also wondered how Waller's efforts on the H^+ problem was going. Ivar Waller, "Der Starkeffekt zweiter Ordnung bei Wassenstoff und die Rydbergkorrektion der Spektra von He und Li^+," *Zeitschrift für Physik*, 38 (1926), pp. 635–646. See also G. Wentzel to Ivar Waller, 27/7 1926, IWA.

32. Ivar Waller to C. W. Oseen, 25/6 1926, CWOA.

33. Ivar Waller to Niels Bohr, 29/12 1926, Niels Bohr Scientific Correspondence, Niels Bohr Archives, Copenhagen, hereafter BSC.

34. W. L. Bragg to Ivar Waller, 26/10 1926, IWA.

35. C. W. Oseen to Ivar Waller, 29/1 1928; Niels Bohr to Ivar Waller, 20/2 1928, IWA.

36. Ivar Waller, "Memories of my early work," pp. 123f. This led to Ivar Waller, "The transition from ordinary dispersion into Compton effect," in *Nature*, 120 (1927), pp. 155–156.

37. Max von Laue to Ivar Waller, 13/10 1926, IWA.

38. Ivar Waller, "Die Einwirkung der Wärmebewegung der Kristallatome auf Intensität, Lage und Schärfe der Röntgenspektrallinien," *Annalen der Physik*, 83 (1927), pp. 153–183.

39. R. W. James to Ivar Waller, 31/8 1926, emphasis in original. See also R. W. James to Ivar Waller, 7/7 1926, IWA.

40. R. W. James, "Early work on crystal structure at Manchester," pp. 423f, 428.

41. Douglas Hartree to Ivar Waller, 1/3 1928, IWA.

42. Douglas Hartree to Ivar Waller, 7/5 1928, IWA. Cf. Froese Fischer, *Douglas Rayner Hartree: His Life in Science and Computing* (New Jersey: World Scientific, 2003), esp. on pp. 38f.

43. W. L. Bragg to Ivar Waller, 4/10 1927, IWA. Jagdish Mehra, *The Solvay conferences on physics: Aspects of the development of physics since 1911* (Dordrecht, 1975), pp. 132–181, esp. on pp. 136ff.

44. R. W. James to Ivar Waller, 26/10, 19/11 1927, IWA.

45. W. L. Bragg to Ivar Waller, 2/11 1927, IWA.

46. R. W. James to Ivar Waller, 5/12 1927, IWA.

47. Ivar Waller and R. W. James, "On the temperature factors of X-ray reflexion for sodium and chlorine in the rock-salt crystal," *Proceedings of the Royal Society*, A117 (1927–28), pp. 214–223 on p. 220.

48. Douglas Hartree to Ivar Waller, 11/9, 7/11, 16/11, 28/11, 2/12, 4/12, 5/12 1928, IWA. The paper appeared in 1929 as: Douglas Hartree and Ivar Waller, "On the intensity of total scattering of x-rays," *Proceedings of the Royal Society of London*, A124 (1929), pp. 119–142.

49. Helge Kragh, "The genesis of Dirac's relativistic theory of electrons," *Archives for the History of Exact Sciences*, 24 (1981), pp. 31–67, on p. 39.

50. Ibid., p. 65.

51. P. A. M. Dirac, "The physical interpretation of the quantum dynamics," *Proceedings of the Royal Society of London*, A113 (1927), pp. 621–641.

52. Wolfgang Pauli, "Zur Quantenmechanik des magnetischen Elektrons," *Zeitschrift für Physik*, 43 (1927), pp. 601–623. On this see B. L. van der Waerden, "Exclusion principle and spin," pp. 221ff.

53. Pauli was the first to admit so, B. L. van der Waerden, "Exclusion principle and spin," p. 223.

54. Arthur I. Miller, *Early quantum electrodynamics: A source book* (Cambridge, 1994), p. 29.

55. Kragh, "The genesis of Dirac's relativistic theory of electrons," p. 50.

56. P. A. M. Dirac, "The relativistic wave equation," in L. Jenik and I. Montvay, eds., *Proceedings of the European conference on particle physics* (Budapest, 1977), pp. 15–34, on pp. 24, 26.

57. Helge Kragh, *Dirac: A scientific biography* (Cambridge, 1990), pp. 57–60; idem, "The genesis of Dirac's relativistic theory of electrons," p. 65.

58. P. A. M. Dirac, "The quantum theory of the electron[, I]," *Proceedings of the Royal Society of London*, A117 (1928), pp. 610–624, on p. 612.

59. Kragh, "The genesis of Dirac's relativistic theory of electrons," p. 63.
60. Oskar Klein, "Elektrodynamik und Wellenmechanik vom Standpunkt der Korrespondens-prinzip," *Zeitschrift für Physik*, 41 (1927), pp. 407–442, on p. 411, n1: "Diese [negative Energie-lösungen] werden wir naturgemäß von der Betrachtung ausschließen."
61. Oskar Klein, "Die Reflexion von Elektronen an einem Potentalsprung nach der relativistischen Dynamik von Dirac," *Zeitschrift für Physik*, 53 (1929), pp. 157–165. Abraham Pais, *Inward bound: Of matter and forces in the physical world* (1986) (Oxford: Oxford University Press, 1988), p. 349.
62. Oskar Klein to Ivar Waller, 10/4 1928, IWA. In October 1928, Klein and Nishina published their work based upon semi-classical assumptions about radiation and on Dirac's theory. O. Klein and Y. Nishina, "Über die Streuung von Strahlung durch freie Elektronen nach der neuen relativistischen Quantendynamik von Dirac," *Zeitschrift für Physik*, 52 (1929), pp. 853–868.
63. Ivar Waller to C. W. Oseen, 3/5 and 13/5 1928, CWOA.
64. Ivar Waller to C. W. Oseen, 3/5 1928, CWOA. The work by Bohr was based on his Como lectures from the Volta conference the previous year and published as: Niels Bohr, "Das Quantenpostu-lat und die neuere Entwicklung der Atomistik," *Die Naturwissenschaften*, 16 (1928), pp. 245–257; idem, "The quantum postulate and the recent development of atomic theory," *Nature*, 121 (1928), pp. 78, 580–590. Published 13/4 1928 and 14/4 1928, respectively.
65. Ivar Waller to C. W. Oseen, 3/5 1928, CWOA.
66. Ivar Waller to C. W. Oseen, 13/5 1928, CWOA.
67. Ivar Waller to C. W. Oseen, 27/8 1928, CWOA.
68. Ivar Waller to P. A. M. Dirac, 28/7 1928, P. A. M. Dirac Papers, Florida State University Libraries, Florida State University, Tallahassee, hereafter PAMDP.
69. Ivar Waller to P. A. M. Dirac, 19/1, 2/2, 18/2 1929, PAMDP and P. A. M. Dirac to Ivar Waller, 5/2 1929, IWA.
70. This paper was to appear as Ivar Waller, "Die Streuung kurzwelliger Strahlung durch Atome nach der Diracschen Strahlungstheorie," *Zeitschrift für Physik*, 58 (1929), pp. 75–94, received 21/7 1929.
71. P. A. M. Dirac to Ivar Waller, 6/3 1929, IWA.
72. Waller, "Die Streuung kurzwelliger Strahlung durch Atome nach der Diracschen Strahlungs-theorie."
73. Wolfgang Pauli to Ivar Waller, 15/11 1929, IWA. The referred work is Waller, "Die Streuung kurzwelliger Strahlung durch Atome nach der Diracschen Strahlungstheorie."
74. Ivar Waller, "On the origin of Dirac's relativistic theory of the electron," tape recorded speech from Linköping Technical University, 24/4 1984.
75. Ivar Waller to P. A. M. Dirac, 2/11 1929, PAMDP.
76. P. A. M. Dirac to Ivar Waller, 18/11 1929, IWA.
77. Ivar Waller to P. A. M. Dirac, 26/11 1929, PAMDP.
78. Niels Bohr to P. A. M. Dirac, 24/11, 5/12 1929; P. A. M. Dirac to Niels Bohr, 26/11, 9/12 1929, BSC. The letters are cited by Kragh, *Dirac*, 90ff; and by Donald Franklin Moyer, "Evaluations of Dirac's electron, 1928–1932," *American Journal of Physics*, 49 (1981), pp. 1055–1062 on pp. 1057f.
79. P. A. M. Dirac to Ivar Waller, 18/11 1929, IWA.
80. P. A. M. Dirac to Ivar Waller, 27/11 1929, IWA.
81. Ivar Waller, "On the origin of Dirac's relativistic theory of the electron."
82. Ivar Waller, "Die Streuung von Strahlung durch gebundene und freie Elektronen nach der Diracschen relativistischen Mechanik," *Zeitschrift für Physik*, 61 (1930), pp. 837–851.
83. Wolfgang Pauli to Ivar Waller, 22/1 1930, IWA.
84. Ivar Waller to P. A. M. Dirac, 5/12 1929, PAMDP.
85. Waller, "Die Streuung von Strahlung durch gebundene und freie Elektronen nach der Dirac-schen relativistischen Mechanik," p. 837.
86. Ibid., p. 843.
87. Ibid., p 844, n. ††.

88. P. A. M. Dirac, "Über die Quantentheorie des Elektrons," *Physikalische Zeitschrift*, 29:16 (1928), pp. 561–563. Waller was invited to the "Leipziger Universitätswoche" by Heisenberg. Werner Heisenberg to Ivar Waller, 17/6 1928, IWA.
89. Werner Heisenberg to Wolfgang Pauli, 31/7 1928, [204], Armin Hermann, ed., *Wolfgang Pauli: Wissenschaftlicher Briefwechsel mit Bohr, Einstein, Heisenberg u.a. Bd. 1: 1919–1929* (New York: Springer Verlag, 1979), p. 468.
90. *Professuren i Matematisk fysik och mekanik vid Stockholms högskola 1930*, O27.

A Nobel Prize for Scientific Revolutions?
Hannes Alfvén and Conflicting Cosmologies

SVANTE LINDQVIST

Japanese Holdouts

At the end of World War II, Japan had 3 million troops overseas, but the war did not end for all of them with Japan's surrender in 1945. Soldiers in isolated regions and on remote islands in the Pacific fought on: some unaware the war had ended, others refusing to believe it.[1] Some hid in the jungles alone, others fought in groups and continued to make guerilla raids—all adhering to the Imperial Army's code of "never surrender." They went into hiding, waiting for attacks that never came and messages from commands that had long since been disbanded. In some cases, it was years before these men realized the war was over.

In Dutch New Guinea, an 18th Army unit took to the jungle when Hollandia fell in 1944. Still not grasping that the war was over, it was not until 1954 that they were brought out of the jungle. On Guam, two Japanese soldiers hid in the jungle for 16 years until 1961. Some ten years later, in 1972, Corporal Shoichi Yokoi was found, also on Guam. He had seen reports of Japan's surrender in leaflets and newspapers scattered about the island but thought that they were American propaganda. On Morotai in Indonesia, Private Nakamura Teruo was found in 1974 when he handed over a well-maintained rifle and his last five rounds of ammunition. The most famous of the Japanese holdouts was 2nd Lt. Hiroo Onoda. For nearly three decades, he remained on the tiny Philippine island of Lubang, hiding in the jungle. Onoda surrendered in 1974, having kept a working firearm all the time.

The number of Japanese holdouts in the Pacific declined—almost exponentially—as the years passed. Are there any left today? Any Japanese veterans in hiding today (2008) would be at least in their early eighties. It is unlikely that any are still alive (but who knows).[2]

What is the relevance of this introduction? The German physicist Max Planck, who received the Nobel Prize in 1918, said:

> An important scientific innovation rarely makes its way by gradually winning over and converting its opponents: it rarely happens that Saul becomes Paul. What does happen is that its opponents gradually die out and that the growing generation is familiarized with the idea from the beginning.[3]

The Biography of a Scientific Theory

Can we write the biography of a scientific theory in the same way as that of a human being? A story with different phases of the life cycle such as birth, adolescence, maturity, declining years, death and—finally—obituaries. Let us examine a case from 20th-century physics, the Klein-Alfvén theory on cosmology.

The birth of the theory took place in the 1950s. Oskar Klein (1894–1977) had studied physics in Stockholm, and he had been an assistant to Svante Arrhenius (1859–1927).[4] He did "important work in quantum mechanics" in the 1920s as an associate of Niels Bohr in Copenhagen.[5] In 1930, he became professor of theoretical physics at Stockholm University.[6] When Klein was appointed, his friend Wolfgang Pauli wrote to him:

> I hope you will now live up to the word Go hence and teach the people. Your great pedagogic powers were always one of your strongest points and a wide field of application awaits them in Sweden. So far there has of course been hardly any theoretical physics in Sweden.[7]

In the 1950s, Klein began to work on cosmology.[8] At an international conference on astrophysics in Brussels in 1958, he presented a cosmological theory "very different from both the big-bang and the steady-state theories."[9] He assumed a dilute cloud of cold hydrogen and electromagnetic radiation as a sub-set of the universe, a "meta-galaxy," which condensed under the force of gravity. As the density increased, the radiation pressure would at some stage balance the gravitational contraction. Once an equilibrium stage was reached, the radiation pressure was assumed to result in an expansion corresponding to the one that had been observed.[10]

We now enter the phase of adolescence, when the theory matured. A few years later, Klein came to the conclusion that a symmetry between matter and antimatter had to be a fundamental assumption in any cosmology, with the universe being composed of equal parts of matter and antimatter.[11] According to Klein, the zero state was still a giant spherical metagalaxy of diffuse, slowly concentrating gas—but one which contained equal amounts of matter and antimatter: protons and antiprotons in equal numbers with a sprinkling of electrons and positrons. When in the course of time the cloud contracted under its own gravitation, the density became sufficiently high for matter and antimatter to begin to annihilate. This radiation halted and eventually reversed the collapse of the remaining matter. The cloud began to expand, and galaxies eventually condensed. In this theory, comparable numbers of galaxies and antigalaxies are expected to exist in the observable universe.

At this point in the adolescence of the theory, our second main character enters the stage: Hannes Alfvén (1908–1995). He is well-known for his work in plasma physics and in particular for his discovery of magneto-hydrodynamic waves ("Alfvén waves"): work for which he was awarded the Nobel Prize in Physics in 1970. Alfvén had studied in Uppsala under the physicist Manne Siegbahn (1886–1978), and in 1940 he became a professor at the engineering school in Stockholm, the Royal Institute of Technology (KTH). Once in Stockholm, Alfvén began to attend Oskar Klein's seminars in theoretical physics regularly and befriended him.

In the spring of 1961, Klein asked Alfvén if he could think of any natural mechanisms that might be able to separate matter and antimatter in the ambiplasma. Such a separation was necessary if the theory was to be credible. Already in the summer of the same year, 1961, Alfvén could suggest a promising separation mechanism, and he became increasingly enthusiastic about Klein's cosmology.[12]

Alfvén—who was a staunch atheist—was in principle favorably disposed to the theory since it did not demand a moment of divine creation (and thus no Creator).

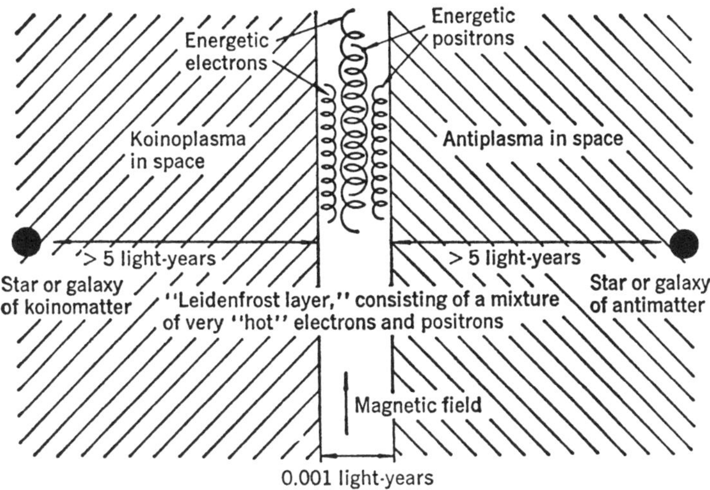

FIGURE 1.   The annihilation sheath separating matter and antimatter according to the Klein-Alfvén theory. From: Hannes Alfvén, *Worlds-Antiworlds: Antimatter in Cosmology* (San Francisco and London: W. H. Freeman, 1966), p. 53.

In 1939, Alfvén had listened to the priest and astronomer Abbé Lemaître explaining his cosmological theory—his "l'Atome Primitive" which was the origin of the big bang theory—at an astrophysics conference in Stockholm:

> To Abbé Lemaître this was very attractive, because it gave justification to the creation *ex nihilo*, which St. Thomas had helped establish as a credo. To many other scientists it was more of an embarrassment because God is very seldom mentioned in ordinary scientific literature. Hence the problem of how the singular state was produced is usually not raised.[13]

To separate matter and antimatter, Alfvén proposed a cosmic variant of the Leidenfrost phenomenon. That is, if a drop of water is placed on a concave and sufficiently hot plate, it may remain there without boiling off for several minutes thanks to a thin layer of vapor that insulates the main body of water from the hot plate itself. Similarly a hot, insulating Leidenfrost layer might be formed at the interface between a world of matter and antimatter. Collisions between protons and antiprotons would produce high-energy electrons and positrons, which would spiral around the magnetic field lines in the layer, emitting gamma radiation.

This was basically the theory that Klein and Alfvén published in 1962 in the journal of the Royal Swedish Academy of Sciences, *Arkiv för fysik*. The theory had now passed from adolescence to maturity.

Arkiv för fysik did not, to say the least, enjoy a wide international readership, but through the second part of the 1960s and up to 1971, Alfvén published articles on the theory in several major international scientific journals: *Review of Modern Physics* in 1965,[14] *Scientific American* in 1967,[15] *Science* in 1969,[16] and *Nature* and *Physics Today*, both in 1971.[17] Lars Bergström, Secretary of the Nobel Committee for Physics, has said that the theory "had a justified place in the physics of the 60s."[18] Per Carlqvist, a long-time collaborator of Alfvén at the Royal Institute of Technology in Stockholm, has said:

> For my own part I long regarded the Klein-Alfvén theory as an interesting and healthy contribution to the cosmological debate. Observations that might corroborate or refute cosmological models were few in those days, but even so the big-bang theory was definitely at the head of the field, particularly after the collapse of the steady-state theory just before 1970. To my mind, it was essential for scientific credibility for there to be more than just a single theory to be discussed at this relatively early stage. But the Klein-Alfvén theory never became, I would venture to suggest, a serious rival to the big-bang theory.[19]

Death, according to most, was instant, with the serendipitous discovery in 1965 of the cosmic background radiation (CBR) by Arno Penzias and Robert Wilson, a discovery for which they were awarded the Nobel Prize in Physics in 1978. To others the theory had its declining years with many illnesses before it finally suc-

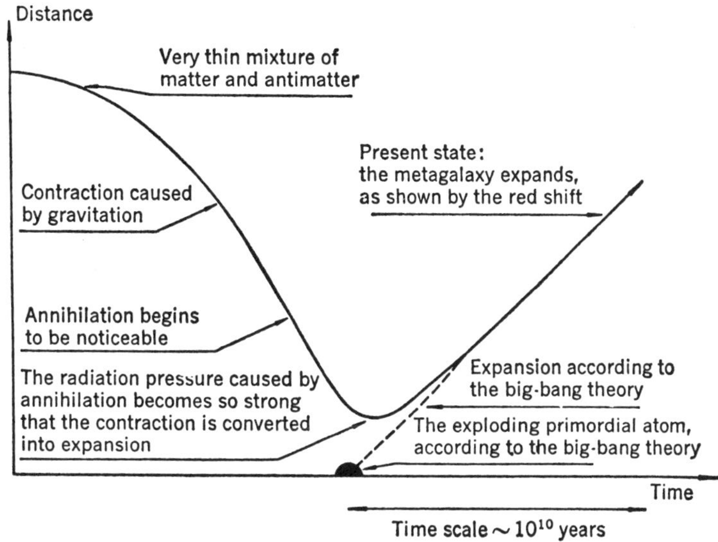

FIGURE 2. The development of the meta-galaxy according to the Klein-Alfvén theory. From: Hannes Alfvén, *Worlds-Antiworlds: Antimatter in Cosmology* (San Francisco and London: W. H. Freeman, 1966), p. 78.

cumbed in the early 1970s. And to some it remained, and to a few still is, alive and vital—but this discrepancy is something to which we will return.

There were five main objections to the Klein-Alfvén theory that emerged from 1965 onwards:

(1) The killer was, as mentioned above, the discovery in 1965 of CBR, which had been predicted by the big-bang theory. Radio antennas had detected microwave radiation of seemingly unvarying intensity pervading the cosmos, and this was widely accepted as the energy echo from the explosion. To many cosmologists this entirely thermal radiation at a uniform temperature (2.73 Kelvin) meant the end of the Klein-Alfvén model, "even if Alfvén could never see it."[20] Per Carlqvist has said that:

Hannes himself did not allow himself to be depressed and suggested that the background radiation might have its origin in electromagnetic radiation reflected by mm-long needles (antennas) in the cosmic plasma. The occurrence of such needles could, however, never be reliably established. With the measurement in about 1990 of the almost perfect black body spectrum of the background radiation this argument ought to have collapsed.[21]

(2) The second objection was that "We now know with even greater certainty that the Alfvén model does not add up, partly because of the absence of X-rays and gamma-rays in the annihilations between matter and antimatter that ought to take

place."[22] But Alfvén saw the absence of X-rays and gamma rays as a result of the Leidenfrost layer, which could efficiently separate matter and antimatter so that only insignificant radiation was released.

(3) One of the great triumphs of the big-bang theory was the confirmation by observations that the universe should consist of 24% helium. This was naturally very inconvenient for the Klein-Alfvén theory, but on this matter, Per Carlqvist has said, "Hannes had, as far as I remember, no comment."[23]

(4) Another objection was the discovery in elementary particle physics of the CP violation, which proved that there is an asymmetry in the universe between matter and antimatter. (This discovery was made in 1964 at the Brookhaven Laboratory by James W. Cronin and Val L. Fitch, and for which they were awarded the Nobel Prize in Physicis in 1980). Per Carlqvist has said:

> I don't think that the discovery of the CP violation upset Hannes so very much. Most high-energy processes still produce roughly as many particles as anti-particles. It has indeed surely been, and to some extent still is, a problem for the big-bang theory to explain why only matter exists in the universe. CP violation has been suggested as offering a solution here, but as yet only on very shaky foundations.[24]

(5) The last objection was that it was difficult to explain the big red shifts that had been observed in the spectra within the framework of the Klein-Alfvén theory. Two of Klein's and Alfvén's pupils and collaborators discussed this problem in a couple of articles, but they found it "difficult (or even impossible) to understand the high flight velocities that had been observed in the galaxies. More recent observations have accentuated the problem."[25]

To Alfvén, all these objections amounted to nothing more than a Near Death Experience for the theory, but that was not the way the scientific community at large viewed the matter. To them, the Klein-Alfvén theory was not a Norwegian Blue with beautiful plumage, pining for the fjords. It was an ex-parrot, pushing up the daisies, filled with sawdust and nailed to the perch.

Obituaries

When did the first obituaries on the theory appear and what did they say? In late 1969, *Nature* published a series of articles by different authors under the heading "Antimatter, Galactic Nuclei and Theories of the Universe."[26] In his article, Gary Steigman at the Institute of Theoretical Astronomy in Cambridge discussed the arguments against the Klein-Alfvén theory, and he concluded:

> Although the Alfvén-Klein cosmology may be appealing, it appears that on close examination it raises more problems than it solves. The above arguments indicate that, at least in its present form, it cannot be an acceptable cosmology.[27]

On another level of the hierarchy of scientific publications—another sort of semi-obituary, giving the theory the benefit of the doubt and thereby postponing its demise, appeared in 1970 in the revised edition of the Indian scientist Jagjit Singh's *Great Ideas and Theories of Modern Cosmology*, published by Dover in New York and aimed at a general audience.[28] Singh said that "the claim has not won general acceptance,"[29] that it "faces serious difficulties right at the outset,"[30] and that "the details of the process have not yet been fully worked out and it is not clear whether it will succeed."[31] He concluded:

> All that may be said in its favor is that it makes our universe of receding galaxies the outcome of a kind of radiation explosion issuing from particle-antiparticle annihilation rather than that of a giant primeval atom as in the big-bang cosmology of Lemaître and Gamow. It therefore postulates a more credible zero state than the big-bang cosmology. But that is about all.[32]

Ten years later, a death certificate was issued with more assurance, but then, not surprisingly, by a proponent of the big-bang theory. In *The Big Bang*, published by W. H. Freeman, Joseph Silk, the renowned Berkeley astronomer, devoted two pages to the Klein-Alfvén cosmology.[33] He said that many objections had been raised to it, and one concerned Alfvén's suggestion of a mechanism to separate matter and antimatter of which "other astrophysicists remain skeptical."[34] Another objection was that copious amounts of high-energy gamma rays, resulting from annihilation of matter and antimatter, had not been observed by space experiments: "Astronomers have indeed measured a diffuse cosmic background of gamma rays, but the flux is relatively low."[35] Silk wrote that the most significant objection to the Klein-Alfvén cosmology stems from the presence of the cosmic background radiation, and that this makes their cosmology "seem extremely dubious."[36]

In 1993, the well-known astrophysicist P. James E. Peebles, Albert Einstein Professor of Science at Princeton University, published his *Principles of Physical Cosmology*.[37] In this major work of 718 pages, Peebles devoted two pages to what has been described as a "referential obituary" on the Klein-Alfvén model.[38] He writes sympathetically that the theory follows an "attractive and sensible path" that keeps to standard physics and minimal extrapolations from what is observed of the universe.[39] "But," he adds, "there is no way the results can be consistent with the isotropy of the CBR and X-ray backgrounds."[40] It was a death certificate stamped with the mark of authority.

In 1996, the Danish historian of science Helge Kragh—already well-known for his scientific biography of Paul Dirac—published his impressive *Cosmology and Controversy: The Historical Development of Two Theories of the Universe*.[41] Here Kragh described how cosmology was transformed into a branch of physics between 1920 and 1970, and he examined how the big-bang theory drew inspiration from and eventually triumphed over rival views, mainly the steady-state theory and its concept of a stationary universe of infinite age. Out of a total of 500 pages, Kragh devoted two pages to the Klein-Alfvén theory. He concluded:

> Although Alfvén's cosmological theory received support from other plasma physicists, it was ignored by most astronomers and cosmologists. It was never developed into a serious alternative to the big-bang theory.[42]

Notice that all the verbs are in the past tense. The Klein-Alfvén theory had now been incorporated into the corpus of scholarship in the history of science. This development could also be illustrated by the declining number of citations to Klein and Alfvén's original 1962 paper in *Science Citation Index*, very similar to the declining number of Japanese holdouts in the Pacific.

In the obituary on Alfvén in the *Biographical Memoirs of the Fellows of the Royal Society, London* (Alfvén had been a foreign member of the Royal Society), the British physicist Sebastian Pease wrote diplomatically that "Alfvén's cosmology has not attracted wide support."[43]

Epitaph

If we were to write an epitaph on the Klein-Alfvén theory today (2008), let us then quote Lars Bergström, Scientific Secretary of the Nobel Committee for Physics:

> I believe that this whole story of the fall of the Alfvén model and the success of the big-bang model are typical examples of how the scientific process works, and should work [. . .] In the early 60s nobody could say that one theory or the other was 'better,' other than possibly on subjective criteria, which varied from person to person. But as the observations became more detailed and new data flowed in, the Big Bang enjoyed more and more successes, while alternative models either gave contradictory predictions or no predictions at all. Of course the scientists of the future (in other words our graduate students) weigh all the models against how they agree with experiments and observations, and in that sense the Alfvén model no longer has any attraction at all. But, as the historical example shows, models which then prove incorrect also play an important part in the complex edifice we call the scientific world view.[44]

Alternative Biographies

This could have been a complete biography of the Klein-Alfvén theory, but the story is more complicated than that. This biography did not capture its entire life: a scientific theory lives many lives, and different ones in different places and on different strata of society. "Revolution" is a concept borrowed from politics. It implies an immediate takeover of power, affecting *all levels* of society at the *same time*. It is usually defined as "The overthrow of one government and its replacement with another" or "a sudden or momentous change in a situation."[45] But other metaphors can be used that may perhaps better describe the scientific enterprise? In an earlier paper, I discussed the theater as a metaphor for scientific controversy, and suggested that the setting of the medieval marketplace, with its many modest stages, may perhaps provide a better parallel than the classical baroque theater.[46] The dividing lines in scientific questions sometimes follow national or institutional boundaries. The rule here is, as the old saying goes, that "what is truth in Berlin and Jena is merely a poor joke in Heidelberg."

The view of what is, and what is not, new scientific knowledge changes with time, but it is quite impossible to establish exactly *when* and *where* this change takes place; it is not something that is agreed upon by everyone at the same time. On the one hand there is a geographical dimension, with different subsets of the international scientific community, and on the other there are a number of non-geographically bounded, overlapping networks of various schools of thought. To this should be added the complexities of the hierarchy of generalization: a hierarchy that stretches from articles in specialized scientific journals, down through more general science journals to textbooks, popular science and newspaper articles. There are time lags and geographical differences within this hierarchy of what is and what is not considered new scientific knowledge.

All this is illustrated by the story of the Klein-Alfvén theory—or rather by the story of its aftermath or life after death. At a time when most astronomers and astrophysicists had already begun to disregard the Klein-Alfvén theory, Hannes Alfvén nevertheless continued to publish new additions and revisions of it—and so he did for the next twenty years until only a few years before his death in 1995. Let us review this publication history briefly!

A popular version by Alfvén on the theory had remarkable success and a widespread international distribution. This short book, only 110 pages, *Världen—spegelvärlden: Kosmologi och antimateria* was first published in Sweden in 1966 (where it appeared in a second edition in the following year). An American and a British edition were published in the same year by W. H. Freeman in San Francisco and London. On the back cover of the U.S. edition, it was advertised as "the first presentation for a general audience of a new theory of the universe," and it was said that "In this book a leading physicist suggests that there are good reasons for believing that half [of the stars and galaxies that we observe] are made of antimatter." In the following years, it was also published in two German editions, a Russian edition, a Norwegian edition, an Italian edition, a Croatian edition, and a Polish edition. The following is a list of the various editions:

Hannes Alfvén, *Världen—spegelvärlden: Kosmologi och antimateria*, Aldusböckerna 161 (Stockholm: Aldus/Bonniers, 1st ed. 1966, 2nd ed. 1967).

————, *Worlds-Antiworlds: Antimatter in Cosmology* (San Francisco and London: W. H. Freeman, 1966).

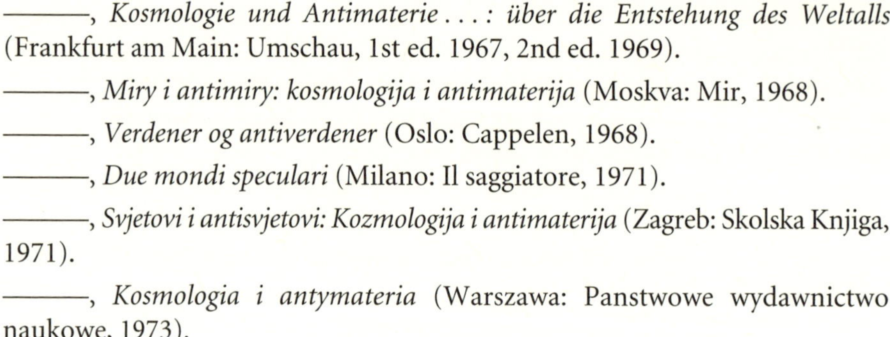
————, *Kosmologie und Antimaterie . . . : über die Entstehung des Weltalls* (Frankfurt am Main: Umschau, 1st ed. 1967, 2nd ed. 1969).

————, *Miry i antimiry: kosmologija i antimaterija* (Moskva: Mir, 1968).

————, *Verdener og antiverdener* (Oslo: Cappelen, 1968).

————, *Due mondi speculari* (Milano: Il saggiatore, 1971).

————, *Svjetovi i antisvjetovi: Kozmologija i antimaterija* (Zagreb: Skolska Knjiga, 1971).

————, *Kosmologia i antymateria* (Warszawa: Panstwowe wydawnictwo naukowe, 1973).

It is interesting that all these editions were published *after* the discoveries of the CP violation in 1964 and of the cosmic background radiation in 1965—but, as said above: the Klein-Alfvén theory still had its rightful place among various cosmologies in the later 1960s.

In 1970, Hannes Alfvén received the Nobel Prize in Physics "for fundamental work and discoveries in magneto-hydrodynamics with fruitful applications in different parts of plasma physics."[47] The Nobel Prize may also have given a boost to Alfvén's confidence and belief in his cosmological theory—not that his confidence needed any boosting, but his followers in Sweden and abroad must have taken heart when he received the Nobel Prize. So many of Alfvén's early predictions—for which he had been scorned by the "scientific establishment" for decades—had, one after the other, been proved right. (This is something which the historian of science Stephen G. Brush has described in an article, "Prediction and Theory Evaluation: Alfvén on Space Plasma Phenomena," published in 1990.[48]) Was it not therefore likely that Alfvén's cosmological theory would also prove correct? The years around 1970 were a crucial time in the development of cosmology: these were the years when the steady state theory finally had to yield to the big bang, and at a time when objections to the Klein-Alfvén theory had been mounting. The Nobel Prize to Alfvén in 1970 may have given the Klein-Alfvén theory an extra credibility through "guilt by association" that enabled it to linger on long beyond its normal life expectancy.

However, in spite of his Nobel Prize, Alfvén sometimes found it hard to have papers on his cosmology accepted by the major science journals.[49] But he kept presenting it "off Broadway," so to speak, in the 1970s in various contexts: in a Swedish popular science journal,[50] at an international conference arranged by Swedish colleagues,[51] in the KTH department's own series of occasional papers,[52] and at a conference arranged by the Armenian Academy of Sciences.[53] At a conference on "Cosmology, History and Theology," staged in Boulder, Colorado, in 1976, Alfvén gave a paper entitled "Cosmology: Myth or Science?" in which he claimed that the big-bang dogma represented "an unscientific attitude and a revival of myth."[54] This had been his opinion ever since he had first listened to Abbé Lemaître at the conference in Stockholm in 1939, thirty-seven years earlier.

In 1980, Alfvén summarized his views in the monograph *Cosmic Plasma* in which he portrayed the universe and the solar system as a system of closed electrical currents.[55] It was published by the respected publisher D. Reidel in the Netherlands in its "Astrophysics and Space Science Library" series, but it was a book that basically was ignored by the scientific community. In the early 1980s, Alfvén did, however, succeed in having some of his cosmological papers accepted by major science journals.[56]

When I interviewed the British astronomer and cosmologist Hermann Bondi in 1996 about Alfvén's work, Bondi made an interesting statement about Alfvén:

> He sometimes gets hold of the wrong end of the stick, but very often others haven't even seen that there is a stick there at all. I'm thinking of his overall work: the importance of electromagnetic forces in cosmology and large-scale astronomy was in many ways introduced by him. That was the stick. And most other people hadn't realized that electromagnetic forces were important. He saw that there was a stick, and then he applied this in many situations when he was very right to apply it but in some where I think that the significance of his concepts is very much in doubt [. . .] He's a monomaniac in some things, but where there was a huge gap in our attitude he corrected it. He overcorrected it in some respects. It is a complex picture. When you examine a student and mark a written exam, and your student gives three excellent answers and one not very good, you don't count the bad answer against the good ones. The good ones are what counts.[57]

The Klein-Alfvén theory was, of course, according to Bondi, the "bad answer" among Alfvén's many successful predictions.[58]

In September 1987, the *Scientific American* published an article on Alfvén and other critics of the big-bang theory under the title "Big-Bang Bashers."[59] The article said that although the big-bang model had resisted all serious challenges:

> Nevertheless, ever since the theory won general acceptance about 20 years ago a few scientists have persistently attacked some of its fundamental assumptions [. . .] These groups, although long relegated to 'fringe' status by mainstream astrophysicists have been invigorated of late by new converts and new findings.[60]

The elder statesman of the plasma dissidents, said *Scientific American*, is Hannes Alfvén. Lawrence M. Krauss of Yale University was quoted as saying that Alfvén and the others who attacked the big bang model did not offer a comprehensive alternative, and that "They're working against something that explains everything very well."[61] This prompted Alfvén—now at the University of California, San Diego, where he had spent his winters since 1967—to write a letter to the *Scientific American*:

Dear Editors:

I was glad to see me listed as a member of the rapidly growing guerilla group of Big Bang Bashers. I was less satisfied with Lawrence M. Krauss' claim that I have criticized only parts of the Big Bang and that I have not offered a comprehensive alternative. In 1960 my senior colleague, or rather teacher, Oskar Klein, introduced me to his way of approaching cosmology. I found his approach very interesting from a scientific point of view, whereas other cosmologies—Continuous Creation and Big Bang—were supported more by megaphones than by scientific arguments.

Since then, I have written the following papers or books in the field . . .[62]

Beginning with his and Oskar Klein's publication in *Arkiv för fysik* 1962, Alfvén then listed no less than 33 articles and books. And he ended his letter in a typical, arrogant Alfvén-way:

May I suggest that Krauss read these papers. When he has thoroughly digested them, I am sure that he will renounce his holy creed which he claims 'explains everything well.' After having done his homework, his renunciation need not take more than three minutes. Perhaps 10^{-43} seconds is enough. (Krauss himself seems to leave that possibility open.)
Sincerely,
(signed)
Hannes Alfvén[63]

On May 30, 1988, Alfvén, now named the "Dean of the Plasma Dissidents" by one of his followers,[64] celebrated his 80th birthday. There was an official celebration at the Royal Institute of Technology (KTH), which was held in the main classroom building (aptly knick-named "Sing-Sing"). Alfvén had a garland of flowers in his silver hair and many speeches were held in his honor. The high point was a song, performed by Professor Göran Marklund (who later succeeded Carl-Gunne Fälthammar, who himself had succeeded Alfvén on the Chair in Plasma Physics at KTH) and Associate Professor Nils Brenning, both dressed in Hawaii shirts, strawhats and accompagning themselves on guitars. To a number of well-known Swedish tunes they had put new words, and the refrain that brought down the house was the following:

"Icke tycka om Big Bang, tycka Big Bang trist"
    (roughly: "Does not like the Big Bang theory, thinks the Big Bang theory pathetic")

We who were there, we all cheered. Because we knew—situated as we were in our little place in time and institutional space—that Alfvén, of course, must be right and all the others wrong. That is not to say that Swedish astronomers and physicists shared his view on cosmology—to them the theory was as dead as to most of their colleagues abroad. But on another level, we the general audience in the Swedish context cheered him on. In some ways, scientific beliefs resembles religion. Your faith is decided by where you happen to be born. Our belief was further strengthened when Alfvén had a long article on his cosmology published in one of the main Swedish dailies in the fall of 1988.[65]

In December 1988, Alfvén was awarded the William Bowie Medal of the American Geophysical Union, its oldest and most prestigious award, for his "outstanding contributions to fundamental geophysics and for unselfish cooperation in research."[66] It was presented to him during the 1988 Fall Meeting of AGU held in San Francisco. In his presentation speech Alex J. Dessler said:

> ... with this award, we honor the underdog, one who has stood up to a powerful scientific establishment, suffered, persisted, and finally prevailed. By making Hannes Alfvén our Bowie Medalist, we do more than recognize a worthy recipient—we make ourselves feel good. Our innate sense of justice is indulged. We bask in the self-satisfied glow that comes from witnessing truth win over the forces that would lead us astray.[67]

Dessler then went through Alfvén's contributions to plasma and space physics, and pointed out the many concepts that today bear his name. He said that most space physicists have learned to listen carefully when Alfvén comes up with one of his ideas. Not everyone of his innovative proposals have proved correct, "but he has almost invariably been right on those that are basic and profound."[68] But what about his cosmology? Well, Dessler said:

> Alfvén has never been content to stay in a field in which everyone agrees with him. He is an iconoclast who prefers the intellectual challenge of scientific debate. He is a sworn enemy of idols of the mind. Thus having gotten space physics straightened out, his activities are now largely directed toward attacking some of the basic beliefs of the astrophysical science establishment. His style has not changed. He plays the role of the iconoclast smashing the most sacred idols of the discipline. For example, he treats the well regarded popular theory of the formation of the universe in a 'Big Bang' as a myth, nothing more.[69]

In his short and arrogant acceptance speech, Alfvén said that he had not solved all of the problems, but had left to the other members of the AGU to solve what he thought the most important of all cosmical problems, namely the question of matter-antimatter symmetry. Oskar Klein's claim that there should be a general symmetry between the amount of matter and antimatter in the universe, Alfvén said, is not generally accepted. Instead it is believed that, for example, one of our closest stars, Alpha Centauri, consists of matter. "The question remains as to whether there exists any scientific reason for this belief." Alfvén then, in a manner very typical of

him, he who had just received a medal from AGU, presented the Society with a new
medal, instituted by himself:

> I think that this problem is of such decisive importance for our views of the
> universe that the one who solves it deserves a medal called the Alpha Centauri
> medal, which in the future may acquire a reputation comparable to even the
> Bowie Medal.[70]

Alfvén then handed over a proposal for the design of what the medal should look
like, and pointed out that "it requires much work by many of you to give it the form
and stature it deserves."[71]

Some two months later, in February 1989, the science section of *The New York
Times* carried a long article entitled "Novel Theory Challenges the Big Bang: Renegade Physicists Say Electromagnetism Shaped an Infinite Universe."[72] The background was the following: Alfvén and some of his followers had met in La Jolla in
the previous week for a three-day workshop on plasma cosmology. The meeting
was held if not "at" at least "near" the University of California–San Diego, and it
was sponsored by IEEE and supported by NASA.[73] *The New York Times* article said
that a group of physicists and astronomers had met to "discuss some recent observations that appeared to contradict fundamental tenets of the Big Bang theory.
Laboratory studies, computer simulations and astronomy's latest findings appear
to buttress the heretical position."[74] Alfvén was quoted in an interview:

> The astrophysicists know they have problems with the Big Bang, and they're
> beginning to study our arguments [. . .] They're cautious, of course. A Christ-
> ian does not go to Mecca and become a convert overnight.[75]

The Boston Globe also carried an article on the meeting, and this widespread
publication of their meeting must have pleased Alfvén and his few followers im-
mensely,[76] and it must have required some skillful lobbying and good contacts with
the press.[77] It is, however, unlikely that the articles changed the views of any professional astronomers and physicists who had already made up their mind on the
Klein-Alfvén cosmology. But again: this illustrates that conflicting theories can remain alive in some institutional settings and on different strata in society long after
they have been proclaimed dead on other levels.

In the late 1980s and early 1990s, Alfvén was able to publish articles on his cosmology in the journals *Laser and Particle Beams* and the *IEEE Transactions*.[78] The
Klein-Alfvén theory was thus able to stay alive on the top—well, almost the top—
of the scientific publication food chain (that is, the refereed, international scientific
journals) some two decades after the theory had been proclaimed dead by most.

And on a lower level in the hierarchy of scientific publications, the Klein-Alfvén
theory lived a vigorous and healthy life. Alfvén's constant criticism of the big-bang
theory, and his claim to have an alternative, received, as Helge Kragh has said,

"considerable attention outside mainstream cosmology."[79] Thus Alfvén's followers were able to publish popular accounts of his cosmology in various popular science magazines.[80]

A main attempt to revive an interest in the Klein-Alfvén theory was the monograph *The Big Bang Never Happened: A Startling Refutation of the Dominant Theory of the Origin of the Universe*, published in 1991 by Times Books/Random House in New York.[81] It was written by Eric J. Lerner, an independent plasma physicist and science writer in Lawrenceville, N.J., who had long been an admirer and follower of Alfvén. Lerner's book was never reviewed in *Science*, *Nature*, or the *American Journal of Physics*. It was, however, reviewed critically in both the popular astronomical journal *Sky and Telescope* and *The New York Times*,[82] and in *The New York Review of Books* the well-known historian of science Daniel Kevles wrote a very critical review.[83] But in another national setting, the well-respected Swedish astronomer and science writer Peter Nilsson (1937–1998) wrote a long and basically favorable review of Lerner's book in one of the main Swedish dailies,[84] which again proves the spatial differences in consensus.

During the last years of his life, in the early 1990s, Alfvén would speak privately of the "Big Bang mafia."[85] His last publication is dated 1992. It was published in the KTH department's series of occasional papers, but it was never published in an international journal.[86] The title was "Can the Big Bang Survive in the Space Age?," and Alfvén died three years later, in 1995, firmly believing that the answer was "No".

What is "the general opinion" among astrophysicists of today, more than ten years after Alfvén's death, on the Klein-Alfvén theory? Lars Bergström, Secretary of the Nobel Committee for Physics, says that "it can now be regarded as scientifically quite dead."[87] Bengt Gustafsson, a professor of theoretical astrophysics at Uppsala University, says:

> As far as I know there is nobody who is still working along those lines today; the theory is in other words stone dead; it is a pity, because it was both simple and beautiful [. . .] The theory also had the merit that it could be refuted.[88]

Per Carlqvist, a long-time colleague of Alfvén at KTH, says that:

> . . . one may say that their theory is almost forgotten today. There are a few, perhaps less well-known, physicists [. . .] who still embrace the theory and work on it. And at the Alfvén laboratory there may be one or two who have been 'brought up' with the theory and who dislike the singularities of the big-bang model.[89]

Can those who still uphold the Klein-Alfvén theory of cosmology today be likened to lonely Japanese soldiers, holding out in the jungle on small islands in the vast Pacific of the scientific community, continuing a war they still have not realized is over? They would, I suppose, resist such a comparison passionately. Perhaps they would rather speak of themselves in terms of courageous dissenters persecuted by the established church: like, for example, the Huguenots in 16th-century France. This at least was the tone of the article "Bucking the Big Bang," which was published in *New Scientist* in 2004, a statement which had been "co-signed by 33 other scientists from 10 counties," all working on alternatives to the big-bang theory (including some of Alfvén's remaining followers).[90]

§ § § §

Finally, let's wind the tape back to the year 1971. In that year, it was fifty years since Oskar Klein had received his Ph.D. from Stockholm University in 1921.[91] He was recognized at the annual degree ceremony as a "jubilee doctor" (as these quinquagenarians are called in Sweden). At the banquet which followed, the 76-year-old Klein had been asked to give a speech. He began by saying that he had come to the conclusion that he should choose "a topic of general interest and at the same time one close to physics, the science to which I have been dedicated. The result was that I chose the topic authority and revolution."[92] In 1971, it was already clear to most scientists (including Klein) that the Klein-Alfvén theory on the origin of the universe would not survive. In his speech, Klein sketched the evolution of physics from Galileo and Pascal to quantum mechanics and relativity. He concluded:

> It is of course natural that things that deviate sharply from what we are familiar with often meet with resistance. Such resistance, when it is not taken too far, may indeed also be justified as protection against casual novelties. [...] After the time of Galileo and Pascal there has naturally often been opposition to new points of view, in physics too. But as far as I know, this has never led to blind faith in authority. As far as physics is concerned, in other words, we have reason to use neither the word 'revolution' nor the word 'authority,' but instead the words 'development' and 'master' and 'teacher.'[93]

This, however, was not an opinion to which Hannes Alfvén would have subscribed. He believed in concepts such as "revolution" and "blind faith in authority" to describe the scientific process, and he never thought that the war was over. Alfvén was a proud member of the guerilla group of "Big Bang Bashers," and like 2nd Lt. Hiroo Onoda, he kept a working firearm and some rounds of ammunition until the end of his life. Bravely, almost single-handedly, Hannes Alfvén took on the whole Pacific fleet.

Notes

An earlier version of this paper was given in the 2004–2005 Thomas Hart and Mary Jones Horning Endowment in the Humanities Lecture Series at the Oregon State University, Corvallis, Oregon, on March 7, 2005. For comments I am indebted to Anders Bárány, Lars Bergström, Per Carlqvist and

Bengt Gustafsson. Their comments were given in Swedish, and the translations of the quotations are mine.

1. The following section is based especially on *http://www.wanpela.com/holdouts/*. There is a huge literature on Japanese holdouts. The most well-known autobiographies are the following: Hiroo Onoda, *No Surrender: My Thirty-Year War* (London: Deutsch, 1975); Itō Masashi, *The Emperor's Last Soldiers* (New York: Coward-McCann, 1967).
2. As late as in 2005, BBC News reported that Japanese officials were investigating claims that two men in their eighties living in jungle in the Philippines were Japanese soldiers left behind after World War II, see: *http://news.bbc.co.uk/2/hi/asia-pacific/4585287.stm*.
3. Max Planck, *Philosophy of Physics* (New York: W.W. Norton, 1936), p. 97.
4. On Klein, see: Oskar Klein, "Ur mitt liv i fysiken," *Svensk naturvetenskap: Yearbook of the Swedish Natural Science Research Council* 26 (1973), pp. 159–172; Inga Fischer-Hjalmars and Bertel Laurent, "Oskar Klein," *Kosmos: Svenska Fysikersamfundets årsbok* 55 (1978), pp. 19–29. Cf. Karl Grandin, *Ett slags modernism i vetenskapen: Teoretisk fysik i Sverige under 1920-talet*, Skrifter Institutionen för idé- och lärdomshistoria no. 22 (Uppsala: Institutionen för idé- och lärdomshistoria, Uppsala universitet, 1999), *passim.*
5. Helge Kragh, *Cosmology and Controversy: The Historical Development of Two Theories of the Universe* (Princeton: Princeton University Press, 1996), p. 223.
6. More correctly: professor of "mechanics and mathematical physics" at "Stockholm University College," only later did Klein become a professor of "theoretical physics" at "Stockholm University"—but in disciplinary and institutional terms this amounted to the same thing.
7. Suzanne Gieser, "Philosophy and Modern Physics in Sweden: C.W. Oseen, Oskar Klein, and the Intellectual Traditions of Uppsala and Lund, 1920–1940," in: *Center on the Periphery: Historical Aspects of 20th-Century Swedish Physics*, Svante Lindqvist, ed. (Canton, Mass.: Science History Publications, 1993), p. 25.
8. Kragh 1996, p. 223. Cf. Oskar Klein, "Some Cosmological Considerations in Connection with the Problem of the Origin of the Elements," in: *Les Processeus Nucléaires dans les Astres*, Mémoirs in 8e de la Société Royale des Sciences de Liège, Quatrième Série, Tome XIII, Fasc. III (1953), p. 42.
9. Kragh 1996 *op. cit.*; Oskar Klein, "Some Considerations Regarding the Earlier Development of the System of Galaxies," in: *Inst. Int. de Physique Solvay, Onzième Conseil de Physique* 11 (1958), pp. 33–51.
10. Kragh 1996 *op. cit.*; Klein 1973, pp. 169–170; Fischer-Hjalmars and Laurent 1978, pp. 26–27.
11. Klein 1973, p. 169–170.
12. Ibid., p. 170; Private communication from Per Carlqvist, February 10, 2005.
13. Hannes Alfvén, "Cosmology: Myth or Science?" in: *Cosmology, History, and Theology*, eds. Wolfgang Yourgrau and Allen D. Breck (New York: Plenum Press, 1977a), p. 7.
14. Hannes Alfvén, "Antimatter and the Development of the Metagalaxy," *Rev. Mod. Phys.* 37 (1965), pp. 652–665.
15. Hannes Alfvén, "Antimatter and Cosmology," *Scientific American* 216 (1967), no. 4 (April), pp. 106–114.
16. Hannes Alfvén and Anna Elvius, "Antimatter, Quasi-Stellar Objects and the Evolution of Galaxies," *Science* 164 (1969), no. 3882 (May 23), pp. 911–917.
17. Hannes Alfvén, "Symmetric Cosmology," *Nature* 229 (1971), p. 184; *idem*, "Plasma Physics Applied to Cosmology," *Physics Today* 24 (1971), pp. 24–33.
18. Private communication from Lars Bergström, January 26, 2005.
19. Private communication from Per Carlqvist, February 10, 2005.
20. Private communication from Lars Bergström, January 26, 2005. He added: "However it is interesting to note that Klein on the other hand seems to have realized the full implications of the CBR discovery."
21. Private communication from Per Carlqvist, February 10, 2005.

22. Private communication from Lars Bergström, January 26, 2005.

23. Private communication from Per Carlqvist, February 10, 2005.

24. *Ibid.* He added: "Klein's interest gradually cooled, however (perhaps as a result of the discussion of CP violation), and Hannes became the driving force behind the Klein-Alfvén theory."

25. *Ibid.* He added: "Bertel Laurent and I discussed this problem in a couple of papers (Laurent, B.E., Bonnevier, B. & Carlqvist, P., 1988, APSS 144, 639; Laurent, B.E. & Carlqvist, P., 1991, APSS 181, 211)."

26. *Nature* 224 (1969), November 1, p. 477.

27. Gary Steigman, "Antimatter and Cosmology," *Nature* 224 (1969), November 1, pp. 477–481, quote p. 480. Cf. *idem*, "Observational Tests of Antimatter Cosmologies," *Annual Review of Astronomy and Astrophysics* 14 (1976), pp. 339–372, esp. 367–368.

28. Jagjit Singh, *Great Ideas and Theories of Modern Cosmology* (1961; New York: Dover, 1970 rev. ed.), pp. 273–277.

29. *Ibid.*, p. 277.

30. *Ibid.*

31. *Ibid.*

32. *Ibid.*

33. Joseph Silk, *The Big Bang* (New York: W. H. Freeman, 1980), pp. 319–321. Cf. the 2nd rev. ed. 1989, pp. 399–401.

34. *Ibid.*, p. 319. In the 2nd rev. ed. 1989, Silk changed this to "*most* other astrophysicists remain skeptical" (emphasis added).

35. *Ibid.*

36. *Ibid.*, p. 320.

37. P. J. E. Peebles, *Principles of Physical Cosmology* (Princeton: Princeton University Press, 1993), pp. 207–209.

38. Private communication from Lars Bergström, January 26, 2005.

39. Peebles 1993, p. 207.

40. *Ibid.*

41. Helge Kragh, *Dirac: A Scientific Biography* (Cambridge: Cambridge University Press, 1990); *idem* 1996. For an interesting comment on Kragh 1996, see a review by one of the cosmologists who took part in this development: Hermann Bondi, "Cosmological Wars," *Nature* 384 (1996), November 28, pp. 323–324.

42. Kragh 1996, p. 383.

43. R. Sebastian Pease and Svante Lindqvist, "Hannes Olof Gösta Alfvén: 30 May 1908–2 April 1995," *Biog. Mems. Fell. R. Soc. Lond.* 44 (1998), p. 12. Pease and I wrote this obituary together, but it was Pease who passed the value judgements on Alfvén's scientific work.

44. Private communication from Lars Bergström, January 26, 2005.

45. *The American Heritage Dictionary of the English Language* (Boston: Houghton Mifflin, 1992 3rd edn.), p. 1545.

46. Svante Lindqvist, "The Spectacle of Science: An Experiment in 1744 Concerning the Aurora Borealis," *Configurations: A Journal of Literature, Science and Technology* 1 (1992), pp. 57–94.

47. *www.nobelprize.org.* The journal *Science* covered the scientific Nobel Prizes that year in their news section in a November issue. Alexander J. Dessler, a space physicist at Rice University, wrote on Alfvén under the title "Swedish Iconoclast Recognized after Many Years of Rejection and Obscurity." Dessler had been one of Alfvén's most loyal admirers since the late 1950s when his teacher S. Chandrasekhar had encouraged him to take a closer look at Alfvén's work. In his article, Dessler mentions most of Alfvén's contributions to space physics, but with one exception: there was not a word on Alfvén's cosmology. See: Alexander J. Dessler, "Swedish Iconoclast Recognized after Many Years of Rejection and Obscurity," *Science* 170 (1970), pp. 604–606.

48. Stephen G. Brush, "Prediction and Theory Evaluation: Alfvén on Space Plasma Phenomena," *Eos* 71 (1990), no. 2 (January 9), pp. 19–33.

49. Personal communication Per Carlqvist, February 19, 2005.

50. Hannes Alfvén, "Big Bang-teorin håller inte: Universum uppbyggt av materia-antimateria," *Forskning och framsteg* 11 (1976), no. 7, pp. 7–13.

51. Hannes Alfvén, "Is the Universe Matter-Antimatter Symmetric?," in: *Antinucleon-Nucleon Interactions*, eds. Gösta Ekspong and S. Nilsson (Oxford: Pergamon Press, 1977b), pp. 261–274.

52. Hannes Alfvén, "Observations and Hypotheses in Cosmology," *TRITA-EPP* 78:11 (1978).

53. Hannes Alfvén, "How Should We Approach Cosmology?" in: *Problems of Physics and Evolution of the Universe*, ed. L.V. Mirzoyan (Yerevan: Publication House of the Armenian Academy of Sciences, 1978), pp. 9–37.

54. Alfvén 1977a, pp. 1–14.

55. Hannes Alfvén, *Cosmic Plasma*, Astrophysics and Space Science Library vol. 82 (Dordrecht: D. Reidel, 1981). Russian edition in 1983.

56. Hannes Alfvén, "On Hierarchical Cosmology," *Astrophys. Space Sci.* 89 (1983), pp. 313–324 (Also published as *TRITA-EPP* 82:03 (1982)); *idem*, "Cosmology: Myth or Science?," *J. Astrophys. Astron.* 5 (1984), pp. 79–98.

57. Interview with Sir Hermann Bondi, Churchill College, Cambridge, March 15, 1996.

58. *Ibid.* When asked directly what he thought of the Klein-Alfvén theory, Bondi said that "I don't think it has found a lot of support. It had great originality in it, but I don't think many people found it very attractive. It is not on the front row."

59. John Horgan, "Big-Bang Bashers," *Scientific American* 257 (1987), no. 3 (September), pp. 18–20.

60. *Ibid.*

61. *Ibid.*

62. Nobel Museum Archives, Hannes Alfvén Collection, Letter to the *Scientific American* by Hannes Alfvén on October 21, 1987, 3 pp.

63. *Ibid.*

64. Anthony L. Peratt, "Dean of the Plasma Dissidents," *The World & I*, vol. 3 (1988), no. 5 (May), pp. 190–197.

65. Hannes Alfvén, "Är universum evigt eller skapat ur intet?," *Svenska Dagbladet* 24 November 1988.

66. Alexander J. Dessler, "Fiftieth William Bowie Medal: Presented to Hannes Alfvén," *Eos* 70 (1989), no. 1 (January), pp. 9–10.

67. *Ibid.*, p. 9.

68. *Ibid.*, p. 10.

69. *Ibid.*

70. *Ibid.*

71. *Ibid.*

72. J. N. Wilford, "Novel Theory Challenges The Big Bang: Renegade Physicists Say Electromagnetism Shaped an Infinite Universe," *The New York Times*, February 28, 1989.

73. Eric J. Lerner, *The Big Bang Never Happened: A Startling Refutation of the Dominant Theory of the Origin of the Universe* (New York: Times Books/Random House, 1991) pp. 273–276. Cf. Anthony L. Perrat, *Physics of the Plasma Universe* (Heidelberg: Springer-Verlag, 1992).

74. *Op. cit.*

75. *Ibid.*

76. The phrase "Alfvén and his few followers" is from Kragh 1996, p. 383.

77. Lerner 1991, pp. 273–275.

78. Hannes Alfvén, "Cosmology in the Plasma Universe," *Laser and Particle Beams* 6 (1988), Part 3, pp. 389–398. (Also published as *TRITA-EPP* 87:07.); *idem*, "Cosmology in the Plasma Universe: An Introductory Exposition," *IEEE Trans. Plasma Sci.* PS-18 (1990), pp. 5–10; *idem*, "Cosmology: Myth or Science?," *IEEE Trans. Plasma Sci.* 20 (1992), 590–600 (Reprint of the article in *J. Astrophys. Astr.* 5 (1984), pp. 79–98).

79. Kragh 1996, p. 383.

80. See, for example: Eric J. Lerner, "The Big Bang Never Happened," *Discover* 9 (1988), no. 6 (June),

pp. 70–79; Anthony L. Perrat, "Not With a Bang: The Universe May Have Evolved from a Vast Sea of Plasma," *The Sciences* 30 (1990), January/February, pp. 24–32.

81. Lerner 1991.

82. E. F. Mallove, "The Big Bang Never Happened," *Sky and Telescope* 82 (1991), no. 5 (November), pp. 492–493; Paul Davies, "20 Billion Years Isn't Forever," *New York Times Book Review* July 28, 1991, p. 20.

83. Daniel J. Kevles, " 'The Final Secret of the Universe'?," *The New York Review of Books* 38 (1991), no. 9 (May 16), pp. 27–32.

84. Peter Nilson, "På väg mot ett annorlunda universum," *Svenska Dagbladet* 20 October 1991 – Nicolai Herlofson (1916–2004), one of Alfvén's closest collaborators and supporters since the 1940s, immediately produced an offprint of the review under the heading of his own one-man publishing company "Pilgrim Press," which he distributed widely within Alfvén's department and to everybody he met.

85. Private communication from Hannes Alfvén on many occasions in the late 1980s.

86. Hannes Alfvén and Carl-Gunne Fälthammar, "Can the Big Bang Survive in the Space Age?," *TRITA-EPP* 90:02 (1990).

87. Private communication from Lars Bergström, January 26, 2005.

88. Private communication from Bengt Gustafsson, January 25, 2005.

89. Private communication from Per Carlqvist, February 10, 2005.

90. Eric J. Lerner, "Bucking the Big Bang," *New Scientist* 182 (2004), no. 2448 (May 22), p. 20. The list of these 33 scientists included Hermann Bondi and Thomas Gold (who together with Fred Hoyle had been the three early proponents of the steady state theory).

91. In the same year Klein published an article on relativity and his metagalaxy, see: Oskar Klein, "Arguments Concerning Relativity and Cosmology," *Science* 171 (1971), January 29, pp. 339–345.

92. Nobel Museum Archives, Hannes Alfvén Collection, Oskar Klein, "Auktoritet och revolution," speech at the post-degree ceremony banquet at Stockholm University on 17 September 1971 (copy of typewritten MS., 3 pp.).

93. *Ibid.*

Re-Reading Bernal

History of Science at the Crossroads in 20th-Century Britain

MARY JO NYE

*D*uring the 1938–1939 academic year, the busy experimental physicist P.M.S. Blackett put together notes for three lectures on the "History of Science" in the Faculty of Arts at the University of Manchester, building upon notes for lectures that he had given in 1936 and 1937, including his May 1937 Presidential Address for the Birkbeck Physics Society. He drafted some further essays, too, all adopting the point of view that would become known as the "externalist" approach in the history of science. By the mid-1940s, Blackett's materials on the history of science included a typescript titled "The Study of the History of Science." This essay is the crux of the article titled "Why Study the History of Science" that was published in an anonymously-authored pamphlet on "The Development of Science" by the British Association of Scientific Workers while Blackett was ASW president during 1943–1947.[1]

The pamphlet became the object of a strongly negative review written in late 1944 or early 1945 by Frank Sherwood Taylor, then the Curator of the Museum of the History of Science at Oxford and president of the British Society for the History of Science between 1951 and 1953. If the author of the article on the history of science had "made a wide study (which I question)," wrote Taylor, "it has not been a deep one." In searching for the important factors in the development of science and scientific discovery, Taylor counseled, "I would urge that no causes be looked for outside the internal logic of science, until those within it have been exhausted."[2]

It would not be surprising that Blackett's approach to the history of science might not be deep or wide. Indeed, it seems unlikely that Blackett might take time to write or lecture at all on the history of science in the 1930s and 1940s. From 1933 to 1937, he headed the physics laboratory at Birkbeck College in London. He moved to Manchester during the 1937–1938 academic year, succeeding William Lawrence Bragg as Manchester's Langworthy Professor of Physics, a position previously held by Ernest Rutherford. Blackett's field of research, first pursued from 1921 to 1933 in Rutherford's Cavendish Laboratory at Cambridge, was particle and cosmic ray physics, and Blackett's continued work in this field would bring him a singly-designated Nobel Prize in Physics in 1947. Blackett had been a young officer in the Royal Navy during the Great War and, as an experienced military man and a prominent scientist, he was a valued member of interwar committees advising on military matters, including the committee chaired by Henry T. Tizard in the late 1930s that recommended that the Air Ministry immediately develop a land-based radar defense against what was anticipated as an inevitable air attack from Germany. During the years of the Second World War, Blackett was scientific adviser to the Anti-Aircraft Command and he became famous for his work in the British Admiralty as one of the founders and directors of operational research.[3]

In the late 1930s and early 1940s, then, Blackett was an active experimental physicist, at the top of his game, running physics laboratories in London and in Manchester, and spending a great deal of time in London for advisory and committee work and, later, for his wartime duties at the Admiralty. Of interest for the historiography of science are the professional and personal reasons that led him to write lectures and essays on the history of science during this period. If his reading was not wide and deep, what did he read and how did it inspire his historical approach in his lectures and contribute to their content? Did Blackett have anything original to say? And, finally, what do we learn from a focus on Blackett about the history of science in Britain at mid-twentieth century and the social-relations-of-science movement known as Bernalism?

Blackett's Introduction to the History of Science and "Bernalism"

In 1969, Blackett presented the introduction to the first J. D. Bernal Lecture, given at Birkbeck College by the x-ray crystallographer Dorothy Hodgkin, who had worked with Bernal in the 1930s at Cambridge. In their lifetimes, Bernal, Hodgkin, and Blackett all had reputations as prominent left-wing British scientists. In his introduction, Blackett recalled some episodes from his longtime friendship with Bernal, including their wartime experiences in operational research and Bernal's leadership in the movement that became known as Bernalism. Blackett noted the impact in Britain of the 1931 Second International Congress of the History of Science and Boris Hessen's widely publicized paper on the "Social and Economic Roots of Newton's *Principia*." Before Hessen, he always had been bored by standard histories of science, Blackett reminisced. Historians described scientists as if they were "living so to speak in a social vacuum." The argument that practical economic needs had driven Newton's revolutionary science, just as they drive modern science, had engaged Blackett's attention to the history of science for the first time.[4]

In his own lectures and essays on the history of science in the 1930s and 1940s, Blackett's starting point was Hessen's point of view, although, as we will see in some detail below, Hessen's historical accounts and conceptions by no means account entirely for Blackett's historical approach. Much attention has been given in the historiography of the history of science to the 1931 congress, and it rightly is perceived as a crossroads in the history and historiography of the discipline.[5]

The Soviet delegation arrived in London in late June of 1931 without having confirmed its plans by the due date, according to the president of the Congress, the British historian of medicine Charles Singer.[6] The Soviet group's eight members included the plant geneticist and President of the Lenin Academy of Agricultural Sciences Nikolai I. Vavilov, the physicist and physical chemist Abram F. Joffe (Yoffe), and the physicist and historian of science Boris Hessen. The leader of the delegation was Nikolai Bukharin, who had helped devise Bolshevik policy during 1925–1927 but lost his position in the Politboro and his editorship of *Pravda* in 1929. In 1931 Bukharin had a minor post as Director of Industrial Research for the Soviet Supreme Economic Council.[7]

Singer arranged a special final Saturday session of ten-minute papers for the Russians, but most of the scientists and the few historians attending the congress from twenty-four countries had already scheduled a sight-seeing trip for the day. At the suggestion of the Marxist mathematician Lancelot Hogben, the Russians launched what the *Manchester Guardian* described as "The Five Days Plan" by turning the Soviet embassy in London into a translating and publishing center, so that copies of all the Russian papers were distributed at the July 4th Saturday morning session at the Science Museum. Three days later a bound volume of the essays appeared as the book *Science at the Crossroads*.[8] All these events were well reported in newspapers and periodicals, including the *Daily Mail*, the *Morning Post*, the *Manchester Guardian*, and the Communist Party *Daily Worker*.[9] Bernal managed to get a review of *Science at the Crossroads* into the conservative weekly *The Spectator* because of his friendship with Celia Simpson, who managed its reviews, and The *New Statesman* published a short article on "Science and Politics in the Soviet Union" written by Bukharin.[10]

Along with Bernal, James G. Crowther, the scientific correspondent of the *Manchester Guardian*, was one of the most influential proponents of the new Russian approach in history of science.[11] Crowther had first visited Russia in the summer of 1929, meeting Vavilov, Joffe and Peter Kapitsa, and making a return visit in November 1930. He published *Science in Soviet Russia* upon his return.[12] Crowther was instrumental in working with the Society for Cultural Relations with the Soviet Union to organize the Russian delegation's presence in London for the Congress, and he organized British trips to the Soviet Union in July and in August of 1931 under the auspices of the SCRSU. The July group included Julian Huxley and Aldous Huxley, and the August visitors included Bernal, the physicist John Cockcroft, and the biochemists Bill and Tony Pirie. In late June and early July of 1931, Crowther acted as guide for the Russians in London, and he took Bukharin to visit Kapitsa in Cambridge.[13]

Kapitsa and Blackett were colleagues in the Cavendish Laboratory. They had collaborated on cloud-chamber work in the early 1920s, but Kapitsa had his own laboratory space in 1931 for his work on very strong magnetic fields.[14] Blackett was just embarking on his work using coincidence Geiger-counters to control cloud-chamber expansion when charged particles pass through the chamber, leading by 1932–1933 to Blackett's description with Giuseppe Occhialini of the existence and properties of the positron or anti-electron.[15] Crowther and Blackett had known each other since at least the summer of 1930 when Blackett and his wife entertained Crowther at one of the Blacketts' dinner parties while they lived for a few months in Berlin, and Crowther was quite interested in Blackett's work.[16]

> At about this time I was much struck by Blackett's use of terms from ball games in his interpretation of his photographs of the tracks of atoms. This made me wonder whether a tradition of ball games had assisted the British in their conspicuous success in particle physics [. . .] the Germans, who have a strong musical tradition, were apt to interpret phenomena in terms of waves.[17]

Crowther would write a great deal in the history of science, as well as in science journalism, inspired, he said, by Boris Hessen's paper. "Hessen's paper revealed to me a method of prosecuting the history of science which was more profound than the conventional one."[18] Crowther conceived the plan of applying the notion of science as a social product to more recent periods in British science, resulting in his publication of *British Scientists of the Nineteenth Century* in 1935, followed by *Famous American Men of Science* in 1937, *The Social Relations of Science* in 1941, and many subsequent volumes.[19] In Crowther's view, Hessen's essay "transformed the history of science from a minor into a major subject," and it did so by transcending the previous "antiquarian" nature of history of science in favor of showing that a knowledge of the scientific past was essential for the solution of contemporary social problems.[20] Crowther recognized that Hessen's ideas were unwelcome among many scientists, as well as among historians, and he witnessed Hessen's scientific colleagues at a banquet in Moscow in 1931 arguing that a differential equation could not be a reflection of social and economic conditions. According to Crowther, when Lev Landau had an opportunity to use the first service for telegraphing pictures between Leningrad and Moscow, Landau transmitted a mathematical diagram caricaturing Hessen's ideas.[21]

Like Crowther, Bernal was enthusiastic both about the Soviet Union and about Hessen's revisionist approach to antiquarian history of science. Returning from his 1931 summer visit to the Soviet Union, Bernal reported that "It was grim but great."[22] In October, Bernal was invited to give a talk on x-ray crystal analysis at the Kapitsa Club, hosted since 1922 in Kapitsa's rooms and then in Trinity College and frequented by Blackett, Cockcroft, Paul Dirac, Ralph Fowler, and C. P. Snow, among others.[23] Snow soon would make Bernal famous by portraying him as "Constantine," a brilliant x-ray crystallographer, in Snow's 1934 novel *The Search.*

It was well-known at Cambridge that the physicists' boss Ernest Rutherford could not abide Bernal on several counts. These included Bernal's open commitment to communism, his ideology of sexual freedom and promiscuity, and his tendency to hand over research projects to students while indulging in highly speculative theorizing.[24] For students, Bernal was inspirational and they were grateful for his insistence that they publish papers in their names alone. From Rutherford's perspective, however, Bernal failed to organize proper research teams.[25] Bernal was assigned to a sub-department at Cambridge where he nonetheless did groundbreaking structural studies of sterols, proteins, and viruses. He worked in what were described as "ill-lit and dirty rooms on the ground floor of a stark, dilapidated grey brick building" with inadequate studentships and technicians.[26] No wonder he addressed himself to the social and economic conditions for scientific work in Britain! With his path blocked at Cambridge, Bernal eventually would succeed Blackett as director of the physics laboratory at Birkbeck College in 1937.

Blackett and Bernal thus had known each other over a long period of time, beginning in Cambridge, and they had many professional and political interests in common. Bernal was among the Cambridge scientists who helped revive the Association of Scientific Workers in 1935, in which Blackett became more and more involved, eventually serving as its president, as noted.[27] As Blackett mentioned in his remarks of 1969, Blackett and Bernal were colleagues in operational research and sided together in the spring of 1942 against the claims by Frederick Lindemann, Winston Churchill's scientific advisor, that evidence collected by Bernal and Solly Zuckerman about German bombing of Hull and Birmingham supported Lindemann's argument for large-scale British bombing of German urban centers in order to undermine German morale. Blackett and Bernal were together at Portsmouth on the eve of D-Day, walking through a maze of tents and huts to look down on the sea, where the boats would launch in the morning, and talking long into the night.[28]

Blackett had moved from Conservative Party sympathies during the Great War to the Labour Party by 1922, but he dated his entry into active political engagement to the 1930s in reaction to the great economic depression, Hitler's rise to power, and the Spanish Civil War. He said that his appearance in 1934 on the BBC radio series "Whither Britain" marked his turn at a crossroads toward political activism and Fabian socialism.[29] The turn coincided, too, with his move from Cambridge to Birkbeck College, near Bloomsbury, a college founded in 1823 as the London Mechanics Institute.[30]

Blackett visited Moscow for physics conferences in September 1935 and September 1937, which increased his interest and sympathy for the Soviet Union's claims to be revolutionizing its nation through applications of science and technology. The 1934 BBC radio talk programs in which Blackett participated tried to make the case for increased public and government support for scientific education and research in Britain as the solution to the problems of the current economic crisis. Julian Huxley moderated a program in which Blackett spoke on "Pure Science," with Huxley taking the view that the line between pure and applied science is arbitrary. Huxley told listeners that they should understand that science is a social activity, and that this is a lesson for scientists to learn as well as statesmen and the public. Blackett agreed and went on to suggest that scientists must step up and make clear to the public the virtues and advantages of science at a time when anti-scientific and anti-intellectual movements were becoming dominant in Germany.[31] Blackett's new scientific and socialist activism soon showed up in the notes that he made for a lecture to be given in Highgate in February 1936. Here can be found his first foray into the history of science as a means of explaining how science works and why it merits public support.

Blackett on the History of Science

Blackett's handwritten notes, headed "Highgate February 1936," consist of four parts, with the second part devoted to a brief résumé of some main lines of development in the history of science. Most of the lecture is drawn from his beautiful published essay of 1933 on "The Craft of Experimental Physics." The Highgate notes provide insight into how Blackett's own laboratory experiences as a physicist played a significant role in his personal reaction to the brouhaha over Hessen's brand of history of science.[32]

Blackett's 1933 essay on the craft of experimental physics was written for a volume that aimed to inform the public about what Cambridge scholars were doing and why they do it. Contributors included R. B. Braithwaite on philosophy, C. P. Snow on chemistry, and C. H. Waddington on biology, among others.[33] For his essay, Blackett drew upon his work in particle physics to explain some of his experimental procedures and results, while putting his work within the larger context of the nature of scientific practice.

Blackett's development into a consummate experimentalist had begun with his naval education at Osborne and Dartmouth naval colleges, a training that gave him probably the most intensive secondary education available in England in the physical sciences and engineering. That naval education and his experiences in the wartime Navy made Blackett a man accomplished in tools, inventions, and instruments before he even entered Cambridge in 1919, where he arguably became the most eminent and the most versatile British experimental physicist of his generation.[34]

Blackett's colorful definition of an experimental physicist is worth quoting in full:

For the experimental physicist is a Jack-of-All-Trades, a versatile but amateur craftsman. He must blow glass and turn metal, though he could not earn his living as a glass-blower nor ever be classed as a skilled mechanic; he must carpenter, photograph, wire electric circuits, and be a master of gadgets of all kinds; he may find invaluable a training as an engineer and can profit always by utilizing his gifts as a mathematician. In such activities will he be engaged for three-fourths of his working day. During the rest, he must be a physicist, that is, he must cultivate an intimacy with the behavior of the physical world [. . .] The experimental physicist must be enough of a theorist to know what experiments are worth doing and enough of a craftsman to be able to do them. He is only preeminent in being able to do both.[35]

Following this passage in the 1933 essay, with which he also began the 1936 Highgate lecture, Blackett remarked in both the essay and lecture on the reciprocal relationship between technical innovations in the laboratory and industry, noting the many years that it may take the physicist to ready an apparatus for the experiments he has in mind and that the physicist gradually develops a hands-on intuitive understanding of how to proceed. The scientific knowledge that results, Blackett continued, is the result of the skill of the hands, as well as the mind.

At this point in the Highgate lecture, Blackett departed from his 1933 essay in order to introduce some historical observations. The handwritten notes suggest Blackett made a transition into his historical comments by observing that science is practical in its origin and that it began flourishing in the Renaissance, with the needs of navigation as one of its prominent spurs. He mentioned astronomy, mechanics, mining, hydraulics, ballistics and other areas of scientific investigation in the seventeenth century, concluding with references to the principle of conservation of energy, the industrial revolution, and the current reaction against science due to the capitalist crisis, a recurrent theme in BBC radio programs and the public media in the early 1930s, as we have seen. The Highgate notes then move back to the text of Blackett's published essay on craftsmanship, with a discussion of how the invention of the wireless valve and the three-electrode thermionic valve, in combination with improvements in the cloud chamber, revolutionized the tracking of nuclear particles. The three-electrode valve, Blackett remarked, was an industrial instrument, used in telephone exchanges to register the number of calls for a subscriber. Toward the conclusion, Blackett again brought together theory and practice, by commenting that the ways of thinking about a simple problem in physics can be divided into two extreme types: the abstract mathematical method and the intuitive methods of everyday life. "Between these extremes lies the method used by the experimental physicist."[36]

The following year, in May 1937, Blackett prepared a Presidential address for the Birkbeck Physics Society. His notes for the 1937 lecture differ substantially from the 1936 Highgate section by including quotations and materials that are developed at even more length in handwritten notes for his history of science lectures at Manchester during 1938–1939. In the lectures and the later handwritten manuscript titled "Science Today" and typescript titled "The Study of the History of Science," we see the enthusiasm with which Blackett took up Hessen's arguments of 1931 and even some of the specific content of Hessen's Newton paper, while combining a close reading of Hessen with reading and references from Crowther, Bernal, R. H. Tawney, Norman R. Campbell, Richard Braithwaite, George Sarton, Sherwood Taylor, Charles Singer, Abraham Wolf, and other sources. In addition, Blackett drafted what he may have intended as an original article studying the "stimulus to scientific discovery that arose from the demands of ocean navigation," consulting books on the marine chronometer and the history of navigation.[37]

Blackett's use of Hessen included the main opening argument of Hessen's 1931 paper, as well as Hessen's examples and quotations. Blackett clearly liked the way in which Hessen posed the problem of the source of Newton's creative genius. Blackett often repeated in his lectures the quotations from Alexander Pope and Alfred North Whitehead that Hessen set up for criticism: (From Pope: "God said 'Let Newton be! And all was light"; and from Whitehead: "'Science and Civilisation'" owes its development to the fact that Newton was born in the very year of Galileo's death. Only think what the history of the development of humanity would have been if these two men had not appeared in the world"). Like Hessen, Blackett promised to offer a "radically different conception" for the origins of Newtonian science than these clichés that the scientific revolution was due to divine providence and personal genius.[38]

Blackett used Hessen's argument that the causes for Newton's invention and discoveries, and for those of his contemporaries in the Royal Society, lay largely in events and values of the English Civil War and Commonwealth. Blackett bolstered Hessen's arguments with his reading of Tawney's *Religion and the Rise of Capitalism* and later, Robert Merton's 1938 *Osiris* monograph "Science, Society and Technology in Seventeenth-Century England." It likely was Crowther who directed Blackett to Merton after Crowther gave a series of lectures in 1937 on the history of science at Harvard University at the invitation of Harvard's President James B. Conant. On his return to England, Crowther, who also lunched with Sarton while at Harvard, pronounced Merton "the most able of all the coming men whom I met at Harvard," praising Merton's new sociological approach to the history of science.[39]

Like Hessen, Blackett taught that the brilliance of scientific achievements during the sixteenth and seventeenth centuries was conditioned by the distintegration of the feudal economy, the development of merchant capital, international maritime ventures, and heavy mining industry. These themes also were ones taken up by Bernal in his *Social Function of Science* of 1939. Blackett offered some of the same concrete examples as Hessen of technical problems that had to be solved: latitude and longitude, the compass and magnetic deviations, the properties of gases and liquids, air pressure, pumps, etc. Like Hessen, Blackett emphasized the importance of the founding of scientific societies outside the structure of the universities and the role of industry in the construction of new instruments that were used by these scientists.[40]

Blackett was not reading economic and social history alone. His familiarity with George Sarton's *The History of Science and the New Humanism* demonstrates how essential Sarton was for scientists such as Blackett who turned their attention to history of science in the mid-1930s. Sarton helped scientists develop the case that scientists should not be blamed for the current economic catastrophe of the 1930s but that the blame lay elsewhere: "The greatest problem which the statesmen of our time have to solve is the humanization of industry and labor, but it will not be easy to undo the evils of a century of technical ruthlessness and unrestrained greed." Most important, Sarton provided an example of how to explain to non-scientists and anti-scientists the need and the pleasure of interesting themselves in "the most exuberant activity" of current times and to understand that the scientific "spirit" is cultural and humanistic at its essence and not "a simple collection of technical and utilitarian recipes."[41]

Sarton's volume, which began as lectures at Brown University in 1930, asked questions not dissimilar from Hessen's introductory problem in his paper of 1931: Are the main events of human history determined by a relatively few men? Sarton's solution was different than Hessen's: "[. . .] as soon as attention is directed upon the deeds which are truly constructive the difference between biographical and non-biographical history dwindles almost to insignificance. The purpose of humanity is to create beauty, justice, and truth, Sarton argued, and these creative activities engage the artist, the social reformer, the saint, and the scientist alike.[42] Of these, it is scientific activity that is cumulative and progressive, so that the history of the progress of civilization is the history of science and the pursuit of truth.[43] In Blackett's view, too, "There is no doubt that in his attitude to his experimental work, a scientist shares much of the feeling of the potter to his clay, the painter to his canvas, or the engineer to his bridge. Far from there being any antagonism between the sciences and the arts, there is in fact no very clear distinction."[44]

While there was a good deal in Sarton with which Blackett could agree, there also was a good deal with which he did not quite agree, as we see in a marginal note in one set of lecture notes ("I would not perhaps go as far as Sarton.").[45] Sarton's statement that "external circumstances" provide only part of the explanation of why and when discoveries are made provided Sarton's rationale for a history of science that emphasizes the history of ideas, not social and economic conditions. "Science develops very much as if it had a life of its own [. . .] it never develops in a political vacuum. Yet each scientific question suggests irresistibly new questions connected with it by no bounds but the bounds of logic."[46] In Blackett's opinion, Sarton's view here was too simplistic, indeed wrong. For one thing, the invention of instruments could suggest entirely new lines of research or it could lead into cul de sacs. In his 1933 essay on the experimental craft, which he quoted in later lectures, Blackett explained that scientific progress often rapidly follows upon the invention of a new method or a new instrument, but "when its limit of usefulness has been reached, investigation may halt, to await the invention of some new contrivance."[47]

Of considerable appeal, however, to Blackett was Sarton's occasional deference to the role of craftsmanship in scientific work and to the importance of technique, if not technology: "The humanity of science is revealed in a humbler manner by the consideration of our instruments. They illustrate that science was not created only by our minds, but to a far greater extent than is usually supposed, by our hands."[48] Blackett quoted Sarton frequently in this vein: "It is as if some scientists (not only inventors) could only think when their hands were busy; it is as if they thought with their hands." And, again: "Much of our experimental knowledge and with a few exceptions the best of it—developed almost naturally with the instruments which human hands had built and used. It is truly essential to exhibit the interdependence of intellectual processes and handicraft, of minds and tools."[49] In his handwritten essay "Science Today," Blackett generalized that "The history of science is to a considerable extent [. . .] the history of instruments."[50]

Blackett also used lines of argument from Bernal's original and controversial book of 1939 on *The Social Function of Science*. Bernal, too, quoted Sarton, "the great historian of science," that science "is essentially human in its origin and growth" and is part of intellectual culture and general humanism, identifying Sarton with the idealist view, dating back to Plato's *Republic*, that science is a kind of pure thought concerned with discovery and contemplation of truth.[51] Another view than Sarton's, wrote Bernal, characterizes science by its utility and its power. Neither view is sufficient: "Modern science derives both from the ordered speculation of the magician, priest, or philosopher, and from the practical operation and traditional lore of the craftsman. Until now, far more attention has been given to the first rather than the second, with the result that progress of science seems more miraculous than it was in fact."[52] The second chapter of Bernal's book is a historical survey, the general lines of which Blackett used in his lectures on history of science, including the description of the rise of agriculture and urban centers, the Hellenistic revival, the birth of science in the Italian Renaissance, the new academies, scientists as craftsmen with patrons, and the development in Leeds, Manchester, Birmingham, Glasgow and Philadelphia of a culture of gentlemen who furthered the science of the industrial revolution.[53]

For Blackett, as for Bernal and other scientists of the social relations of science movement in the 1930s, it was imperative to explain to students and the public what science is, in order to undercut anti-intellectualism and anti-scientism, but also in order to get public support for investment and commitment to scientific research and development in Britain.[54] The best way to understand science, or any organism or activity, as Blackett put it to his Manchester Arts students, was to know science's history, not to "lay down a system of abstract definitions of its scope, methods, and conditions" about which it is hard to find agreement anyway. Talk about the invention of the method of experimental discovery was unsatisfying to Blackett. Rather, scientific method is only a more sophisticated version of "systematized common sense, as applied in all ages by practical people who have a job to do."[55]

Blackett's most finished essay having to do with the history of science is the typescript dating from 1943 or 1944, which appears to have been the core of the anonymously authored essay titled "Why Study the History of Science" published by the Association of Scientific Workers. Blackett's typescript answers the question "Why study the history of science?" with a completely utilitarian and policy-centered argument.[56] In the past, Blackett argued, the growth of science has been uneven and this cannot be risked in the future. In order to maximize efficiency in the war against fascism and in postwar reconstruction, it is important to understand the past causes of growth and decline in scientific advance. The answer does not lie in the relative difficulties of scientific subjects nor in an internal logic of science, but can be found in the historical relations between economic development and technology, for instance, how rapid economic expansion in the sixteenth and seventeenth centuries presented a host of technical problems that stimulated the various scientific fields that answered to these practical needs. This sort of advance happened again in the eighteenth century, and in the late nineteenth century. New sciences developed as the "brewers wanted better methods of brewing, the landowners longed to be rid of the scourge of diseases that ruined their silkworms and sheep, and they got science to solve their problem through bacteriology."[57] There is little here of Sarton's beauty, justice, and truth.

Blackett did not mean to imply that every scientist is consciously subjected to the will of captains of commerce. In his early years (at least before he moved to London), Newton had little interest in practical problems, but his friends such as Hooke, Wren and Halley were deeply interested in practical problems and brought pressure to bear on Newton, leading him back from his pursuits of alchemy and theology to the *Principia*. (Note Blackett's well-informed appraisal of Newton's interests.) The scientist does not choose his work solely in a social vacuum, Blackett explained, but he chooses from problems that are suggested from the outside world or that come from the work of other scientists. Fascist Germany now presented the example of the diversion of German scientists into ideologically-motivated scientific research, such as race theory, and into the demands of the German war machine, suppressing scientific freedom. The challenge for British policy lay in developing planning of science for future needs, but without abandoning fundamental research and scientific freedom in favor of purely technological work.[58]

Why study the history of science, then? Blackett's answer in this essay centers on the needs of contemporary science and society: "The study of the history of science is of value only in so far as it provides conclusions on which to base present and future action." This is Bernalism plain and clear. Further, writes Blackett, the moral lesson of a study of the history of science is that science is so closely linked with social life that any scientist who cares for the progress of his subject must be ready to take action outside his immediate scientific sphere in order to further his own scientific work.[59] Bernal was more specific in *The Social Function of Science*, warning of the danger for scientists of behaving like their colleagues in Germany who had employed the rhetoric of independence and chosen not to meddle in political and social matters.[60]

Elsewhere in his writing on history of science, Blackett warned that science "for its own sake" can only be pursued when the major social problems of the world have been solved. "Each scientist has to make his own decision as to the orientation of his interests, how much he [. . .] adds to knowledge [. . .] and how much of his energies to divert to the contemporary social and political problems which are so much the result of science."[61] This was the program of the social relations of science movement and of scientists' responsibility to both their science and to society.

Bernalism and Some Paradoxes for the History of Science

Frank Sherwood Taylor was not the only scientist or historian who disputed the aims and content of the kind of history of science offered by Crowther, Bernal, Blackett and those who argued for the social responsibility of the scientist in Britain and the historical role of social and economic needs in scientific advance. By the time that the British Society for History of Science was founded in 1947, liberal humanists, many of whom were historians and philosophers, had come to dominate teaching programs established earlier at University College London, Oxford, and Cambridge, and most of these scholars oriented themselves firmly toward Sarton's new humanism, the history of scientific thought, and the independence and neutrality of science from external influences. At the first ordinary meeting of the BSHS in October 1947, Benjamin Farrington, the Marxist classicist and author of histories of Greek science, questioned the astrophysicist and philosopher of science Herbert Dingle about Dingle's June 6th Inaugural Lecture at University College London. Dingle was the first full-time professor and head of the UCL Department of History and Philosophy of Science, a department that included as part-time or full-time members Singer, the philosopher Abraham Wolf, Douglas McKie, who had been educated in chemistry, and Angus Armitage, whose degree was in astronomy.[62] Farrington, who soon would become an administrator in the UNESCO Commission for the History of the Social Relations of Science, criticized Dingle for considering only the "thought aspect" of science and failing to see science "as an integral part of human history in general" in which scientific advance is the result of the quest for the "practical mastery of nature."[63]

Singer, who had chaired the 1931 Congress, became the first president of the new British Society for the History of Science. His presidential address of 1948 in part constituted a response to Farrington. Singer identified William Whewell, the nineteenth-century Master of Trinity College at Cambridge, as the nineteenth-century pioneer of the modern history of science and Sarton as its twentieth-century leader. The role of the history of science today, said Singer, is the same as that of humanism five hundred years ago: to heal the great discontinuities and disruptions that threaten civilization by broadly surveying the history of science, which is the most truly humane and international of all studies.[64]

Singer's definition of the history of science in his presidential address specifically included in its purview a study of the "conditions under which such men have laboured and lived, the examination of their training and mental history, of their ways of life, and the human setting," along with study of social, economic, and philosophical conditions that "fail to produce effective scientific fruits, or that yield only bizarre or deformed products."[65] Crowther later claimed that he had a conversation with Singer in 1937 in which Singer told him that the approach Crowther was taking in history of science was making Singer's own work out of date, and that if he were only now beginning his studies, he would adopt Crowther's point of view.[66] Singer, however, like many of his colleagues, was more likely to invoke social and economic history in order to explain failures or perversions of science than its successful advances. Singer stated in his presidential address that the true scientific attitude demands "independence of all other judgements that influence mankind—judgements, for example, based on fashion or tradition or taste or passion or class."[67]

Anna-K. Mayer has analyzed the development of the history of science discipline in Britain, including the fortunes of the Cambridge History of Science Committee. Needham and Walter Pagel, a German refugee pathologist working at the Papworth Institute near Cambridge, founded the Cambridge History of Science Committee in 1936, setting up a lecture series and an exhibition celebrating Cambridge science. Needham's *History of Embryology* had appeared in 1934, and Pagel had written monographs on Van Helmont and philosophical medicine and on religious motivations in seventeenth-century biology, rejecting, as Mayer notes, the positivist protocol of separating science from magic and advocating a deeply historicist approach to the history of science.[68] Following the disruptions of the war, when Needham and Pagel both were absent from Cambridge, the History of Science Committee was taken over by the historians Herbert Butterfield and Michael Postan, the paleobotanist Hugh Hamshaw Thomas, the modernist theologian Charles Raven, and the professor of English literature Basil Willey. The historiographical approach shifted to the celebration of empiricism over mysticism and to the history of thought with strong emphasis upon "experience" and "virtue" in science.[69] As Mayer argues, the birth of modern science (a theme taken up by Alexandre Koyré in France) was an intellectual revolution, and scientific work was "a distinterested journey of the mind."[70]

Two other themes were dominant in the immediate postwar approach to the history of science in Britain: the theme of anti-Marxism and the theme of the freedom of science. Both Bernal and Crowther recognized that Marxist ideological jargon alienated most British readers, even though Bernal still succumbed to it. As Charles Singer's wife Dorothy Waley Singer wrote Joseph Needham regarding the 1931 Congress, "I listened hard and with open mind, and the impression gained was that they adopted toward Marx exactly the medieval attitude towards Aristotle—that nothing could be right unless it could be traced to his words."[71] Bernal conceded that the appeal to Marxism likely immediately influenced British academics "not to listen to the arguments which followed."[72]

Among the most widely discussed of negative reactions to Bernal's sometimes overtly Marxist *Social Function of Science* was a review by the Hungarian-born German refugee physical chemist and philosopher of science Michael Polanyi, a close, if highly argumentative, friend of Blackett's in Manchester's Science Faculty. Polanyi was increasingly alarmed in the late 1930s at talk about the need for planning and organization of science in Britain along models similar to the Soviet Academy of Sciences or the emerging CNRS in France. Polanyi also objected to the formation in 1938 of a Division for the Social and International Relations of Science in the British Association as the result of efforts by Bernal, Needham, Haldane, and Richard Gregory, the influential editor of *Nature*. Blackett was a member of the Division's subcommittee that was to address issues of scientific and technological development.[73]

Only a minor part of Bernal's *Social Function of Science* actually had to do with the history of science. Its great originality and influence lay in its formidable presentation of information and statistics about scientific research in Great Britain and the proliferation of scientific work throughout universities and government-sponsored agencies, many of them oriented toward industry, agriculture, and military defense. Bernal proposed a major reorganization of research, including the formation of a central endowment based in government and industrial funds that would be distributed by a council on the basis both of the need for internal development of the sciences, according to scientists' estimates, and the need for particular developments in individual sciences on account of national needs.[74]

In a chapter on international science, Bernal praised Marx, Engels and Lenin for their understanding of science and lauded Soviet achievements since 1917, detailing ten key problems identified by the Soviet Academy of Sciences in the Third Five-Year Plan. Bernal also described dialectical materialism as a useful tool for Soviet science, co-existing with traditional methods of induction and proof. Incredibly, Bernal offered the example of Nikolai Vavilov and his institute for plant research as a "beautiful" instance of the integration of fundamental discoveries in "genetic principles" with useful purposes in agriculture.[75] Vavilov would disappear from public view in 1940.

For Polanyi, and others, it was infuriating that Bernal failed in any way to criticize the Soviet Union at a time when it was well-known that trials and arrests had been made of many Soviet scientists and that no Soviet scientist was free to cross the border. As to dialectical materialism, "Not one in a hundred physicists believed this nonsense, but no one could dare to contradict the statement publicly."[76] Soviet planning of science and denial of the distinction between pure and applied science was both erroneous and dangerous to the functioning of science as an "organism of ideas" that "cannot be deflected from its internal necessities by the prospects of useful applications."[77]

Polanyi collected a series of essays about planning, the Soviet Union, and Bernalism into a book titled *The Contempt of Freedom: The Russian Experiment and After* (1940), which he sent to Blackett in 1941. Blackett wrote his friend that he found in the work a hostile attitude not toward the Soviet Union alone, but, as Blackett put it, "to all that is generally called progressive politics—'progressive obsessions,' in your words."[78] During the war years, Polanyi, Taylor, the Oxford zoologist John Baker, and the retired Oxford botanist George Tansley were among those who argued publicly against government planning, linking their objections to a rejection of economic interpretations of the history of science.[79] In 1941 the group founded the Society for Freedom in Science, with a membership that included by 1947 the historians and philosophers of science William Dampier, Gavin de Beer, Herbert Dingle, Alexander Findlay, Michael Postan, Charles Singer, Sherwood Taylor, H. Hamshaw Thomas, E. T. Whittaker and, surprisingly, Joseph Needham.[80] (Needham, it should be remembered, shared many views with these humanist scholars, especially in Needham's Christian thinking that he sometimes expressed in sermons as Master of Caius College at Cambridge.[81])

In the late 1940s, Blackett's interests turned toward new researches in physics, moving away from particle physics to the Earth's magnetism and eventually toward investigations of magnetic evidence for continental drift. He became widely known and politically suspect in Great Britain and the United States for his opposition to American atomic weapons policy and to British development of nuclear weapons, although his political views were sufficiently respectable and valuable that he became an advisor to Harold Wilson's Labour government in 1964. For the previous ten years, he met with a group of scientists including Bernal and Snow that advised Hugh Gaitskell, as leader of the Labour Party, and then Wilson and his Shadow Minister for Education Richard H. S. Crossman.[82]

Bernal, who continued to do laboratory work at Birkbeck College, on the structure of water and other liquids, took on the task of writing a large survey of the history of science, which appeared in 1954. Bernal's *Science in History* grew from a single volume of 867 pages in 1954 to four volumes of 1,328 pages in the fourth edition of 1968, with many revisions, including the removal of Trofim Lysenko's name from later editions of the book.[83] The book received much favorable notice, including a review of the last edition by Loren Graham, who praised Bernal's illustration of "the role of artisans and technology in scientific development as well as the impact of economic differences in society upon science."[84] In explaining his reasons for spending so much time on history of science in the 1950s and 1960s, Bernal told Aaron Klug, who was working at the time in Bernal's laboratory at Birkbeck College, that he hoped by his omissions and errors to stimulate others who were more qualified and leisured to take up the task. Bernal also remarked that he thought his experiences as a scientist who had participated in some critically important work might give him an advantage over professional historians.[85]

For all the alleged radicalism of Bernal and the Bernalists, many of their experiences and views about the nature of science corresponded with assumptions and descriptions of the British historians of scientific thought. Bernal or Blackett no more than Sarton or Singer doubted that scientific development is advancement in knowledge that gets closer to the truth about the natural world. Nor did Polanyi doubt the truth value of scientific laws and theories, despite his criticisms of philosophical claims for scientific objectivity and his attempts to define scientific truth in terms of what he called personal knowledge. One of the primary differences between the Bernalists and historians of ideas lay in the emphasis by Bernal and Blackett (and equally Polanyi) on understanding how science works, not just what theories say. For Blackett, as we have seen, science is craftsmanship, much of it developed gradually into a kind of intuitive feeling for the apparatus.

For the Bernalists, more significantly than for historians of thought, science is a social network and a social activity. Things get done because "a handful of the more important scientists in the country know one another," as Bernal put it, and workers in a field "arrange among themselves, when they are on friendly terms, the kind of work each of them intends to pursue and the relation of one man's work to another." "To some it is a game against the unknown where one wins and no one loses, to others, more humanly minded, it is a race between different investigators as to who should wrest the prize from nature."[86] How similar all this is, too, to Polanyi's emphasis on tacit knowledge, apprenticeship, science as social practice, the competitiveness and obsession of scientific work, and the intimacies of the republic of science.[87]

On the matter of the kinds of moral and ethical values and conduct that result in scientific knowledge, the Bernalists and historians of thought also fundamentally agreed, although the Bernalists employed a language that was more sociological in character. In what Gary Werskey characterizes as the most widely quoted passage in all of Bernal's writings, Bernal wrote in 1939 of what Robert Merton would define in 1942 as the scientific norms of universalism, communism, disinterestedness, and organized skepticism.[88]

> In its endeavour, science is communism. In science men have learned consciously to subordinate themselves to a common purpose without losing the individuality of their achievements. Each one knows that his work depends on that of his predecessors and colleagues, and that it can only reach its fruition through the work of his successors. [. . .] Each man knows that only by advice, honestly and disinterestedly given, can his work succeed, because such advice expresses as near as may be the inexorable logic of the material world, stubborn fact. Facts cannot be forced to our desires, and freedom comes by admitting this necessity and not by pretending to ignore it.[89]

The Bernalists were most radical and most controversial in their advocacy of scientific planning, scientific dependence on social needs, and scientists' responsibility to the public. For Bernal, freedom of science meant freedom from want, just as surely as it meant freedom from police supervision or from bureaucratic administration. In Bernal's view, guarantees of scientific freedom can be achieved only if there is public confidence that public self-interest lies in scientific advance.[90] In contrast to the current rhetoric of "pure" science, Bernal argued, scientists had long been accustomed to taking for granted the usefulness of their work, whether that utility lay in the glory of God or the benefit of mankind or personal satisfaction.[91] More tendentiously, Bernal and his associates, Blackett among them, offended or alienated many intellectuals by making an implicitly moral critique of scientists, historians of science, and philosophers of science who could be accused of complacency in their profession of commitment to the purity of science.

In this regard, Werskey has characterized Bernal's *Social Function of Science* as a book that was "monumentally indiscreet."[92] Indeed it was indiscreet on the same scale as James Watson's *Double Helix* or C. P. Snow's *Two Cultures*. Bernal quoted a passage at length from one of the characters in Aldous Huxley's *From Point Counter Point*:

> I perceive now that the real charm of the intellectual life—the life devoted to erudition, to scientific research, to philosophy to aesthetics, to criticism—is its easiness. It's the substitution of simple intellectual schemata for the complexities of reality [. . .] Till quite recently, I must confess, I too took learning and philosophy and science—all the activities that are magniloquently lumped under the title 'The Search for Truth'—very seriously [. . .] Shall I ever have the strength of mind to [. . .] devote my energies to the more serious and difficult task of living integrally?[93]

Bernal then continued, indelicately, "This attitude, though rarely admitted, is actually extremely widespread among scientists, particularly those in the safer and more comfortable positions [. . .] 'Whenever I look out at the world,' a professor once remarked, 'I see such misery and mess that I prefer to bury myself in my own work and to forget about things that in any case I could do nothing about.'"[94] In his review of Bernal's book, Polanyi took special offense at Bernal's statement that the ideal of pure science "was a form of snobbery, a sign of the scientist aping the don and the gentleman" and that no economic system would pay a scientist just to search for truth.[95]

In the 1930s and 1940s, the Bernalist approach to history of science in the social-relations-of-science movement had the potential to invigorate history of science as a part of general historical practice. In the short run, it largely failed because its opponents loathed its Marxist ideological jargon, pro-Soviet politics, and proposals for central planning and funding of science and technology. In the longer run, the history of science profession took a turn back toward Bernalism in the 1970s, institutionalizing a new historiographical approach in 1975 in the founding of the Society for Social Studies of Science and in the endowment of the 4-S Bernal Award in 1981. The practitioners of science studies dismissed not only the notion that the essence of science is its internal thought and logic, but that scientists are objective and disinterested in their scrutiny of the natural world and in their relationships with each other as members of the scientific community.[96] Scientific practice and laboratory instruments operating in the interstices of science and technology in local cultures and trading zones became the objects of study. By the 1980s, as argued by Paul Forman in a recent essay, science studies had reformulated the relationship between science and technology into one of technoscience, in which science is the gift of technology rather than the reverse, and in which physics is one of the "technical sciences." Forman identifies this historiographical turn with postmodernism.[97]

Paradoxically for the legacy of the Bernalist scientist-historians, the practitioners of science studies, who appear to be Bernalist offspring, have by and large rejected the assumption of the privileged status of scientific knowledge that the Bernalists, the historians of thought, the scientist-philosopher Polanyi, and the sociologist of science Merton all took for granted. More to the point with respect to Bernal, Blackett, Crowther, and other members of the science-and-social-relations movement, postmodernists in the 1980s and 1990s often adopted a skeptical attitude toward Bernalist confidence in the benefits of science for humanity and the Bernalist campaign on behalf of the scientists' responsibility to society.[98]

Among the founders of the social-relations-of-science movement, C. P. Snow's *Two Cultures* was one of the last restatements of the call to arms begun in the 1930s. In the last section of his 1959 lecture in the Senate House of Cambridge University, Snow warned that "[. . .] people in the industrialized countries are getting richer, and those in the non-industrialized countries are at best standing still [. . .] This *disparity* between the rich and the poor has been noticed. It has been noticed, most acutely and not unnaturally, by the poor."[99] A "scientific revolution on the world-scale," Snow challenged his audience, was the responsibility of the scientists, intellectuals, and political leaders of Great Britain.

The challenge was not issued by Snow alone, however. Blackett took it up, as well, in a new set of lectures on the history of technology that he gave in 1958 and 1959 in Calcutta, Nottingham, Manchester, and London. Much of the content of these lectures is similar to Blackett's history of science lectures from a decade earlier, augmented by new reading in economic history, particularly W. W. Rostow, and by use of current economic statistics. The message of the lectures had changed, however. It was no longer only the need for the protection and stimulation of the sciences in Great Britain, but the imperative necessity for aiding India and other third-world countries in the development of science and technology. The material level of civilization in East and West had been comparable in the sixteenth century, said Blackett, before an industrial and economic take-off occurred in Great Britain, which was built upon world trade, slavery, naval power, imperialism, and India. Western successes coincided with the worsening of living standards of laborers worldwide. Like Snow, Blackett emphasized the widening gap in income between rich and poor. Modern science and technology are our greatest achievement, Blackett concluded in a November 1959 lunchtime lecture at Imperial College, but if the benefits are not spread, they will have led to our greatest failure.[100] The theme of the widening gap between the rich and the poor was the theme again of Blackett's own Rede Lecture at Cambridge in May 1969.[101] Blackett passed away in 1974, still faithful to the themes and goals of the social-relations-of-science movement and constructing a history of science and technology in its behalf.

Notes

It is a great pleasure to write this essay in tribute to Tore Frängsmyr, a much-valued colleague and friend. Tore has written of the influence of George Sarton on the history of science in Sweden and Sarton's friendship with Johan Nordström, the first Professor of the History of Ideas and Learning (Idé- och Lärdomshistoria) and history of science at the University of Uppsala in 1932. Of Nordström, Tore writes that Nordström used science as one way of understanding the intellectual history of humanity and that he was an externalist in the best meaning of the history of science as history of culture. This has been Tore's approach in history of science, too, studying the discoveries and theories of science in "their philosophical, organizational, and practical context and implications" (Tore Frängsmyr, "The New Academies and the Scientific Climate of the Eighteenth Century," *Proceedings of the American Philosophical Society*, 143 (March 1999), pp. 109–115, on p. 109. On Sarton and Nordström, "Pioneer George Sarton Inspired Swedish History of Science," *Uppsala Newsletter History of Science*, 38 (Fall 2006), pp. 1–2. On the history of science as a discipline in Denmark, Sweden, Finland, and Norway, Tore Frängsmyr, "History of Science in Scandinavia," *Archives Internationales d'Histoire des Sciences*, 35 (1985), pp. 400–407; and in Sweden, Tore Frängsmyr, ed., *History of Science in Sweden: The Growth of a Discipline, 1932–1982*. Uppsala Studies in History of Science, Volume 2 (Uppsala, 1984).

In writing this essay, I am grateful to Erwin N. Hiebert for giving me photocopies of Blackett's lectures in the history of science from the Blackett Papers at the Royal Society. I hope Erwin will like the use to which I have put them. I thank Giovanna Blackett Bloor for permission to consult and quote from the correspondence and papers of Patrick Maynard Stuart Blackett at the Royal Society, where I have enjoyed the privilege of doing research on several occasions, and I thank Robert A. Nye for his criticisms of an earlier draft of this essay.

1. Files F1–F8 in The Papers of Patrick Maynard Stuart Blackett, OM FRS, Baron Blackett of Chelsea (1897–1974), Library, Royal Society of London, hereafter abbreviated BP. All materials are written in Blackett's hand unless otherwise noted. F1 includes "Highgate February 1936," 1 page; Birkbeck Physics Society Presidential Address, May 1937, 1 page; untitled, undated ms. of 8 pages with 1st page missing. F2 includes "History of Science" with notation "Faculty of Arts 1938–1939?" (three lectures), 14 pages. F3 includes "Draft for article" 1938"?, 11 pages; "Science Today," 16 pages, n.d.; "History of Science," 5 pages, n.d.; F4 includes "Baconian Society St. Albans, January 1941," 2 pp.; "The Study of the History of Science," n.d. but notated as probably 1943–1944, 11-page typescript. F5 includes "The History of Science from Columbus to Newton," Manchester University Faculty of Arts 1947, two lectures, 5 pages and 7 pages; also, F5 includes three pages with notes taken for lectures of May 1951 and an undated sheet headed "Properties of Matter 1948. History of Science 2 lectures" with a bibliographical list that includes the names Hessen, Pledge, Crowther, Whitehead, Wolf, Sherwood Taylor, Hogben, Singer, Braithwaite, Lilley, Gordon Childe, Tawney, Norman Campbell, and Bernal, among others. F8, F9, and F10 include notes and bibliographical references as well as a penciled graph appearing to correlate numbers of scientists, "treasure" and prices and wages from 1400–1900. F6 and F7 include notes for lectures on the history of technology given at Calcutta in 1958, Nottingham, Manchester, and Imperial College in 1959 (see endnote 100). The anonymously authored "Why Study the History of Science?" appears as pages 2–5 in *AScW Memorandum: The Development of Science*, 8 pp. (London: Association of Scientific Workers, 73 High Holborn, 8 November 1943).

2. F. Sherwood Taylor, "Is the Progress of Science Controlled by the Material Wants of Man?" *Society for Freedom in Science Occasional Pamphlet*, no. 1 (April 1945), 15 pp., on pp. 10, 7.

3. For Blackett's life and work, see Mary Jo Nye, *Blackett: Physics, War, and Politics in the Twenti-eth Century* (Cambridge, MA: Harvard University Press, 2004); Peter Hore, ed., *Patrick Blackett: Sailor, Scientist, Socialist* (London: Frank Cass, 2002); and Sir Bernard Lovell, "Patrick Maynard Stuart Blackett, Baron Blackett, of Chelsea," *Biographical Memoirs of Fellows of the Royal Society*, 21 (1975), pp. 1–115. For a brief biographical sketch, Nye, "The Most Versatile Physicist of His Generation," *Science*, 296 (5 April 2002), pp. 49–50.

4. Blackett's Introduction to the first J. D. Bernal Lecture, which was given by Dorothy Hodgkin, 23 October 1969, at Birkbeck College, 8 half-pages, on p. 4 (BP: H.142). And, Boris Hessen, "The Social and Economic Roots of Newton's *Principia*," in *Science at the Crossroads* (London: Kniga, 1932), pp. 147–212.

5. The year 2006 marked a 75th anniversary that provided the occasion for conferences at Prince-ton University and the Science Museum in London; on this, see Helena M. Sheehan, "J. D. Bernal: Philosophy, Politics and the Science of Science," *Journal of Physics: Conference Series*, 57 (2007), pp. 29–39. In general, Robert Fox, "Fashioning the Discipline: History of Science in the European Intellectual Tradition," *Minerva*, 44 (2006), pp. 410–432, on p. 412. Fox includes some of the relevant historical literature on the Congress in note 6, p. 412, including articles by Anna-K. Mayer, "Fatal Mutilations: Educationism and the British Background to the 1931 Interna-tional Congress for the History of Science and Technology," *History of Science*, 40 (2002), pp. 445–472 and "Setting Up a Discipline, II: British History of Science and the 'End of Ideology,' 1931–1948," *Studies in History and Philosophy of Science*, 35 (2004), pp. 41–72. Fox also notes a forthcoming Oxford D.Phil. dissertation by Christopher Chilvers, "Something Wicked this Way Comes: The 1931 Congress and the Russian Delegation." Also see Loren R. Graham, "The Socio-political Roots of Boris Hessen: Soviet Marxism and the History of Science," *Social Studies of Sci-ence*, 15 (1985), pp. 705–722.

6. Fox, "Fashioning the Discipline," p. 413.

7. See C. A. J. Chilvers, "The Dilemmas of Seditious Men: The Crowther–Hessen Correspondence in the 1930s," *British Journal for the History of Science*, 36 (2003), pp. 417–435, pp. 422–426) and Andrew Brown, *J. D. Bernal: The Sage of Science* (Oxford: Oxford University Press, 2005), pp. 105–106.

8. See Gary Werskey, *The Visible College: A Collective Biography of British Scientists and Socialists of the 1930s* (London: Free Association Books, 1988; first published by Allen Lane, 1978), pp. 138–140; and William McGucken, *Scientists, Society and State: The Social Relations of Science Move-ment in Great Britain 1931–1947* (Columbus: Ohio State University Press, 1984), pp. 72–73.

9. Chilvers notes the articles "Bukharin Shows His Colours: Propaganda and Hate," *Daily Mail*, 6 July 1931: p. 7; "Soviet Delegates Pulled Up," *Morning Post*, 6 July 1931, p. 14; and "Only a Few Minutes Each: How Soviet Scientists Were Treated at Congress," *Daily Worker*, 6 July 1931, p. 1 (in Chilvers, "Dilemmas of Seditious Men," p. 426, notes 62 and 65). On *The Guardian* and *The Spectator*, Werskey, *Visible* College, p. 140.

10. Werskey, *Visible College*, p. 140 and note on pp. 140–141.

11. Chilvers, "Dilemmas of Seditious Men," p. 426; and Hilary Rose and Steven Rose, "Red Scientist: Two Strands from a Life in Three Colors," in *J. D. Bernal: A Life in Science and Politics*, ed. Brenda Swann and Francis Aprahamian (London: Verso, 1999), p. 143.

12. J. G. Crowther, *Science in Soviet Russia* (London: Williams and Norgate, 1930).

13. Chilvers, "Dilemmas of Seditious Men," pp. 421–422; and Brown, *J. D. Bernal*, p. 109.

14. J. W. Boag et al., *Kapitza in Cambridge and Moscow* (Amsterdam: N. Holland, 1990), pp. 151, 156, 159–160, 169. And J. G. Crowther, *The Cavendish Laboratory 1874–1974* (New York: Science History Publications, 1974), pp. 190–191.

15. P. M. S. Blackett and G. P. S. Occhialini, "Some Photographs of the Tracks of Penetrating Radi-ation," *Proceedings of the Royal Society of London*, A139 (1933), pp. 699–726; and P. M. S. Black-ett, "The Positive Electron," *Nature*, 132 (16 December 1933), pp. 917–919.

16. J. G. Crowther, *Fifty Years with Science* (London: Barrie and Jenkins, 1970), p. 62.
17. Ibid.
18. Crowther, *Fifty Years with Science*, p. 79.
19. J. G. Crowther, *British Scientists of the Nineteenth Century* (London: K. Paul, Trench, Trubner & Co., 1935), published in the United States as *Men of science: Humphry Davy, Michael Faraday, James Prescott Joule, William Thomson, James Clerk Maxwell* (New York: W. W. Norton, 1936); J. G. Crowther, *American Men of Science* (New York: W. W. Norton, 1937); J. G. Crowther, *The Social Relations of Science* (New York: Macmillan, 1941).
20. Quoted from Crowther, *Social Relations of Science*, p. 432, in Chilvers, "The Dilemmas of Seditious Men," p. 428.
21. Crowther, *Fifty Years with Science*, pp. 86–87.
22. Quoted in Brown, *J. D. Bernal*, p. 110.
23. Brown, *J. D. Bernal*, p. 111.
24. David Wilson, *Rutherford: Simple Genius* (Cambridge, Mass.: MIT Press, 1983), p. 565; and Werskey, *Visible College*, p. 81.
25. Eugene Garfield, "J. D. Bernal—The Sage of Cambridge," *Essays of an Information Scientist*, 5 (1981–1982), pp. 511–523, on pp. 515 and 518, drawing from Max Perutz, "A Sagacious Scientist," *New Scientist*, 90 (1981), pp. 39–40, and a personal communication from W. Traub, 29 March 1982.
26. Werskey, *Visible College*, p. 81.
27. Werskey, *Visible College*, p. 235.
28. See account in Solly Zuckerman, *From Apes to Warlords: The Autobiography of Solly Zuckerman* (New York: Harper and Row, 1978), pp. 140, and 139–148; and Paul Crook, "The Case against Area Bombing," in *Patrick Blackett*, ed. Peter Hore, pp. 167–186. On D-Day, J. D. Bernal, "D-Day Diaries," in *J. D. Bernal*, ed. Swann and Aprahamian, pp. 196–211, on pp. 196–197, 198.
29. P. M. S. Blackett, "Interlude on Politics," ms. (BP: A10A).
30. See Eric Hobsbawm, "Bernal at Birkbeck," in *J. D. Bernal*, ed. Swann and Aprahamian, pp. 235–254, on p. 237.
31. P. M. S. Blackett, "Pure Science: Discussion with Professor P. M. S. Blackett," in *Scientific Research and Social Needs*, ed. Julian Huxley (London: Watts, 1934), pp. 203–224; and Blackett, "The Frustration of Science," thirteen-page typescript (BP: H.1) that appeared as Chapter 7 in *The Frustration of Science*, ed. Sir Daniel Hall, J. G. Crowther, and J. D. Bernal (London: Allen and Unwin, 1935). See McGucken, *Scientists, Society and State*, pp. 77–80.
32. On Blackett as a scientific leader, see Nye, Chapter 6 on "Scientific Leadership: Recognition, Organization, Policy, 1945–1974," in *Blackett*, pp. 142–168.
33. Harold Wright, ed., *University Studies: Cambridge, 1933* (London: Ivor Nicholson and Watson, 1933).
34. On Blackett as a physicist, see Sir Edward Crisp Bullard, "Patrick Blackett [. . .] An Appreciation," *Nature*, 250 (1974), p. 370. On his naval education, see Nye, *Blackett*, pp. 15–24; Evan Davies, "The Selborne Scheme: The Education of a Boy," in *Patrick Blackett*, ed. Hore, pp. 15–37; Geoffrey Sloan, "One of Fisher's Revolutions: The Education of the Navy," in ibid., pp. 38–54.
35. P. M. S. Blackett, "The Craft of Experimental Physics," in *University Studies*, ed. Wright, pp. 67–96, on p. 67.
36. Blackett, "Craft of Experimental Physics," quotation from p. 89. Blackett, "Highgate February 1936," 1-page ms notes (BP: F1).
37. Quoted from P. M. S. Blackett, "Draft for article? 1938?" (BP: F3.). This essay on navigation includes references to Rupert Thomas Gould, *The Marine Chronometer: Its History and Development* (1923), Frédéric Philippe Marquet, *Histoire générale de la navigation* (Paris, 1931), and Bensaude, *L'astronomie nautique au Portugal à l'époque des grandes découvertes* (Paris, 1912).
38. Hessen, "Newton's 'Principia,'" p. 1.

39. Crowther, *Fifty Years with Science*, p. 178.

40. See Hessen, "Newton's 'Principia,'" pp. 5, 7–16, 19–20.

41. See Sarton, *History of Science and the New Humanism*, pp. 184, 56–57 for quotations, and Blackett, "History of Science, Faculty of Arts 1938–9?" (BP:F2).

42. Sarton, *History of Science and the New Humanism*, pp. 3, 6, 9, 10.

43. Sarton, *History of Science and the New Humanism*, pp. 10, 12–14.

44. Blackett, "Science Today" (BP:F3).

45. Blackett, in "History of Science, Faculty of Arts 1938–9?" (BP:F2) with marginal notes in brackets: "I would not perhaps go as far as Sarton."

46. Sarton, *History of Science and the New Humanism*, pp. 21, 177.

47. Blackett, "Craft of Experimental Physics," p. 74; and Blackett, "Draft for article ? 1938" (BP:F3).

48. Sarton, *History of Science and the New Humanism*, p. 28.

49. Blackett, "Science Today" (BP:F3).

50. Blackett, "Science Today" (BP:F3).

51. Bernal, *Social Function of Science*, p. 5 (quoting from Sarton, p. 68); and p. 4.

52. Bernal, *Social Function of Science*, p. 13.

53. Bernal, *Social Function of Science*, pp. 20, 25; and Blackett, "Science Today" (BP:F3).

54. Probably the most widely quoted example of anti-scientism in Britain in the inter-war period was the sermon preached by E. A. Burroughs, the bishop of Ripon, in the improbable setting of the 1927 meeting of the British Association, when he suggested "that the sum of human happiness outside scientific circles would not necessarily be reduced if for ten years every physical and chemical laboratory were closed." Crowther claimed in his 1941 book on *Social Relations of Science* that the sermon at Leeds played a causal role in the rise of the social-relations-of-science movement in the 1930s. See Anna-K. Mayer, "'A Combative Sense of Duty': Englishness and the Scientists," in *Regenerating England: Science, Medicine and Culture in Inter-War Britain*, ed. Christopher Lawrence (Amsterdam: Rodopi, 2000), pp. 67–106, on pp. 73, 89.

55. Blackett, "History of Science. Faculty of Arts 1938–9?" (BP:F2).

56. Blackett, "The Study of the History of Science," typescript, on p. 1 (BP:F4).

57. Blackett, "The Study of the History of Science," typescript, pp. 1–4, 6 (BP:F4).

58. Blackett, "The Study of the History of Science," typescript, pp. 9–11 (BP:F4).

59. Blackett, "The Study of the History of Science," typescript, pp. 10, 11 (BP:F4).

60. Bernal, *Social Function of Science*, p. 395.

61. Blackett, untitled, undated ms. of 8 pages with 1st page missing, in Section 6 (BP: F1).

62. On history of science at University College London, William A. Smeaton, "History of Science at University College London: 1919–47," *British Journal for the History of Science*, 30 (1997), pp. 25–28.

63. Quoted from the report in the *Bulletin of the British Society for the History of Science*, 1 (1949), pp. 6–7 of Farrington's paper "What Must We Include in the History of Science?" and Dingle's response, in Fox, "Fashioning the Discipline," on p. 422. Among Farrington's writings on Greek Science are *Science in Antiquity* (1936); *Science and Politics in the Ancient World* (1939); *Greek Science: Its Meaning for Us, Part I* (1944); and *Greek Science: Its Meaning for Us, Part II* (1949).

64. Charles Singer, "The Role of the History of Science," reprinted in *British Journal for the History of Science*, 30 (1997), pp. 71–73, from *Bulletin of the British Society for the History of Science*, 1 (1949), pp. 16–18.

65. Singer, "The Role of the History of Science," p. 71.

66. Crowther, *Fifty Years with Science*, p. 178.

67. Singer, "The Role of the History of Science," p. 71.

68. Joseph Needham, *A History of Embryology* (Cambridge: Cambridge University Press, 1934). Anna-K. Mayer, "Setting Up A Discipline: Conflicting Agendas of the Cambridge History of Science Committee, 1936–1950," *Studies in the History and Philosophy of Science*, 31 (2000), pp. 665–689, on p. 672.

69. Mayer, "Setting Up A Discipline," pp. 676–682.

70. Anna-K. Mayer, "Setting Up a Discipline, II: British History of Science and 'The End of Ideology,' 1931–1948," *Studies in the History and Philosophy of Science*, 35 (2004), pp. 41–72, on p. 41.

71. Quoted in Werskey, *Visible College*, p. 145, from Letter from Dorothy Singer to Joseph Needham, 10 August 1943.

72. Quoted in Werskey, *Visible College*, p. 145, from J. D. Bernal, "Science and Society," *The Spectator*, 11 July 1931, reprinted in J. D. Bernal, *The Freedom of Necessity* (London, 1949), p. 338.

73. McGucken, *Scientists, Society and State*, p. 132; and Rose and Rose, "Red Scientist," pp. 140–141.

74. Bernal, *Social Function of Science*, especially pp. 310, 317.

75. Bernal, *Social Function of Science*, pp. 223–231.

76. Michael Polanyi, "The Rights and Duties of Science," *Manchester School of Economics and Social Studies*, 19 (1939), pp. 175–193, reprinted in *The Contempt of Freedom: The Russian Experiment and After* (New York: Arno Press, 1975; originally London: Watts, 1940), pp. 22–23.

77. Polanyi, "Rights and Duties of Science," p. 8.

78. See Nye, *Blackett*, p. 38.

79. Michael Polanyi, in *Manchester Guardian*, 7 November 1942; and Polanyi, "Autonomy of Science," *Memoirs and Proceedings of the Manchester Literary and Philosophical Society*, 85 (1943), pp. 19–38; John Baker, "Counterblast to Bernalism," *The New Statesman and the Nation*, 29 (1939), pp. 174–175. Also see McGucken, *Scientists, Society and State*, pp. 266–275.

80. Society for Freedom in Science, List of Members (June 1947), photocopy from Museum of Comparative Zoology Library, Harvard University, accession date October 17, 1949. Thanks to Kristin Johnson for this photocopy.

81. Mansel Davies, "Joseph Needham," *British Journal for the History of Science*, 30 (1997), pp. 95–100, on p. 99.

82. Nye, *Blackett*, pp. 100–142, 158–168.

83. Brown, *J. D. Bernal*, pp. 363–366.

84. Loren Graham, Review of *Science in History*, in *New York Times Book Review*, 1 August 1971, pp. 1, 18, quoted in Garfield, "J. D. Bernal," p. 516.

85. Brown, *J. D. Bernal*, p. 364.

86. Bernal, *Social Function of Science*, pp. 36, 113, 97.

87. In particular, see Michael Polanyi, *Science, Faith and Society* (Oxford: Oxford University Press, 1946), *Personal Knowledge: Towards a Post-Critical Philosophy* (Chicago: University of Chicago Press, 1958), and *The Tacit Dimension* (Garden City: Doubleday, 1966). On Polanyi, see Mary Jo Nye, "Michael Polanyi (1891–1976)," *HYLE*, 8 (Fall 2002), pp. 123–127; and Mary Jo Nye, "Science and Politics in the Philosophy of Science: Popper, Kuhn, and Polanyi," in *Science as Cultural Practice/Science in an Age of Extremes*, ed. Moritz Epple and Claus Zittel (Akademie Verlag Berlin), in press.

88. Werskey, *Invisible College*, p. 189. See Robert K. Merton, "A Note on Science and Democracy," *Journal of Legal and Political Sociology*, 1 (1942), pp. 115–126, reprinted as "The Normative Structure of Society," in Merton, *Social Theory and Social Structure: Toward the Codification of Theory and Research* (Glencoe, Ill.: Free Press, 1949), pp. 307–316. A recent analysis of Merton's norms is found in Stephen Turner, "Merton's 'Norms' in Political and Intellectual Context," *Journal of Classical Sociology*, 7 (2007), pp. 161–178.

89. Bernal, *Social Function of Science*, p. 416.

90. Bernal, *Social Function of Science*, pp. 321–323.

91. Bernal, *Social Function of Science*, p. 95.

92. Werskey, *The Visible College*, p. 155.

93. Quoted in Bernal, *Social Function of Science*, p. 97, from Aldous Huxley, *From Point Counter Point*, pp. 442–444.

94. Bernal, *Social Function of Science*, pp. 97–98.

95. Polanyi, *Contempt of Freedom*, pp. 13, 14.

96. An early example is S. B. Barnes and R. G. A. Dolby, "The Scientific Ethos: A Deviant Viewpoint," *European Journal of Sociology*, 11 (1970), pp. 3–25.

97. Paul Forman, "The Primacy of Science in Modernity, of Technology in Postmodernity, and of Ideology in the History of Science," *History and Technology*, 23 (2007), pp. 1–152. Among many others, Forman discusses the views of Bruno Latour in *Science in Action* and Otto Sibum in "What Kind of Science is Experimental Physics?" *Science*, 306 (1 October 2004), pp. 60–61. Also see two sets of essays on "Technoscientific Productivity," ed. Ursula Klein, *Perspectives on Science*, 13 no. 2 (Summer 2005) and 13 no. 3 (Fall 2005).

98. On scientists, responsibility, and political activism, see Mary Jo Nye, "What Price Politics? Scientists and Political Controversy," *Endeavour*, 23 (1999), pp. 148–154.

99. C. P. Snow, *The Two Cultures* (Cambridge: Cambridge University Press Canto Edition, 1993), pp. 41–51.

100. P. M. S. Blackett, "Aspects of the History of Technology," Three Lunch-Time Lectures at Imperial College, October–November 1959, 12 handwritten pages with 3 typewritten pages of quotations (BP:F7); and "Historical Trends of Technology," Calcutta, 1958, 1 handwritten page (BP:F6); untitled lecture notes, Nottingham, 2 pages (BP: F6); and "Reflections on History of Technology since 3000 BC," Mond Lecture, Manchester, 1959, 3 pages with 2 pages of quotations (BP: F6). On Blackett's visits to India and his role as an advisor to the Indian government, see Nye, *Blackett*, pp. 162–167.

101. P. M. S. Blackett, "The Gap Widens," Rede Lecture, Cambridge, 15 May 1969 (Cambridge: Cambridge University Press, 1970).

History of Science in the Age of Policy

SVEN WIDMALM

Introduction

In current European research policy, "innovation" and academic "excellence" are both emphasized. Ideally, it is often implied, both should occur together, as in some American universities and institutes of technology, where success is simultaneously achieved by academic and economic criteria. Especially in biotechnology, there is an emphasis on the simultaneous achievement of academic excellence and commercial success. This is according to the common conviction that the distance between fundamental research and commercial application is especially short in this field.[1] The idea is by no means new. In the wake of World War II, the discipline of physics was viewed similarly, setting a Big Science pattern that influenced policy thinking for decades. Today, however, the importance of short lead times is emphasized more. It is thought that it is possible to force the development of inventions by researchers through management and policy means.

Another recent tenet of policy thinking, at least in Sweden, is that resources must be concentrated in areas of strength in order to achieve world leadership in at least a few fields. The logic seems indicative of a kind of globalization *angst*. It is not enough that Swedish science scores very well according to academic indicators relative to population or GNP. In order to survive in the global competition, Sweden should score well also in absolute numbers in at least a few fields. Only then might research efforts translate into broad economic success, and perhaps Nobel Prizes. Implicitly or explicitly it is recognized that such a policy would demand the abandonment of research in many other fields.

These are two examples of changes in the last couple of decades that indicate a growing sense of urgency among policy makers and politicians regarding the importance of knowledge production, and especially science and engineering, in society. The phenomenon is reflected also in the many claims by observers in universities as well as in the political sphere, in media and so on, that we live in a world where advanced knowledge is not only the engine of economic progress but a defining characteristic of our culture. Popular concepts used to bring home this point of view are "knowledge society" and "knowledge economy" (ubiquitous), "learning economy" (Bengt-Åke Lundvall), "information society" (Manuel Castells), "risk society" (Ulrich Bech), "mode 2 society" (Helga Nowotny et al.), and "knowledge politics" (Nico Stehr). Under such conditions, one might expect that the relevance of history of science for not only policy but also the general understanding of our society should be growing. But this is hardly the case. Contemporary discussions about the importance of advanced knowledge seem to be lacking historical perspectives. Using a term from cybernetics and the sociology of science, history is often *blackboxed* in current policy and policy-related research.

In the following, I will attempt to expand this diagnosis somewhat, discussing examples from policy and policy-related research in Sweden, and commenting on the perils of blackboxing history when analyzing the knowledge society from a policy perspective.[2]

A Brief Background

In the 1990s, Swedish research policy went through convulsions, the effects of which are still with us. In a shrinking economy, with regard to research financing, the Social Democratic Ministry for Education made a concerted effort to break down academic hierarchies—e.g. by upgrading many teachers to the professorial level and by breaking male dominance on the professorial level. As part of the same attempt to modernize the Swedish system for research and higher education, attempts were made to make academic culture and research more oriented towards broad social needs. Thus a "third task" was invented for the university that would henceforth, besides carrying out research and education, interact on several levels with the "surrounding society."[3] The Ministry in the second half of the 1990s has been described as "activist"; confrontations between the minister Carl Tham and the Swedish professoriate, particularly in the sciences, were legion in this period. In 1997, the government appointed an investigation into the future needs of Swedish research, which in 1998 resulted in the report "Research 2000" (*Forskning 2000*). The working group responsible for the report had been asked to, for example, come up with suggestions about how research could become better at fulfilling social and economic needs. The climate of distrust between academics and the Ministry was, however, such that the report more than anything else became a pamphlet defending "traditional" academic values—resulting in more heated debate, but perhaps not a great deal of action.[4]

Eventually, however, new reforms were implemented, importantly the abandonment of traditional "sectorial" research in favor of a policy aimed at bridging the perceived distance between research, business and other "stakeholders" in a more general sense. In 2001, the different research councils for sciences, social sciences and the humanities were merged into one council (the Swedish Research Council, *Vetenskapsrådet*) and in the same year a new council-like organization, Vinnova, was created. The task of the latter—a government agency (*myndighet*)—was among other things to support collaboration between academic research and the business sector, which it currently does with an annual budget for research financing of about SEK 1.5 billion (more than $200 million).

Vinnova is possibly unique among funding agencies and such organizations in that it is named after a particular social-science theory, namely that of Innovation Systems (Vinnova translates as the "Swedish Governmental Agency for Innovation Systems"). It has lived up to its name by tirelessly promoting a peculiar vision of how innovation should be supported that is founded on the neo-Schumpeterian innovation systems approach and also on economic-geographic cluster theory and the ideas of Henry Etzkowitz and Loet Leydesdorff, the so-called Triple Helix model. This vision is ambitious to say the least, depicting a national innovation system that encompasses all of society. This is not original in itself, as matters having to do with knowledge production are seen by many as defining society in the West and increasingly elsewhere—as the expressions listed above ("knowledge society," etc.) bear witness to. Vinnova is peculiar, however, in that it is a government agency that does fulfill its mission to support economically oriented research but which also promulgates an ideology which, as we shall see, gives the agency itself a central position in a technocratic scheme to promote social, economic, and intellectual development.

I will not speculate on how typical the Swedish case is in a European context—the ideology of Vinnova obviously has many siblings in countries like Finland (often held forth as a shining example by Swedish innovation enthusiasts), Germany, or Britain—the OECD has been important in fostering enthusiasm for ideas about an innovation-driven economy in Sweden and other industrialized countries.[5] The Swedish case is of interest in itself, however, as a case study of something that should be of general concern for historians of science: the emergence of a new (or at least modernized—much of this thinking is rather 19th century) kind of ideology which gives science a central role in the development of society, while at the same time attempting to subjugate it to technocratic manipulation.

This development got off the ground in the 1990s, behind our backs so to speak, when many historians of science and others in the Science and Technology Studies (STS) field, and also a few scientists, were busy with the Science Wars and other cultural debates. It offers us an interesting challenge, as innovation thinking represents a particular view of what science is, what it should be, and not least what it has been, in the past. In brief, this view implies that science is changing from an undesirable state of (relative) unproductiveness and (undemocratic) standoffishness to a state of transparency and socioeconomic productiveness. Furthermore, it is implied that this change can be hastened and facilitated by policy means. Last but not least, it is implied that such policies would benefit all of society, as innovation is the central means of fostering increased productivity these days. This policy paradigm is indeed technocratic, because it suggests that experts will be able to decide what is best for science, businesses, and society at large.

Science According to Policy

The linear model, according to which fundamental research would generate economically productive knowledge, implied that the romantic notion of free research was possible to combine with demands of the welfare state and capital that science should produce technology, growth and national security.[6]

Vannevar Bush's *Science—The Endless Frontier* from 1945 is often seen as an especially important statement concerning the linear model and the importance of autonomous research. Bush spoke about the importance of science for technological and economic development as glowingly as any modern innovation researcher, but—unlike many of his heirs—he strongly underlined that research must be autonomous:

> Scientific progress on a broad front results from the free play of free intellects, working on subjects of their own choice, in the manner dictated by their curiosity for exploration of the unknown. Freedom of inquiry must be preserved under any plan for Government support of science [...] Support of basic research in the public and private colleges, universities, and research institutes must leave the internal control of policy, personnel, and the method and scope of the research to the institutions themselves. This is of the utmost importance.[7]

It has often been implied that Bush's ideals set the agenda for post-war research policy; at the same time, much research funding was forcefully directed towards specific goals—defense, different branches of industry, and so on. The relationship between ideology and practice was far from unequivocal.[8] In general, academic ideals like those of the linear model, the Humboldtian university or the Mertonian norms should be seen as ideological constructs and their relation to academic practice as something to be investigated and not taken for granted. This does not diminish their importance, their value, or their perniciousness, however. This essay will deal with current policy mainly on the ideological level, whereas its practical effects will only be touched upon lightly.

Towards the end of the 20th century, policy in the United States and Europe became more interested in holding science economically, and possibly democratically, accountable. The linear model was now largely abandoned in favor of structural modes of thinking involving entities such as systems, clusters, and networks. At the same time, interest grew in the social sciences to study the relationship between academy and industry and to offer recipes for how to make the "system" run more efficiently. This kind of research—in business studies, policy studies, economic geography and so on—tends to give an economistic view of science simply because it focuses on its economic effects. In particular, economistic thinking tends to get the upper hand when this kind of research becomes intimately involved with the policy process.

Policy-related research of this kind is exemplified by studies of the so-called "Triple Helix" (fusing academic, government and business interests),[9] "clusters"[10] and "systems of innovation."[11] The well-known books by Michael Gibbons, Helga Nowotny and others on Mode 1/Mode 2 are broader but have a similar tendency.[12] In Sweden, these models of thinking have gained wide acceptance. The rhetoric of Triple Helix is routinely used when collaboration between universities or colleges with local municipalities or government and businesses are discussed;[13] innovation systems and clusters are frequently invoked in these discussions as well. Cluster research is strong at several Swedish universities, innovation systems studies have been lavishly funded at Lund University and, as has already been pointed out, it has been used in the christening of the funding agency Vinnova. Mode/Mode 2 is not popular in the same sense but seems to be more frequently used at higher levels of policy. It does have strong links to the Swedish policy community, however, as the two books promoting these concepts were initiated and financed by Swedish funding agencies (the now defunct Board of Research Councils and Council for Research on Higher Education, and also the Bank of Sweden Tercentenary Foundation).

Policy actors have recently made a fetish of "innovation," to such an extent that "research policy" is today gradually transmogrifying into "innovation policy."[14] A good example of how research policy is regarded among innovation enthusiasts are the ideas of Thomas Nordström, Rector at Kristianstad "University"[15] for a brief period of eight months, when he quit to become project leader at Vinnova, because the academic environment could not cope with the pace of change he thought was necessary.[16] In an interview in 2006, after he had moved to Vinnova, Nordström declared that the problem with Swedish research was that it was not organized along the lines of major corporations like IKEA or Volvo, and that researchers had too much latitude to pick problems as they choose. Academic freedom should still be considered a priority, but it ought to consist of freedom to develop solutions to problems defined by policy. Furthermore, Sweden could not afford to fund "world-class" research in many areas—hence it would be wise to concentrate resources on a few elite universities and a few research areas where one could hope to become competitive globally.[17]

These were perhaps consciously provocative statements, but as we shall see, they reflect a policy ideology promulgated by Vinnova and high-level representatives of business, the engineering community, and some unions. It is, at least on a discursive level, closely connected to a European rhetoric of sustainable growth and global competition, as laid down in the Lisbon strategy from 2000. At the same time, it is thoroughly nationalistic in that "Sweden" and "Swedish" research or industry are focused and described as something separate from, and in fierce competition with, "global" actors. This way of thinking has often been cloaked in terms of something called the "Swedish paradox."

The Swedish Paradox

The notion of a Swedish paradox in research policy seems first to have evolved in the early 1990s within the emerging paradigm of innovation-systems research where it was claimed that Sweden, though a lavish spender on research and development (R&D) performed badly in high-tech branches such as electronics.[18] Discussions similar to this one had been going on earlier in the UK and became a staple of EU-policy rhetoric from the mid-1990s, there labeled the "European paradox."[19] (Here I will discuss the Swedish case, but it should be remembered that discussions in Sweden closely paralleled policy developments in the EU and the OECD.)

Originally the Swedish paradox was primarily seen as a structural problem within Swedish industry. Sweden has, to an unusual extent for such a small country, long been dominated by a few large corporations such as Ericsson, Volvo or ABB. The fact that Sweden seems to spend comparatively large amounts of money on research is explained by the high R&D spending of these corporations. Hence, a leading innovation-systems proponent like Charles Edquist could plausibly claim that the relative lack of growth in the high-tech area was a "market failure," giving cause for political interference in the market economy.[20]

Soon, however, a new interpretation of the paradox emerged among a heterogeneous group of policy makers and policy wonks at government agencies, foundations, universities, (mostly) big business, and so on. It had been noticed that Swedish scientists, perhaps especially in the life sciences, were fairly successful in publishing internationally.[21] Hence the paradox was now reinterpreted to mean: if Swedish scientists are so smart, how come we're not rich? The market failure identified by innovation research was more and more seen as a kind of academic failure. The conclusion was that researchers, no matter how brilliant, have to be firmly guided in order to produce more useful knowledge, or innovations. In April 2004, representatives for three large unions, the CEOs of the Confederation of Swedish Enterprise (*Svenkt näringsliv*) and the Royal Swedish Academy of Engineering Sciences (Lena Treschow Torell—a major Swedish policy ideologue), and the chairpersons of several large research foundations published an appeal for a new research policy in the leading Swedish daily newspaper, *Dagens Nyheter*. Referring to the Lisbon strategy, the signers demanded that policy had to guide research much more directly towards areas of economic interest for Sweden:

> We [...] want to underscore that these demands do not imply that we question the quantity of fundamental research undertaken at Swedish universities and colleges or that we want to decrease demands for high scientific quality in research—rather fundamental as well as applied research should be more geared towards areas where businesses and other parts of society experience urgent needs, needs that are of key importance for an active growth policy.[22]

Hence, like Thomas Nordström, the signers of this appeal suggested that academic research should really go on as before but with a slightly different agenda—oriented towards innovation and other useful things rather than the somewhat disorganized quest for academic excellence as such that seemed to have guided it earlier.

Clearly, researchers cannot be trusted to decide what to work on; this had become a cornerstone of research policy since the debacle of Research 2000. In these discussions, it is always unclear exactly what "researchers"—almost always this term refers to persons in natural science or engineering—had been up to in the past. Obviously, Swedish scientists were up to *something*, as shown by their excellent track record in publishing. But exactly *what* is, in the kind of broad policy discourse considered here, something of a mystery. Often it is suggested they had been locked up in an ivory tower. Now they should be forced out, into the market or perhaps the Agora (though the interest in democracy, accountability and transparency in science has been somewhat weak in Sweden compared to other countries in the EU, like Denmark or the Netherlands). This was emphasized in the Government Bill (*propostion*) on research in 2004, where it was pointed out that fundamental and applied research were really very much the same thing these days, as the step from discovery to application had become so short as to be almost imperceptible.[23] Again the same logic reappeared: changing research policy into innovation policy would not really make a great difference; scientists should be able to go on much as before, if they could only concentrate on more important things than they had before (whatever that was).

I have tried to imply that policy discourse in Sweden over the last few decades has not been very much informed by knowledge of history of science or any other kind of history (with the exception of quantitative economic history focusing on broad trends—there has been an interest in that). In fact it uses historical straw men rather freely, as in the recurring complaints about ivory-tower science. This is true not only of policy bureaucrats, politicians, and so on. Policy-related research seems to suffer from the same malaise, as has been pointed out now and then by irritated historians, and as I will come back to in the final section of this essay.[24]

In my view, this is a serious problem, and not only because an apparent disregard of history is always distasteful to a historian. The lack of historical sense becomes politically more problematic as research policy becomes politically more important, and ideologically more ambitious. Innovation systems theory is a perfect

example of this. The idea that science, technology and so forth should be understood from a systems or network perspective is in itself perfectly sound from a historical or STS perspective—it has been a staple of such research for many years. But
as the innovation-systems model recognize no natural limits to the system, its political implications likewise become limitless. The heart of an innovation system
may consist of things like research institutions, high-tech firms, and policy actors.
But as the system also includes broader phenomena—taxation, communications
and so on—the theory really amounts to a complete vision of society, from the perspective of its innovative capacity.

In short, innovation-systems theory provides a vision of an *Innovation Society*—
a technocratic dream as comprehensive as any of its historical forbears. In Sweden,
this vision has its ideological center in the government agency Vinnova.

The Innovation Society

In Sweden, there has recently been a debate concerning the fact that some government agencies have become self-appointed promulgators of ideology in their various spheres of action.[25] This is certainly true of Vinnova. In the many reports and
other texts produced by the agency, a broad vision of society is delineated that is
said to be nonpolitical but which is nevertheless ideological. If implemented, this
vision would have repercussions for Swedish society that go far beyond Vinnova's
official task, to promote sustainable growth by facilitating technology transfer, innovation, learning and so on.[26]

Vinnova's intellectual point of departure is an amalgamation of the innovation
systems model, the idea of clusters, and the Triple Helix model, resulting in a broad
conception that technological development and growth are the products of local-
ized and systemic interaction between research, business, and the public sector. It
sees as its task to help guide such interaction, and the agency aspires to be a vital
component in a broad national struggle to survive economically in the face of hard-
ened global competition and various domestic problems (environmental, demo-
graphic, etc.).[27]

Vinnova frequently makes use of the rhetoric of the Swedish paradox and provides various explanations for the relative lack of high-tech industry in Sweden despite high R&D expenditure.[28] When it comes to academic research, which is our
main interest here, Vinnova has identified a need for restructuring and redirection
of funding towards applied research. In order to achieve this, policy must become
more energetic and compel universities and funding agencies to rethink their priorities. This problem is discussed in terms such as "critical mass," "focusing," "concentration of resources" (*kraftsamling*), and "prioritizing." The goal is that Sweden,
in a few years, shall be able to make economic investments in some areas that are
among the highest in the world, also in absolute numbers. Unlike the United States,
a small economy like the Swedish one cannot afford to be a world leader in many

areas of R&D. But, according to Vinnova, the goal of achieving sustainable growth in a world marked by global competition demands that Sweden attain world leadership at least in *some* areas:

> To achieve international competitiveness and economic growth the absolute size of R&D activity is more important than relative size. [...] An efficient Swedish national innovation system and sustainable growth in Sweden absolutely demands that Swedish R&D-environments, that are concentrated geographically and as to subject matter, have critical mass. [...] The creation of such environments put great demands on the prioritizing of public research funding. [...] The creation of strong R&D environments also demands that those who do R&D—that is to say, businesses, research institutes, and universities—collaborate in an efficient manner.[29]

It is interesting to note how frequently words like "Swedish" and "national" are used in this sort of text. Vinnova sees itself (indeed it is its official task) as upholding a national interest in a world where business, science, and also politics are becoming increasingly international or even global. To uphold such a position leads to a kind of political schizophrenia. High-quality science is obviously international or global, but it should be forced to fulfill national economic interests. The fact that successful enterprises tend to become international (or American) gives cause for reflecting on exactly whose interests heavily concentrated funding of research by the Swedish state serve. Vinnova and others belonging to the "innovation movement"[30] tend to favor adjusting taxation (downwards) and salaries (upwards) so that successful innovators and businesses will not flee the country; other aspects of society are taken into consideration when evaluating the innovation system as well.[31] It is therefore easy to see that all of society, directly dealt with by the model or not, will be implicated through an implementation of Vinnova's policies.

The national task of policy is sometimes emphasized in a way reminiscent of how the "yellow peril" was discussed about a hundred years ago. The present minister of research, Lars Leijonborg, while visiting a Swedish university, expressed such sentiments: "In China the students have a strong driving force. They work hard in order for China to become as strong as the Western nations. You feel a little pain in the stomach when you visit the universities in Beijing and Shanghai and see how hungry they are. Swedish students also have strengths and you will do all right, but study hard, because they do."[32] While revising this paper, I listened to the transmissions from the Nobel festivities where the leader of the Liberal party, Jan Björklund, was interviewed. He experienced similar feelings as his party-comrade Leijonborg when reflecting on the fact that most Nobel laureates had, for some time, been American and that in a couple of decades they might instead be Chinese or Indian: "we [i.e. the Swedes] must improve ourselves" was his conclusion.[33] Similar views were recently expressed in unusually stark language in a newspaper article signed by leading representatives of several large Swedish companies and universities, and also the CEOs of Vinnova and the Royal Swedish Academy of Engineering, Per Eriksson and Lena Treschow Torell: "Global competition never sleeps. The race is on around the clock, from the west to the east, from the leading American universities to the growth-hungry [sic] students in China and India."[34] Considering the frequency with which references are currently made to the growing competition from Asia, it seems appropriate to speak of a "knowledge scare" that fosters a climate in which science is seen more and more as a means to ascertain national economic (or environmental) survival, and where other purposes of research are seen as trivial in comparison. In this discourse Asia—China and India in particular—are described as threats, whereas the United States is seen as the glowing example that should be emulated.

As for scientific research, Vinnova recommends putting all eggs in only a few baskets.[35] Policy should promote a few areas that will give the greatest possible dividends in the next decade or so. In order to do this, policy makers need to be able to look into the future, they need some sort of crystal ball. According to Vinnova the crystal ball is innovation research—not free innovation research, naturally, but internationally prominent innovation research that is guided by policy.[36] Hence Vinnova has put a good deal of money into funding "centers of excellence" in this area—research units working on innovation systems, clusters, and so on. This sort of closes the circle of the logic of Vinnova's vision of the Innovation Society: the agency promotes innovation and technology transfer in key areas of R&D; simultaneously it promotes research on what those key areas are, thus hoping to produce knowledge that will help direct policy. Vinnova's vision starts and ends with the notion that innovation is the central phenomenon in research and in society as a whole. This is because without innovation the Swedish nation will perish in the global struggle for economic survival. Vinnova gives itself a key role in this strangely 19th-century technocratic vision of social and economic development.

A Technocratic Policy Paradigm

I have used the word "technocratic" in order to describe the ideology of Vinnova and like-minded policy actors. I will finish this discussion of policy by delineating what I will call the "technocratic paradigm" of current policy thinking—by way of summarizing and also to set the stage for the final brief discussion about the possible roles of history in the policy context.

An important figure of thought in current policy thinking is the idea that academic research leads the way and that business follows—in the sense that business is depending on new knowledge produced in academe, but also in the sense that business is keen to invest where there is a strong research environment. The "clustering" of academic and commercial entities is seen as an important factor behind economic growth. Policy thinking in Vinnova's vein also presumes that developments in science and to some extent business must be *directed* in order to achieve maximum efficiency in an innovation system where costs are escalating and competition is global. This is, it is often implied, a matter of national economic survival. It is assumed that government should probably increase research funding and it is always stressed that funding must be directed towards economically productive fields (in effect making the national system for research into a money-laundering business, transferring tax payers' money into support for local industries).

Hence, there are enough doctrine-like ideas like these around to make it proper to talk of a technocratic paradigm in research policy, which was originally inspired by interpretations of the great successes of physics during the war and since the 1980s by developments in life science. An outline of the current form of this paradigm, focusing on the central questions of research and commercialization, would contain the following elements:

- Policy should promote innovation *and* academic excellence because . . .
    - There is only a short step from research to commercial application;
- Resources for research must be concentrated in order to achieve world leadership in some areas because of . . .
    - Fierce global competition and . . .
    - Escalating costs;
- Business wants to cluster with excellent science;
- Future developments in science and business can be directed by government and policy experts;
- Government investments in research and technology transfer serve a *national* interest.

This line of thinking contains elements of a modernized planned economy, where government agencies attempt to construct competitive research environments and to direct future developments in R&D and business. Vinnova has suggested a 10-year plan for biotechnology, which should lead to, for example, a broad public acceptance of this somewhat controversial area, and to a doubling of successful biotech start-ups as well as of national biotech export incomes.[37]

The technocratic paradigm is not uncontested, but it has gained a strong foothold among policy actors, and it is in line with thinking in the EU or the OECD. It is applied rather indiscriminately and is affecting Swedish universities on a broad scale—not only the sciences but all academic research. Rectors are more and more acting as CEOs, attempting to implement government policies in order to improve chances of gaining economic resources that have become scarcer (relatively speaking) since the 1990s, when the university system in Sweden expanded rather dramatically.[38] Universities are now competing for resources by implementing reforms that are thought to improve their chances of achieving excellence and economic productivity in areas that are considered relevant from a policy perspective.

Though it rests on some sort of academic foundation, Vinnova's ideology may nevertheless be seen as totalitarian, from an academic perspective: policy-defined priorities are supposed to elevate a few areas (like biotechnology and innovation research) to high levels of funding, whereas others will be left with little or no support. The vision is supported by influential actors like Lena Treschow Torell, already mentioned, and her husband Michael Treschow, chairman of the boards of Ericsson and Unilever. According to the latter, relations between academy and industry should be characterized by "nearly complete convergence as to goals and driving forces."[39] This vision of academic research, and of academic life in general, makes a truly radical break with Humboldtian as well as Mertonian ideals.

Back to the Past

Vinnova is representative of a fairly broad movement among influential actors in Swedish and European policy, business, and academic research. For someone belonging to the latter category, but working in the history of science rather than innovation studies, Vinnova becomes a kind of funfair mirror in which certain elements of science are recognizable whereas others are alarmingly distorted.

Vinnova's view of science is instrumental and strangely millennial: science is a tool for achieving the Holy Grail of innovation that will save the nation. Other purposes or effects of scientific research are not so much criticized (other than for costing too much) as ignored, left out of the picture, made invisible. The same is true for science and scholarship that cannot easily be associated with the innovation concept, such as history or particle physics (the latter discipline has suffered severe cutbacks in Sweden lately, the number of national laboratories having been reduced from four to two).

The role of history in an academic landscape where Vinnova-style policy is gaining in influence is worth pondering. History and the humanities in general are affected by the broad adoption of technocratic management principles, like quantitative evaluation. But that is not my concern here. Rather I want to speculate on what the role of historical thinking is in policy, and what it could be. The following are three notable characteristics of policy thinking with regard to history:

- First, policy and policy-related research tend to "blackbox" history. That is, it is seen as something given, a finished product that does not merit renewed scrutiny.
- Second, the blackboxed version of history tends to be used as a straw man when defining our present predicaments and suggesting remedies for the future. Usually past science is seen as academically introverted, perhaps locked up in the proverbial ivory tower. This is true of rhetorical mission statements from Vinnova and its allies, but to some extent also of more serious scholarship, as in the discussion regarding mode 1/mode 2.
- Third, the relevance of history is diminished precisely because it deals with the past, whereas policy and policy-related research presumes to deal with the future.

Together these three points illustrate the historical fallacy of policy thinking: the idea (if that is the right word) that political action has no need for historical analysis as such, but that it does need a pre-packaged view of the recent past which can be used pedagogically, or propagandistically, in order to make clear why radical actions are necessary. In Sweden, this discursive use of history has aptly been termed the "rhetoric of system change" (*systemskiftesretorik*), as it tends to be used when larger social systems are subject to drastic reform.[40] Such rhetoric tends to produce historical caricatures like the straw men just mentioned.

It does seem that current system-changing trends in science policy constitute a formidable challenge to a subject area like history of science. First, because policy defines the value of research in a way that tends to diminish the importance of the humanities in general, and of much social science and natural science, too. Second, because it promotes a view of science that is rhetorical and unhistorical, which implies a further denial of the value of the historical study of science.[41]

I prefer to look at these tendencies, not as a sign of active hostility against the humanities and historical research but rather as a result of innovation fetishism and consequent tunnel vision. I also regard them as extremely dangerous, as they tend to negate many important functions of knowledge production that cannot be associated with innovation or sustainable growth—for example the critical tradition in social science and the humanities. Furthermore, technocratic science policy is democratically unsound in that it seems to presume that the interests of politics, knowledge production and corporate capitalism are identical (cf. the quote from Michael Treschow above).

What, then, could history of science provide to remedy the situation? Historians of science will probably not be able to offer the kind of policy prescriptions that are currently provided by innovation researchers. I think our kind of advice, if we find it in us to offer advice, will be of a more negative kind. I suggest the following three examples of policy relevance for history of science:

- First, history must never be blackboxed. Historians of science have often participated in STS campaigns against false historical consciousness among scientists. The traditional blackboxed version of history of science as being progressive and cumulative *per se* has been deconstructed since the days of Thomas Kuhn. This work must be extended to include the social sciences, in particular policy-oriented social science.
- Second, historians should engage in the perhaps thankless task of debunking straw men in policy. A good one to start with would be that of the ivory tower. The fact that academic science has always, though often under a cloak of otherworldly detachment, been engaged in socially relevant activities should be pointed out to policy's mythmakers.
- Third, and in a more constructive spirit, historians should be able to demonstrate that historical (as well as sociological, anthropological, and so on) understanding gives a qualitatively better understanding of science than that provided by innovation perspectives where technical as well as socio-cultural aspects of

science are disregarded. Why not start by reminding innovation thinkers that the resilience of science has something to do with the fact that it has been multifunctional, embedded in a number of activities (education, technological development, culture), with a number of different kinds of "stakeholders," and that unintended effects (like the death of democracy as we know it) might result from turning research into a commercially oriented monoculture?

These are only a few suggestions regarding the possible uses of history in general and history of science in particular in a policy context. As none of them promise to bring short-term economic returns, however, they will probably not be heeded by policy makers intent on building the Innovation Society. They may be of interest to policy thinkers and makers who are critical of innovation fetishism, however. Such people do exist. Some pioneers of policy-oriented innovation studies have lately made comments suggesting feelings akin to those of Dr. Frankenstein contemplating his monster:

> As a kind of countervailing power to the colonizing tendency emanating from market-oriented innovation policy, we see a need to develop a wider field of politics—*knowledge politics*—that covers all aspects of knowledge production and takes into account that the production of knowledge has much wider scope than just contributing to economic growth. This includes of course knowledge necessary for social and ecological sustainability but not only that. In rich societies, it should be possible to afford culture, ethics and knowledge for its own sake, not only knowledge that promotes innovation and economic growth.[42]

Notes

1. This idea is promoted e.g. in *Strategi för tillväxt—bioteknik: En livsviktig industri i Sverige*, VINNOVA policy 2005:2 (Stockholm: VINNOVA, 2005), p. 15.
2. For a more detailed presentation of some aspects of this problem area, see Sven Widmalm, "Innovationssamhället," in Mats Benner and Sverker Sörlin, eds., *Kunskap—Sveriges framtid?* (Stockholm: SNS förlag; forthcoming).
3. On developments in general during this period, see Mats Benner, *Kontrovers och konsensus: Vetenskap och politik i svenskt 1990-tal* (Nora: Nya Doxa, 2001).
4. Magnus Eklund, *Adoption of the Innovation System Concept in Sweden* (Uppsala: Acta Universitatis Upsaliensis, 2007), ch. 4. Cf. Peter Schilling, *Research as a Source of Strategic Opportunity? Re-Thinking Research Policy Developments in the Late 20th Century* (Umeå: Umeå University, 2005), ch. 5.
5. Eklund, pp. 11–14, 36–43, 108, 116.
6. On the linear model, see Karl Grandin et al., eds., *The Science-Industry Nexus: History, Policy, Implications* (Sagamore Beach, MA: Science History Publications/USA, 2004).
7. Vannevar Bush, *Science—the Endless Frontier* (1945; Washington, DC: National Science Foundation, 1960), pp. 12, 33. Bush's book was immediately recognized as an important policy document in Sweden. See Sven Widmalm, "The Svedberg och gränsen mellan teknik och vetenskap,"

in idem, ed., *Artefakter: Industrin, vetenskapen och de tekniska nätverken* (Hedemora: Gidlunds, 2004), pp. 149–87, on p. 179.

8. Cf. Daniel S. Greenberg, *Science, Money, and Politics: Political Triumph and Ethical Erosion* (Chicago: The University of Chicago Press, 2001), ch. 3.

9. Henry Etzkowitz and Loet Leydesdorff, eds., *Universities and the Global Knowledge Economy: A Triple Helix of University-Industry-Government Relations* (London: Continuum, 1997).

10. For an overview of cluster research, see Harald Bathelt, Anders Malmberg and Peter Maskell, "Clusters and Knowledge: Local Buzz, Global Pipelines and the Process of Knowledge Creation," *Progress in Human Geography*, 28:1 (2004), pp. 31–56.

11. For a recent overview, see Lars Coenen, *Faraway, So Close: The Changing Geographies of Regional Innovation* (Lund: Circle, 2006), ch. 3.

12. Michael Gibbons et al., *The New Production of Knowledge: The Dynamics of Science and Research in Contemporary Societies* (London: Sage, 1994); Helga Nowotny et al., *Re-Thinking Science: Knowledge and the Public in an Age of Uncertainty* (Cambridge: Polity, 2001).

13. At Umeå University there is a Triple Helix Hall in the "house of collaboration" (*Samverkanshuset*); at Linköping University there is a "HELIX VINN Excellence Centre."

14. Merle Jacob and Luigi Orsenigo, *Från forskningspolitik till innovationspolitik: Högskolans och näringslivets samverkanspolitik i Sverige 1990–2005* (Stockholm: Studieförbundet Näringsliv och Samhälle, 2007).

15. A comment regarding the terminology of the Swedish university system is called for here. In Sweden, establishments of higher education are separated into two main groups: universities and "colleges" (*högskolor*). The latter do not have the same status and privileges as the former—e.g. they do not have the right to bestow doctorates. Kristianstad's establishment for higher education is not a university by this definition, but when its name is translated, the term university is used. See: http://www.hkr.se/Default.aspx. Many of the Swedish colleges want to attain university status, and this is a perennial cause of grievance and debate.

16. Per-Olof Eliasson, "Kulturkrock på högskola medförde att rektor avgick," *Universitetsläraren* (2005). http://www.sulf.se/templates/CopyrightPage.aspx?id=2785.

17. Per-Olof Eliasson, "Ny strategi behövs för nyttoperspektiv," *Universitetsläraren* 9 (2006), http://www.sulf.se/templates/CopyrightPage.aspx?id=4975.

18. Charles Edquist and Maureen McKelvey, "Högteknologiska produkter och produktivitet i svensk industri," in *Forskning, teknikspridning och produktivitet*, Expert report no. 10 in SOU 1991:82 (Stockholm: Allmänna förlaget, 1991), pp. 119–81. Cf. Charles Edquist and Maureen McKelvey, "High R&D intensity without high tech products: A Swedish paradox?" in Klaus Nielsen and Björn Johnson, eds., *Institutions and Economic Change: New Perspectives on Markets, Firms and Technology* (Cheltenham: Elgar, 1998), pp. 131–149.

19. Giovanni Dosi, Patrick Llerena, Mauro Sylos Labini, *Science-Technology-Industry Links and the "European Paradox": Some Notes on the Dynamics of Scientific and Technological Research in Europe*, LEM Working Paper Series 2005/02. (http://www.lem.sssup.it/WPLem/files/2005-02.pdf).

20. Charles Edquist, *Innovationspolitik för Sverige—mål, skäl, problem och åtgärder*, VINNOVA Forum Innovationspolitik (VFI) 2002:2 (Stockholm: Vinnova, 2002), pp. 16–17. Cf. Charles Edquist and Maureen McKelvey, "Introduction," in Edquist and MacKelvey, eds., *Systems of Innovation: Growth, Competitiveness and Employment*, Vol. 1 (Cheltenham: Elgar, 2000), pp. xi–xx, on p. xx; Charles Edquist, "Systems of Innovation Approaches—Their Emergence and Characteristics," in ibid., pp. 3–37, on p. 21.

21. Talk about a Swedish paradox has been criticized by some scholars. Staffan Jacobsson and Annika Rickne describe it as "a 'dominant belief,' or conventional wisdom" in Swedish policy. According to their analysis, the Swedish research system is comparatively successful not because it is particularly well-funded but because it is efficient. Staffan Jacobsson and Annika Rickne, "How Large is the Swedish 'Academic Sector' Really? A Critical Analysis of the Use of Science and Technology Indicators," *Research Policy* 33 (2004), pp. 1355–1372.

22. Anders Narvinger et al., "Låt näringslivets behov styra forskningen," *Dagens Nyheter*, April 11, 2004: "Vi vill [. . .] understryka att dessa krav inte innebär att vi ifrågasätter omfattningen av den grundforskning som bedrivs vid svenska universitet och högskolor eller att vi vill göra avsteg från kraven på hög vetenskaplig kvalitet i forskningen—vad det handlar om är att tydligare inrikta såväl grund- som tillämpad forskning mot de angelägna behovsområden vi ser i näringsliv och övriga samhället, behov som utgör nyckeln till en aktiv tillväxtpolitik."

23. *Forskning för ett bättre liv*, Regeringens proposition 2004/05:80, pp. 10, 140, 151.

24. Dominque Pestre, "The Production of Knowledge Between Academies and Markets: A Historical Reading of the Book The New Production of Knowledge," *Science, Technology, and Society* 5 (2000), pp. 169–181; John Krige, "The Industrialization of Research. Commentary: In Praise of Specificity," in Grandin et al., eds., pp. 133–37; Terry Shinn, 2002, "The Triple Helix and New Production of Knowledge: Prepackaged Thinking on Science and Technology," *Social Studies of Science* 32/4 (2002), pp. 599–614. Cf. the special issue on "Mode 2 revisited," *Minerva*, 41:3 (2003).

25. See for example opinion pieces by Tobias Krantz and Henrik von Sydow, and Jacob Palme in *Dagens Nyheter* July 22, 2006 and *Svenska Dagbladet* July 28, 2006.

26. For a presentation of Vinnova, see http://www.vinnova.se/. Vinnova's official "vision" is: "VINNOVA makes a clear contribution to Sweden's development as a leading growth country." For a more detailed description of vision and goals, see http://www.vinnova.se/In-English/About-VINNOVA/. Another visionary statement appears in *Behovsmotiverad forskning och effektiva innovationssystem för hållbar tillväxt*, Vinnova Policy VP 2003:3 (2002), pp. 161. Chapters 1–2 of this interesting document gives a useful exposé of the ideological foundations of Vinnova and will here be used in order to capture its character and flavor.

27. *Behovsmotiverad forskning*, ch. 2.

28. Ibid., pp. 36–37.

29. Ibid., p. 29: "För internationell konkurrenskraft och ekonomisk tillväxt är den absoluta storleken på FoU-verksamheten viktigare än den relativa. [. . .] För effektiviteten i det svenska nationella innovationssystemet och för hållbar tillväxt i Sverige är det av avgörande betydelse att geografiskt och ämnesmässigt koncentrerade svenska FoU-miljöer har kritisk massa. [. . .] Skapandet av sådana miljöer ställer stora krav på prioritering av offentliga forskningssatsningar. [. . .] Skapandet av starka FoU-miljöer kräver också att utförarna av FoU, det vill säga företag, forskningsinstitut och högskolor, samspelar effektivt." Cf. ibid., pp. 46–47.

30. There is an evangelical tone in some of Vinnova's publications indicating that members of what is sometimes called the "innovation movement" see themselves as missionaries for a common cause that bridges political and other dividing lines. A good example is the publication Tor Bonnier and Per-Olof Berg, eds., *Svensk innovationskraft: "Visionen måste vara starkare än motståndet"*—the title of which translates "Swedish innovative power: 'The Vision must be Stronger than the Resistance'." The publication even contains a manifesto.

31. Thomas Andersson, Ola Asplund and Magnus Henrekson, *Betydelsen av innovationssystem: Utmaningar för samhället och för politiken*, VFI 2002:1 (Stockholm: Vinnova, 2002), pp. 73, 82; Charles Edquist, "Innovationssystem och innovationspolitik," in Bonnier and Berg, pp. 17–23, on p. 22.

32. The comment was made by Leijonborg during a talk to students at Linköping University, Nov. 21, 2007. http://www.liu.se/liu-nytt/arkiv/nyhetsarkiv?newsitem=12041: "I Kina har studenterna en stark drivkraft. De jobbar med målet att Kina ska bli lika starkt som västländerna. Man får lite ont i magen när man besöker universitet i Beijing och Shanghai och ser hur hungriga de är. Svenska studenter har också styrkor och ni kommer att klara er hyggligt, men plugga på, för det gör de!"

33. This might sound like festive whimsy, but as a matter of fact the Liberal Party demanded, in the last election campaign, that Swedes should become better at research so that they could win more Nobel Prizes. The transmission was on the channel P1 of Swedish public radio around 7 P.M., December 10, 2007. Björklund in fact said the Swedes must "*spotta upp oss*," which literally

translates as "pluck up courage." Nowadays, however, the expression more often seems to mean "improve oneself," or "to become better at something."

34. Charlotte Brogren, Birgitta Böhlin, Per Eriksson, Peter Holmstedt, Anders Narvinger, Göran Sandberg, Jan-Erik Sundgren, Lena Treschow Torell, Pia Sandvik, "Regeringen saknar forskningsvision," *Svenska Dagbladet* Oct. 13, 2007: "De globala konkurrenterna sover aldrig. Tävlingen pågår dygnet runt från väst till öst, från toppuniversiteten i USA till de tillväxthungriga studenterna i Kina och Indien." The expression "growth-hungry" (*tillväxthungriga*) is a Swedish neologism, at least when applied to a category like "students."

35. *Behovsmotiverad forskning*, ch. 8.

36. Ibid., ch. 12. Cf. Edquist, *Innovationspolitik för Sverige*, pp. 46–47.

37. *Strategi för tillväxt*, p. 11.

38. For a perceptive and thorough case study of Lund University from this perspective, see Fredrik Melander, *Lokal forskningspolitik: Institutionell dynamik och organisatorisk omvandling vid Lunds universitet* 1980–2005 (Lund: Statsvetenskapliga institutionen, Lunds universitet, 2006).

39. Mikael Treschow, "Eftersträva excellens och samsyn," in Bonnier and Berg, pp. 41–44, on p. 43: "i det närmaste hundraprocentig konvergens i mål och drivkrafter."

40. Bo Malmberg and Lena Sommestad, "Ljus över landet: Om samhällsvetenskapens potential i en krisfylld tid," in Jonas Anshelm, ed., *Skall vetenskapen rädda oss?* (Stockholm/Stehag: Brutus Östlings Bokförlag Symposion, 1996), pp. 67–152.

41. In fact, there is some interest in the history of science, as shown by the enormous attention surrounding the tercentennial of Carl von Linné this year. But manifestations of this kind seem to have more to do with the growing interest in marketing and branding of universities or even branches of science.

42. Bengt-Åke Lundvall, "Innovation Systems between Policy and Research," paper presented at the Innovation Pressure Conference, Tampere, March 2006, pp. 15–16. Cf. Bengt-Åke Lundvall, *The University in the Learning Economy*, DRUID Working Paper No. 02-06 (2006); Richard R. Nelson, "The Market Economy, and the Scientific Commons," *Research Policy* 33 (2004), pp. 455–471. Thanks to Lundvall for allowing me to quote the first, and sending me the second of these papers.

A Note on the "Integrity" of the University[1]

SHELDON ROTHBLATT

The university's corporate structure, according to one historian, is the key to its long-run survival.[2] And that survival is filled with visual and tactile memories and reminders expressed as symbols, idioms, emblems and iconographic references to an ancient guild or special past. Distinctive costume—hats, badges, awards, rings, even neckties or scarves—are nearly universal. By color and markings, gowns signify discipline, rank and distinction. Seals, scepters, academic chains mark off one university from another. Ceremonies, processions and rituals define matriculation and graduation, and in some countries accession to a professorial chair is accompanied by a special lecture and occasion. Examinations, which are various, have been conducted over the centuries with an acute sense of theater.[3] Both the professorial world and the student estate are substantially defined through ritual and custom. Fraternities and sororities in the United States, student societies in nineteenth-century Germany, the medieval "nations" still in existence at Uppsala or Lund University, or the springtime student songfests held on the university bridges at Dorpat in Estonia provide sub-corporate identities in a world once singularly hierarchical. And there are special recurring ceremonies, or shall we say, folk customs whose origins are probably obscure and are created by students at a time in their lives when pomp and circumstance often appear pleasingly absurd.[4]

The array of customs and symbols required for corporate self-recognition provide universities with what John Henry Newman, whose lectures on the meaning of a university have created a scholarly industry, once called "integrity," a word that he said was taken from Aristotle. Actually, Newman applied the word to a college rather than a university. It was the college or the college idea that provided universities with a certain necessary ballast, a solidity connecting past and present, because universities, by their nature, were otherwise restless and "progressive." Integrity is a stronger word than "identity," reaching towards a loftier sense of itself, pointing to a special inner coherence and wiring, a particular kind of cultural glue. Newman did not use these words, however. He said that integrity transformed "being" into "well-being."[5]

The numerous symbolic references that have marked the history of universities are also captured in "placemaking." The arrangement and styles of buildings and spaces on a given territory are cultural glue. They attract and fix student and faculty loyalties. In fractured disciplinary environments, they create a sense of the whole being greater than any of its parts. They provide continuity: emotional and sentimental anchors in a world where experiences are fleeting. They unite the generations, keeping graduates close to the institution. And they have contributed to the formation of social and political elites, a consequence that has not escaped the criticism of more democratic ages that only see "elitism" as unwarranted privilege. But these remarks require substantial qualification. What buildings, symbols and ceremonies actually "mean" surely must depend upon viewers and historical circumstances. That is the rub, for to be effective, to promote a common sense of the importance of attending a place, meanings must be universal. At this time, I will leave that issue partially open while I select examples of placemaking traditions, survivals and innovations from various countries. There is no question but that Americans have always been attracted to universities and colleges as "places" for undergraduates.[6] The issue may be more problematic with respect to the corporate identity of the instructional classes in the era of government funding, market discipline and research missions. And it may also be more problematic for the types of higher education that cannot be called "undergraduate."

In the United States, the corporate integrity of the university lies in the college idea, as Newman said, which is itself expressed as a territorial entity denominated the "campus." The campus contains the indispensable innumerable symbols and structures: buildings, gardens, bridges, walks, avenues, glades, statues, plazas, fountains, statuary, towers, gateways and also follies, quirky leftover inheritances in the form of inscriptions, unlikely structures and almost unusable ones, such as a log cabin on the University of California, Berkeley campus. The word "campus" originated in the late eighteenth century according to Paul Venable Turner's landmark study of its origins and evolution. He notes the departure from European antecedents. "The American campus has been shaped less by European precedents than by the social, economic, and cultural forces around it. As a result, it has been the laboratory for perhaps the most distinctively American experiments in architectural planning."[7] The American campus has a defined perimeter. It is true that at present the edges are often ragged, as expansion into circumjacent neighborhoods has blurred the lines that once announced a separation. In particular the research university's imperial need for room in which to carry on the ever-accumulating disciplinary specialties has led it to invade spaces once the sole preserve of cities. The actions are not always appreciated by municipal authorities and neighborhoods.

The late historian Robert Brentano once made the point that a distinctive university architecture, or an interest in having one, did not appear until about the fifteenth century.[8] The Italians wanted universities to resemble the structures to be found in their cities. The English were more interested in developing the quadrangular and cloistral form as a means of creating a separate university city. While an analogy with monastic cloisters immediately flits to mind, these, which usually encircled a burying ground, were not the only sources. The fifteenth-century manor house and outlying buildings have been claimed as the likely origin of Cambridge collegiate courts.[9] But Continental universities also made use of the court, if not always to define separate colleges (although these were certainly abundant in the earlier centuries of French, German, Belgium or Spanish universities). In all instances, courts and quadrangles turn activity away from the urban street, narrowing and controlling access in both directions. In Florence, the entrance to some of the courts are almost hidden from the passerby; but the great screen at King's College Cambridge, which is parallel to the street, does the opposite by grandly and teasingly announcing the existence of a privileged sanctuary within.

The colonial American university college was a descendent of Cambridge collegiate architecture, particularly the revisions introduced at Emmanuel College during the Puritan hegemony of the later seventeenth century, possibly building upon an earlier precedent at Gonville and Caius. The courts around which teaching, eating, sleeping, study and administration were grouped and the sense of separation from the surrounding community (whether or not that was the intention[10]) were transformed in the passage across the Atlantic. The availability of land and the absence of competing towns or cities led to the introduction of a more open and elongated quadrangular area, often only a suggestion of the prototype. Plentiful land and wood encouraged a spatial design that allowed for buildings to be separated as protection against the spread of fire. The loose or partially enclosed space, first used at Harvard (the Yard), has a lively and lasting history and has never truly disappeared. It is in fact a staple of American collegiate architecture to be found almost everywhere. However, since the court does not ordinarily separate street from campus interior, it has assumed the form of an inner park crossed by paths with easy access from all sides. In some cases, the park has become an unofficial botanical garden planted up with specimen trees. It is difficult to find an American liberal arts college or university that has a fundamentally different core pattern or does not paraphrase the form of an inner park.[11]

The open campus was a concession to democratic taste, a belief that even if universities became private (a post 1820-development), they should in some sense be public. A next step was to read the spaces with the aid of a new aesthetic that had grown up in the eighteenth century alongside an older understanding, more neoclassical, that buildings and expanses represented an "idea" such as virtue, beauty, courage, knowledge, masculinity. French Enlightenment utopian city planning, as in the designs proposed for Chaux in Burgundy, made full use of such notions. The newer aesthetic argued that in determining whether a particular building or space was significant, associations supplied the criteria. Moods, personalities, events, the sublime—these could be evoked by the right combination of symbols and designs. A building did not therefore have to be absolutely beautiful or perfect to be important. (Hence, of the three visual responses as defined by writers like Edmund Burke, the picturesque or "irregular" was sufficient.) Associations provided all the necessary meanings and strengthened the inherited understanding of the college as an isolated but self-contained beacon of civilization on a continent inhabited by indigenous populations. Associations also allowed for history to become a factor in a university's cultural life since history supplied all the necessary ingredients.

By the early nineteenth century, the elements comprising the Campus Visible were now in operation and constituted an essential aspect of the Romantic Movement. The elements were picturesque landscaping, based on the English estate garden, buildings that were functional but could also be follies, and a way of understanding aesthetic arrangements and signs that tied undergraduates, their tutors and professors to the institution's historical past. Furthermore, the integrity of the Romantic campus was itself a vital part of an undergraduate's liberal education, a factor in the process of character shaping essential to that conception and an active player in the assigned collegiate mission of turning callow youths into mature graduates. Youth was defined as a "dangerous age," as a vulnerable stage in the life cycle, one that required continual adult superintendence. Residence was of course salient, for otherwise the effect of the campus on the minds and feelings of undergraduates would be diluted through competition with other influences. Once again we find that John Henry Newman understood the development exceedingly well, not only the importance of the campus as living space but also as a performer in education's intense and unpredictable drama. His novel of the 1840s, *Loss and Gain*, was the first fictional representation of the university as actually alive. His view, that universities are places or milieux, templates for rites of passage and consequently contributors to the education of students, has not always been adopted, but it persists in much new campus planning and architecture and remains prominent in the literature. In 1989, the Arizona State University Design Review Board produced a description of a campus as "culturally instructive, introducing the individual to the rich set of information, values, principles, and experiences which art, landscape architecture, and architecture are capable of embodying [. . .] Buildings should be designed as cultural artifacts in their own right, rich in allusion, metaphor, symbol, and ordering: they should stimulate engagement and reward contemplation."[12]

Historical associations abound at Sweden's original university, Uppsala, whose buildings old and new weave in and out of the town's center and provide ample postcard depictions of a university integrated with its city. However, the central composition of Historicum, bishop's palace, medical amphitheater modeled on Pavia, and the imposing early twentieth-century Aula Magna grouped around a park with three cathedral towers within the visual perimeter provides a set of picturesque associations that cannot be surpassed anywhere. The splendid castle on the hill (used by the University) and magnificent university library in every way dwarf the shopping streets and offices. So despite its dispersal patterns, Uppsala always presents itself as above and greater than the town that jostles it. Uppsala belongs, in feel, to the older Lutheran Establishment and to royal Sweden.

But the symbols arranged around and within the Royal Institute of Technology (KTH) in Stockholm are in yet a different way similarly astonishing declarations of an institution with a special identity and position within the kingdom. The sculptor Carl Milles' "Well of Industry" greets visitors as they move through architecturally arranged trees, severely pruned in the French style, and ascend the great width of stairs leading to the first set of courts. His reliefs of "Man's Struggles with the Four Elements" appear in the next set. Atop the buildings are Ivar Johnsson's allegorical figures of Architecture, Shipbuilding, Chemistry and Civil, Mining, Electrical Engineering and Mathematics. Further on Stig Blomberg's "Thor and His Goats" leap determinedly from a wall into the Student's Courtyard. And there are many more gods and spirits depicted as overseers of the useful arts and granting their approval to the harnessing of nature so crucial to the KTH mission. Eller Lallerstedt's deliberately archaic building style of 1917 is suggestive of the Nordic past and the solid strength of the Swedish national commitment to technology, yet the entire composition, just like Stockholm's famed City Hall, is strangely modern, a very original re-working of architectural grammar.[13] Do the industrious engineering students notice the heavy symbolism? The architecture students won't, since they have been banished to a depressing building of postwar vintage some streets away.

In the trio of mid-nineteenth-century buildings comprising the University of Oslo, we find some fascinating symbolism as the new University grew with the new city. (The University dates from 1811, or almost from the time that Denmark ceded overall control of Norway to Sweden.) The buildings are a knock-off of Schinkel's designs for the University of Berlin and are longitudinally placed to express the University's association with the sources of Norwegian social and political authority. On one side lies the royal palace, on the other the Storting, Norway's parliament, as if to say that the professors were not taking chances. Edvard Munch's murals wrapped around the walls of the Aula Magna of Oslo are thrilling and fortunately in an uplifting style for the painter that gave us *The Scream*.

The pleasure of associations and symbols, or possibly postmodern jokes and teasing that capture the folly styles of the past, are evident in the otherwise functional architecture of the University of Twente at Enschede on the Dutch-German border. Built as a campus, the University has a collection of towers and ponds, one of them containing a drowned building. But Robert Venturi's postmodern allusions are offered more seriously. His extension to the Bard College library in Annandale-on-Hudson in the state of New York has been praised by the architectural critic of *The New York Times* as an "honor to the classical canon" and a "contribution to the Great Books debate." This critic finds the debate captured in the annex's various sides, each of which is different. A tour around the structure reveals a surprising complexity. The interior is described as soaring, light-splashed and "benignly labyrinthine: a gentle maze for the pursuit of knowledge."[14] Not all post-modern features are so fulsomely honored. "Most," writes a disgruntled commentator, are "a necrophilic disturbing of the graveyard."[15]

Associations of another kind are important to Stockholm University as relocated from its sometime central urban location to the more suburban Frescati. Seen from the air, Frescati is linked to the Royal Institute of Technology and to the capital itself through a network of spectacular waterways and woods. This is a genuinely exciting prospect, but an aerial view is not typical, so the question is how the complex is viewed from the ground.

The University lies within easy walking distance of museums and famous scientific societies to help form an intellectual, scientific and cultural complex. The great glass classroom and office slabs of the University are not to everyone's taste, especially as the interior of the buildings are hard to negotiate. Entries are mid-way between floors, and the long corridors, all identical with low ceilings, provide a cramped and unfriendly feeling. The confusion is similar to that of Odense University in Denmark, now renamed the University of South Denmark, whose inner thoroughfares, despite the aid of signage, do not provide an easy way to read the connected buildings. But while Odense sits alone in the countryside, a huge exterior lawn between the slabs and newer opposing structures provide Frescati with a campus sense. Filled with students on a warm day in late May, the overall effect is merry, giving some credence to the view that provided with sufficient space, students will invest it with their own meanings. Further on, one of those leftover folk structures from the past, the "Villa Bellona," is a lovely red house serving as a tiny reminder of an older Sweden. Newer buildings, several by the distinguished architect Ralph Erskine, are very fine indeed: an Aula Magna with numerous wood touches in the Nordic vein and a grand and lofty library. Dining and attractive outdoor seating areas abound, and there are landscaped interior courts between the modern buildings.[16]

Postwar architects and planners around the world were cognizant of American campus planning traditions and symbols however they chose to use or ignore them. The designers of Sunshine Coast University College in Sippy Downs, Queensland, Australia were impressed by the axial arrangement of Thomas Jefferson's thrilling conception for the University of Virginia (the "Lawn") where two Palladian pavilions face one another across the long ends of the greensward. Other Palladian buildings address one another across the longer sides. This idea of opposing structures received a different treatment at Samford University because that institution, unlike Mr. Jefferson's "godless university," is a Baptist foundation. Samford is located in the suburbs of Birmingham, Alabama. It moved from its original location in the country after the Second World War and was rebuilt largely in the neo-Georgian style. At either end of a great central park are two churches, one grandly Baroque, the other a Wren-like chapel. In-between and in the middle, approached orthogonally by a fine set of stairs, is the library. The ensemble boldly declares that learning (the library) is supported on either side by religion. A bit of fun in the form of a bronze statue of its seated and bemused eponymous pint-sized donor sits at the bottom of the stairs leading to the library above.[17]

While Uppsala dominates its town, and the original buildings of Oslo are very much a union of university, city and government, the American campus idea contained a decidedly anti-urban strain, although so attenuated at present that its shadows are barely perceived.[18] There is also the fact that campuses that were once solitary are now surrounded by towns and cities, so that part of the later story of the American campus is the old issue of town versus gown. The anti-urban strain in the earliest decades of American campus planning is, comparatively speaking, an historical anomaly. The European university was really a profoundly urban institution, if not always first located in the largest municipalities.[19] This made perfect sense. Population centers provided the markets for educational services, the housing and economic support. Furthermore, in the Victorian medievalist Frederic Maitland's sharp colloquialism—I hope I recall correctly—some ideas only come when people are "tightly packed." Peter Hall's prose is more direct but no less certain: "every great burst of creativity in human history" takes place in an urban setting.[20] But cities of significant size were few in the early periods of the Republic, and the conspicuously plural aspects of American society encouraged the proliferation of institutions serving the needs of diverse religious societies which were expected to nurture and socialize the youths in their temporary care. Eventually competition between institutions for students, professors, income and reputation further stimulated the tendency to use signs, structures and spaces that called attention to the virtues of a particular campus. To this day, students use campus appearance as a factor in choosing where to apply. The campus "place," promoting history, beauty and associations, has been a selling point in a business environment.

An explosion of campus construction occurred in 1960s Britain in the era denominated "Robbins" because of the expansion recommendations made by the committee that this famous economist chaired. "Greenfield" sites were chosen very much in the American vein, and new campuses were built expressing almost every planning idea, campus or urban, that had ever been advanced. Stefan Muthesius, in a comprehensive analysis of 1960s construction, calls the experiments "utopianist" to capture the sense of daring (if not always successful) solutions to learning and living undertaken by distinguished architects. Part of the daring was to essentially reject the neo-historical styles of pre-1945 architecture.[21]

The American campus idea was not much in evidence in Britain before the 1960s, although the foundation of the University of Birmingham in neo-Byzantine style around 1900 is an exception. However, whereas Birmingham is sited in a park, it did not adopt the court or quadrangle so much in evidence at the civic university colleges of Britain and so striking a feature of American campuses. But it is replete with an array of symbols and references, statues of Darwin, Shakespeare, Virgil, Beethoven and other paragons, and visual notes on the economy of the industrial midlands. Birmingham also possesses a central defining tower (modeled on the Torre di Mangia at Siena)[22] such as those evident at Berkeley (the Campanile copied but in concrete not stone from the one in the Piazza di San Marco in Venice) and Stanford. All the other many Victorian institutions were city university colleges essentially without campuses, formed into federations with a central examining and degree-granting core on the model of the University of London. Yet their architecture and symbolism are full of interest; and notwithstanding the many critics who called them "unlovely," their styles are on a par with that of Victorian Oxbridge.[23]

President Clark Kerr of the University of California, a graduate of the elite Quaker College of Swarthmore in Pennsylvania, was certainly not one to disparage rich legacies. He contributed from his own pocket in order to make the Berkeley campus appear friendlier, more respectful of its natural site and more comfortable to students. His dream was also to create a collegiate university within the family of University of California campuses, which in fact he did in the 1960s near the seaside town of Santa Cruz. Critical attention was captured by the building of colleges as villages, no two alike, in a redwood forest overseeing the Pacific Ocean. Hoping to capture for public higher education the spirit of Swarthmore, and true to the history of the Campus Romantic, the paths, individualized colleges, gardens, enclosures, spaces, open farmland and even a restored gatehouse farm building together met the ideal of the picturesque campus, the campus of experience. The architects "discarded the traditional hierarchy of spaces: an academic and administrative core with a periphery of residential and social areas."[24] One of the colleges, in fact, was built as a miniature Mediterranean town with streets and open-air cafes.

The seaside town of Santa Cruz is small, a bit isolated and a tourist spot. It does not qualify as an urban area, and the residents were probably of two minds about having a major university nearby, pleased by the economic advantages and possibly the prestige but also concerned about automobile traffic, large numbers of young people seeking amusements, and the inflation of land costs for housing, as well as new roads. So the situation was typically American of the earliest periods, colleges and universities either distant from towns and cities or poised nervously on the edge.

At virtually the same time that Kerr invested an uncommon amount of personal emotion in the college campus idea as applied to public education, he described the arrival of a new type of university that, he said, did not represent Newman's idea of what a university should be. In his extraordinary lectures on what he called "The Research Grant University," the Godkin Lectures delivered at Harvard in 1963, Kerr announced the birth of the "multiversity," so-called because a new educational reality demanded "a new word to [. . .] make the point that Alma Mater was less an integrated and internal spirit, and more a split and variable personality." "Environment" had replaced "community." No purpose was gained by denouncing the newcomer (which in fact had been growing for some time) since it was "rooted in the logic of history."[25] Santa Cruz was a revolutionary departure for publicly-assisted higher education. Although Kerr realized that his Santa Cruz dream incorporated into a multi-campus system of public higher education was at odds with the realities that he described, he clung to the belief that the two could co-exist within a differentiated system. But faculty and students combined to destroy his hopes. The faculty wanted the Research Grant University, except for those swept up in the counter-culture of the 1960s, and the activist undergraduates of the 1960s took the notion of a Campus Romantic well beyond the respectable, serious, liberal arts inheritance that Kerr desired. The net effect was to undermine the assumptions upon which the Campus Romantic was based and even to introduce new buildings into virgin forests that were quite at variance with the original conception of collegiate villages. Overall, in keeping with what appeared to be a less hierarchical spirit emanating from the student disturbances and protests of the 1960s, the relations between undergraduates, graduates and professors relaxed. They appear to have relaxed everywhere. First names often replaced titles, and in some cases, notably at Berkeley, students refused to wear gowns during graduation ceremonies in order to stress their independence, individuality and defiance of the corporate authority of the institution. Professorial dress became informal, not only in America but elsewhere, and suits and ties were less in evidence, although gowns made a reappearance later on.

Serious challenges of a different kind to the Campus Romantic and its integrity had in fact been voiced at the beginning of the twentieth century but were never strong enough to overcome the taste for disguising new universities as ancient institutions whose prestige relied upon historical references and styles. Thorsten Veblen, an American *enfant terrible*, excoriated university boards of trustees for following in the wake of early builders. He declared the habit of building in historical styles in order to identify with the university's glorious past generic history to be fakery. Buildings were pretentious, representing the desire of new wealth to commemorate itself.[26] Since so many American architects of the *fin de siècle* were trained in the Beaux Arts school in Paris, a ready supply of talent kept the historical styles alive. The Futurists and those who espoused the twentieth-century International Style were unsentimental and quick to denounce apostates like Eero Saarinen, whose design additions for Yale University did not fit the Modernist mold.[27] Other critics, following Veblen's lead, regarded the mass of campus decorative and symbolic elements, the rituals and ceremonies with their hierarchical features as class-reinforcing, reflecting centuries of privilege and exclusion. Education expressed by coats of arms in stained-glass windows was misleading. Education was a social good dispensed functionally like other social goods such as health, transportation and housing. Careers and labor markets mattered. In any case, the stress had to be on the product and not on the institution generating the product.

A late twentieth-century critic provided a withering denunciation of the Campus Romantic, declaring it to be luxurious, backward-looking, and of course "bourgeois." Campus architecture should never be cluttered or pretentious. It should never attempt to impose upon the student whose freedom to roam needed to be respected. It should be modest, nakedly contemporary, reflecting a no-nonsense industrial civilization. At best the buildings should be shells whose interiors allowed for continual reconfiguration as befit the times, rather like a railroad station filled with kiosks. If monotony and impermanence were the price, so be it. At least, in the language of the day,[28] the appearance was "honest." After the Second World War, the literature on planning seemed to turn against the suburban campus style and towards an interest in the revival of the university as a city or as city-connected. One heard about "precincts," "linked nodes," "linear planning" and "networks." The vocabulary was top down, technocratic, instrumental and impersonal.[29]

The rise of the research university and the "rediscovery" in Anglophone countries that the earliest universities were founded for the purpose of professional education, the teaching of the liberal arts being merely preparatory,[30] suggested that the Campus Visible was not necessarily indispensable to the higher learning. The university might be a "place," but it was also a collection of disciplinary specialties and subjects, to include some not usually classified as liberal, such as high-level engineering (introduced at Cambridge after a bitter fight[31]), that were oriented towards careers and problem-solving in nations undergoing industrialization and urbanization. Academic reformers were concerned with the competition and examples emanating from Germany, whose military and naval successes were regarded as security threats. German *Technische Universitäten* and innovative research captured the imagination of scholars and scientists who, frankly, were bored with collegiate forms of education designed for late adolescents. In Civil War era America, Congress legislated the "land grant university," also called the "people's university," requiring such institutions to offer instruction in agriculture and the mechanical arts (engineering), and thus was born the designation "A & M," as in Alabama A&M. While the genius of the legislation was that it did not forbid the teaching of any other subjects, nor did it prescribe the settings in which instruction and research were to take place, it appeared that the liberal arts campus would no longer be a primary object of higher education.

Another countertrend was the coming of mass higher education, defined by Martin Trow as the condition reached when 15% of the available student cohort is enrolled in post-secondary education.[32] An institution defined as mass-access loses much of the focus, the intellectual integrity, typical of the Campus Romantic; and as the twentieth century continued, especially after 1945, campus populations became increasingly composed of students who were not callow youth in need of careful counseling and moral superintendence but "mature" or adult and part-time degree students, many with families, homes and work responsibilities that competed with the attention customarily demanded of the historic campus. Furthermore, spaces once carefully arranged to suggest an integrated college experience were now, especially on the largest campuses, invaded by giant structures such as stadia, Olympic-sized swimming pools, exercise rooms, concert and dance halls, cinemas, art galleries, huge libraries and research laboratories in profusion. Smaller campuses had versions of these, appropriately scaled. On the large campuses, quaint architectural oddities that escaped demolition often appeared engulfed, fading memories of a commitment to a culture of the liberal arts designed for the few. Where medical schools made an appearance, the Campus Visible had a rival indeed as parking lots and clinics competed for space with the inherited spectra of liberal arts buildings. A visit to Vanderbilt University in Nashville, Tennessee or the University of California at Los Angeles adequately demonstrates the dominance of the giant structures.

The interesting point is that despite some major challenges to the conception of the campus as a microcosm of its corporate history, the Modernists, rejectionists and the arrival of the mass-access university did not altogether mean that the battle over whether universities should continue to find tactile ways of expressing their integrity was lost. New collegiate universities, for example, were built, not only at Santa Cruz, or at another campus of the University of California at San Diego, but in the nation that inspired the college idea that led to the campus. This should not be surprising, except that Oxford and Cambridge had no cluster college imitators in the higher education building boom of the Victorian era, and Scotland, which once had collegiate universities, had long since given up on the idea. Among the experiments were colleges at the universities of York and Canterbury, albeit less distinctive or separate than at Oxford and Cambridge. But otherwise, the Robbins era universities were campuses, if not exactly romantic. In some cases, roadways extended beneath or around the campus perimeter, classrooms and residence halls were intermixed with offices. A quadrangle appeared at Sussex. Lakes or ponds and watercourses were created; gardens proliferated. Locations were chosen near small historic towns with their own array of picturesque legacies. Besides York and Canterbury, there was also Norwich. Sussex University was sited on downs, a short train ride from the seaside resort of Brighton.

A later higher education expansion occurred at a time when government funding for higher education could not be so generous. Instead of greenfield sites, brownfields were chosen, land that was polluted or had been used for other purposes. Instead of making students travel and reside on country or small town campuses, urban universities would be brought to them. Subsequently, "eco-architecture" was given more prominence. Planners became far more conscious of environmental issues. "Energy efficient" buildings appeared using lichen and moss roofing, photovoltaic cells, and recycled materials were featured prominently. Nottingham's Jubilee campus and East London's Docklands are two examples of institutions that were born this way, returning to the city idea once the natural partner for universities.[33]

Fascinating eco-friendly ideas were employed in locations far from Europe and America. For example, the landscapers and architects of Sunshine Coast University College were very cognizant of the location's tropical ecology and were eager to preserve its plant life. They chose growth appropriate to the site and also took advantage of the available light and winds by using verandas, breezeways, screens, filters and overhangs. Many of these features were incorporated into the library, described elatedly by a reporter as having "timber-louvered screen walls and a wackily multi-pitched sawtooth roof." The structure was "articulated by steel columns, flanked by hardwood seating in the Doric manner and punctured by palm trees." The library was "soft but dignified, warm but grand, friendly but also uplifting" and expressed an overriding humanist conviction." The drawings do indeed make the campus appear friendly and appealing in line with a concern to avoid the lofty and arid towers of Corbusier (and versions thereof, as at the Catholic University of Nijmegen in the Netherlands) or the monumental, overwhelmingly self-important facades of earlier generations of Beaux Arts influences so oppressive to some observers, so impressive to others. The intriguing designs for Sunshine Coast University College include landscaping directions for "platonic shapes" accompanying the campus spine. "Plato believed that the way to perfection and truth was through the eternal and constant principles of mathematics." [34] Here was symbolism carefully thought through. Whether students understood the message is another matter altogether.

Nordic respect for a pristine environment is evident at new universities such as Karlstad in Sweden. The buildings, on the edge of the city, look out upon forests of pine, so that the university is in fact dramatizing through its own location the transition from wild to cultivated or the interpenetration of the two. But America too has a legacy of respect for nature, despite many outrages. The grand and the monumental figure, but also the more intimate landscapes favored by the so-called Hudson River school of nineteenth-century painters. The architects of Parkland College in Champagne, Illinois, a community college, rejected the familiar plan of a single structure in order to create villages, the overall footprint capturing the spirit of the surrounding "checkerboard of farmsteads." [35] But in Sweden, and elsewhere, the urban element was not abandoned as planners reflected the postwar social democratic policies of ministries. Universities could be used to revitalize regional economies, but also to provision far-flung communities with cultural amenities in short supply. The subject has occasioned a considerable amount of international comparative analysis on the strengths and drawbacks of various schemes. [36] The history of Linköping illustrates an early stage of the new policies. Its site was a farm offered to the state by the local community. Little effort was expended on placemaking. The strongest considerations were to integrate the university's studies with the dominant economic interests of the city, notably Saab industries, and to provide the region with a trained labor supply. An additional consideration preventing a distinct identity from forming was the initial proposal, soon abandoned, to join Stockholm University as a southern outpost. But in any case, Linköping did not receive full university status until 1975.

The new university was clearly intended to be a mass-access institution, efficient in the production of graduates and providing training relevant to local needs. That conception produced a view of the university as mainly utilitarian, an institution without a past and unaligned with traditional European landmarks (such as placement near a cathedral, as in the case of Uppsala in the seventieth century or University of Kent in the twentieth). One building was wryly described by undergraduates as a hangar for Saab aircraft. Yet a change of mood occurred. The newer structures were softer, faced with brick and less industrial in aspect. More interior space was provided for meetings and social occasions and more effort was taken to hide views of the parking lot. The composition termed "Zenith," which houses the bookshop and restaurant, was of a different and friendlier order altogether. With the construction of the classically symmetrical administrative building called "Origo," the University acquired a place that satisfies some of the purposes of a symbolic center.[37]

Göteborg, the second city of the kingdom, has a university with a hybrid aspect. Founded in the nineteenth century when it was a more modest enterprise, Gothenburg University essentially conforms to the dispersed model of the European university, its buildings spread about the city. Two new structures of the 1990s, the School of Economics and the Faculty of Social Sciences, are integrated into the city fabric. The first, attempting to establish a distinct identity, confronts the streets and "communicates through its façade." The second strives to integrate itself with the general housing plan, wishing to be closer to the idea of a city.[38] A newer complex comprising a humanities and music building almost edges towards the campus style as a loose cluster around a large area of grass. The university has grown along with its bourgeois city set on the trading routes to Britain and northern Europe. Its associations are with that city; its buildings join the parade of city symbols.

Södertörn in south Stockholm represents yet a different type of urban university, one designed to aid in social renewal. Easily accessed by train, it is not a campus but shares a site with various businesses and is placed in an industrial and ethnically-diverse neighborhood where land is scarce. Yet those who were instrumental in forming the spaces, meaning many of the administrators and faculty themselves, were eager to find a distinctive means of expressing an identity that captured some of the mood of a more traditional institution, or at least respected the right of students to enjoy pleasing and comfortable surroundings. A select group traveled to the United States to visit a number of different campus settings. On a tight site and as in the case of Karlstad, multi-story buildings were designed with large atria, providing a generous sense of space and allowing the strange and wonderful Nordic light to pour into the interior cavities. In the case of Södertörn, the openings also serve as performance halls and venues for dinners and receptions.[39]

The new campus university of Umeå in the north of Sweden, but not the far north, is also part of the social democratic plan of regional development. Buildings are placed freely in the open landscape and vary according to height. The choice of yellow brick as the dominant façade material, as at Århus in Denmark, was intended to soften or domesticate the overall effect. The University has grown substantially in the intervening years and now displays an increasing density, the city away from the city. The center symbolism of the planned American campus—a tower, church or library—is missing, as if it were a pretentious and unnecessary statement and out of keeping with Swedish egalitarianism, but the newer Universum or campus administration building is in fact a conspicuously large structure with prominent glass features. However one may feel about giving such prominence to an administrative structure (albeit one far more pleasing than the threatening postwar vulgarized classicism of Berkeley's partial equivalent), the leadership of Umeå was always interested in creating an attractive campus for a part of Sweden far from the traditional university centers of prestige and rank such as Uppsala and Lund. In this sense its mood was different from that of the early Linköping. Umeå challenged the functionalist's credo that a university should dissociate itself from patterns of building and display that suggest privilege.

After much of the massive and anonymous architecture of the 1960s and later, buildings such as those of the University of Montréal at Québec, stark functionalism (but with a touch or two of the past buried in the interiors) and swallowed up by the city, the old places shine even more brightly. Yet Québec in the Latin Quarter is easily accessible by a train that runs beneath; and if Stockholm University is no longer in the middle of the capital, it too has its own underground train station that meets the edge of the greensward. Almost the furthest imaginable step towards the anonymous campus with no visible integrity were the structures designed by postwar French planners, for until the 1990s these universities were not given names but were known only by numbers. Momentarily enthralled by the American campus idea and eager to locate universities in the environs surrounding Paris, planners ringed the capital with new universities. However, a misreading of American campus planning, a failure to notice how self-contained were the colleges and universities, resulted in structures that lacked the numerous amenities of American colleges and universities. The apparent assumption was that the new sites were still part of Paris, and it was expected that students would still have conventional recourse to the cultural and recreational advantages offered by a great metropolis. But in truth, the students were isolated. By the 1990s, the original policy of dispersal was being rethought.[40]

The placemaking of the original Oslo University has been mentioned. But the new campus is of a wholly different order and conception. It lies in an immediate suburb of Blindern also reached by its own train. It is mainly a postwar composition, functional and straightforward, a series of large parallel buildings lining a wide, central and stepped axis. However, the new library bespeaks more warmth and interest. The old physics building has a nice interior entry, and there remain a few oddities of the site: a grouping of prewar Palladian-style buildings around an open court more typical of Oxbridge colleges, obviously self-conscious imitations, and a lovely craft-style period house containing student services.[41]

A campus may be compared to a city. It may in its range of amenities and functions imitate a city or it may partake of the city, becoming part of its morphology and elevation and linked to its transportation system. Rarely do we find a new university that is actually a city built from scratch. But such is Louvain-la-Neuve in Wallonia, which, however, pays some homage to pre-existing communities on its edges. The linguistic fracture of Belgium, more pronounced after the war than earlier, resulted in the original Leuven University becoming wholly Flemish. Banished from the ancient site, French-speaking Louvain is built atop a great platform in the countryside with trains running beneath. Also underneath are three stories of underground parking. On top is a city intermingled with a university. But it is a city in the picturesque style, with narrow streets, a Place des Wallons encircled by shops and a Place de l'Université. There is a lake. The entire conception is breath-takingly bold.[42]

Louvain benefits from not having to compete with a city, since it is specifically designed to be one. A different and interesting example of a university in a city that has in a sense struggled to declare its independence, a university that is in fact the archetypal European university writ particularly large, is the University of London founded in the 1820s to be a metropolitan university serving catchment areas about three or four kilometers distant. Beginning with a single university college and a medical school (its principal *raison d'être*), London today is a collection of quasi-independent schools, colleges and institutes everywhere dispersed. The first structure, now called University College London, presents itself as a considerable neo-classical façade fronting a more or less routine city thoroughfare. The second foundation, King's College, presents a less imposing aspect because its central axis, given the nature of the original site, is perpendicular to the street. Entry is required to view the façades. Far off to the south lies the Royal Holloway College in a fantasy chateau style surrounding a court. It was founded as a college for women by a wealthy manufacturer of quack medicine who was acquainted with Matthew Vassar, the American donor of Vassar College for women (no longer single-sex). Associated hospitals are to be found everywhere in the capital, although amalgamations have occurred. Amongst a South Kensington complex of museums lies the Imperial College of Technology, now also a medical school, which was a late addition to the galaxy of famous high-level engineering establishments that includes KTH, the Massachusetts Institute of Technology, the Charlottenburg and others in Switzerland, Austria and Denmark. We may pause for just a moment before the Imperial College, whose overbearing façade declares that this is an important building, but the crowded surroundings disturb any attempt to take in the presentation. Within there is an enormous court, but it is lined with some indifferent International Style Buildings. A great tower in the imperial style of British India lies within the court, guarded at its base by two lions. In short, identifying symbols are present but hard to detect; and this is true of much of the dispersed University of London, which does not possess a coherent architectural place within the vast spread of a world capital. Located near University College, the Senate House with a large tower was meant to provide a central symbol, but it too must compete with the formidable and plentiful iconography and emblems of the metropolis.

I began this discussion with references to the special corporate identity of the university, its guild heritage as it emerged from the particular institutional organization of the medieval period. I explained that the campus was an American territorial expression of that self-referencing identity, that it represented integrity precisely because it possessed a visible boundary and plentiful symbols anchoring the institution to its past. It was not exactly what sociologists have called a "total institution" because it lacked the final coercive authority of a hospital, prison, barracks or, an example once used, of an English boarding school. But as long as the campus did not have to compete with the city, it was a successful container of all kinds of customs, rituals and symbols. Thus, degree-granting ceremonies possessed a certain finality: not only was a particular course of studies being validated, but the recipient was being sent away into exile, as it were, the Garden of Eden with its guardian angels now closed to them. Commencement speakers in America conventionally spoke of the need for graduates to go into the real world, there, it was hoped, to apply the principles and ethics learned during a sojourn at the Campus Romantic.

Time has eroded the becoming simplicity of this scenario. The American campus, I have argued, was designed and intended to be a romantic experience for young people coming of age. But I cannot be certain whether the campus has the same meaning for the research student, the mature student, or for the intellectual life of the university. For postgraduate students, the rite of passage is socialization into professional careers. The campus idea, while it may in fact be relevant to that process, is not a character in the romantic drama.[43] The dispersed urban campus, or even the compact, high-rise New York University, which has no campus but shares a small amount of public space in the form of Washington Square, has not been severely handicapped intellectually. Manhattan supplies its own excitements, especially as so many students either work or live at home. Medical schools prefer an urban location; and it is no accident that law schools in England, the inns of court, are not at Oxford or Cambridge but in the center of London where barristers can plead. We can generally speculate that beauty and historical references are appealing and desirable. The distinguished Harvard professor Henry Rosovsky writes that "the physical setting in which we labor matters enormously." He contrasts the "urban squalor" of Harvard Square, with the "oasis" that is Harvard Yard. "I think with pity of my neighbors arriving in downtown Boston to inhabit all day the synthetic atmosphere of yet another glass-lined tower."[44] But glass-lined towers certainly occupy territory on many world campuses.

Public exposure of the financial scandals that occasionally beset universities, and just closer reporting in general, removes some of the mystery and magic of academic life, whose corporate practices were once not widely understood. A certain egalitarianism has undermined the distance between teacher and taught, also contributing to a decline in the enchantments of the university. On a different point altogether but even more relevant, recent commentators have suggested that universities have lost their integrity because outside pressures have shifted control of the university away from collegial governance, or even governance shared between boards and senates, to markets and national governments, or state governments in the case of the United States. The large numbers of part-time faculty, a very rapid growth in America since 1975, means that institutions cannot count on their loyalty or participation in governance.[45] And what is true of the United States is perhaps even truer of Europe, where universities are heavily dependent upon state largesse. Private research universities in the United States, which also depend upon external financing, have more integrity in Newman's sense, but they too are not exactly free agents. Smaller colleges, which recruit faculty from the same research universities, have a stronger sense of their historical identity but are not immune from the interior fracturing caused by disciplinary specialization. Several scholars have noted how "donnish dominion" has declined in Britain, even at the most collegial of institutions, Oxford and Cambridge.[46] Everywhere, it appears, universities are struggling to define the areas of academic freedom and institutional autonomy that remain. Furthermore, never before in the long history of universities have the "invisible colleges," the national and international networks of scholars and scientists, been so prominent and important to careers. The frequent absences of research faculty from campuses have been much commented upon. National and international markets for professorial talent inevitably dilute the loyalty that academics may have for their home institutions. It was once suggested that relationships are more based on invisible colleges than on home institutions.[47] We may then well ask why universities should any longer bother with symbols, ceremonies and rituals or why they should even bother to commission structures that are not straightforwardly functional if there is nothing intact to symbolize, and the inner spirit, once captured by the plentiful associations of the Campus Visible, has faded.

Many more types of higher education institutions could be mentioned as examples of campus and non-campus institutions. However, this contribution to the festschrift is not a comprehensive survey but a note towards the definition of integrity, a brief reflection upon the use and meaning of the campus, the urban university, the associations, locations and proximities of structures and the symbols and customs surrounding the principal missions of instruction, training and research. Older notions keep reasserting themselves even as critics find some fault in them; and newer notions about democracy (attractive locations cannot be "democratic" unless everyone has access to them, impossible under mass higher education circumstances) or utility and buildings suited to them do not always succeed in capturing attention or loyalty. The vast scale of modern building is challenged by picturesque touches—a student street of shops at Nijmegen, for example, or an interior tropical garden; attractive seating areas in the long corridors of Odense; the atria of the newer Swedish universities; the efforts to correct some of the earliest expressed forms at Linköping; the splendid library building of New York University tucked in amongst the sometime office buildings converted to classroom use, delightful mews nearby and the acquisition of low-rise houses from the past lining Washington Square. Picturesque touches appear where least expected, and at the same time the traditional campus is invaded by the large structures needed for science and technology missions. Aesthetic opinions will differ on the success of the imports, whether they are slavish, outmoded and tired uses of older planning ideas (or inappropriate for damp climates, such as the open spaces of the University of Essex) or insensitive functionalism, a disregard for what makes students and teachers comfortable in a busy competitive world. However, there has been no lack of trying on the part of architects, planners and university committees.

What this survey reinforces is the common view that a complex congregation of higher education institutions reflects the complexity of modern life, its variety, markets and assumptions about the different forms of learning and education. But we are still somewhat up in the air about the connection between the settings that I have described and the different kinds of education required of contemporary society. I am more confident in believing that the Campus Romantic worked best for elite liberal arts education precisely because its targets were clear: impressionable post-adolescents making the transition from school to university, from youth to maturity. Architects and planners of the postwar era thought continually in terms of communities, but that also became communication, getting students to and from universities and colleges.[48]

Muthesius makes the point that utopianist campus planning shared many of the ideas informing New Town planning in the era of Britain's Welfare State.[49] At first, the Campus Romantic rejected the city, but over time the two drew closer together, the university rejoining the city, merging with it and making itself as accessible as necessary for an era of mass higher education. This is undoubtedly a gain for democracy, but does it mean a loss of the integrity and well-being so intimately connected to real and metaphorical boundaries? The subject of the relation of environment to learning is serious enough to warrant some speculation about whether a walk along paths strewn with platonic shapes has any meaning in our global, professional, bureaucratic, frantic and uneasy world.

Notes

1. Sources that are not referenced derive from personal knowledge.
2. J.K. Hyde, "Universities and Cities in Medieval Italy," in *The University and the City: From Medieval Origins to the Present*, ed. Thomas Bender (New York and Oxford: Oxford University Press, 1988), p. 20.
3. Hilde de Ridder-Symoens, "Management and Resources," in *A History of the University in Europe*, II, ed. Hilde de Ridder-Symoens (Cambridge: Cambridge University Press, 1996), pp. 205–208. William Clark, *Academic Charisma and the Origins of the Research University* (Chicago and London: University of Chicago Press, 2006), pp. 54, 105, 117, 360, 406, 442.
4. Sheldon Rothblatt, *The Modern University and Its Discontents: The Fate of Newman's Legacies in Britain and America* (Cambridge: Cambridge University Press, 1997), pp. 81–82. For the purposes of this essay, I will eschew technical discussions of the symbols and forms of communication found in the anthropological literature. Essentially, I am primarily referring to those visible signs that reinforce a sense of participating in a larger shared experience, signs that dramatize participation, establish order and continuity and replace the profane by the sacred, the humdrum by the holy. See Barbara Myerhoff, "Rites and Signs of Ripening: The Intertwining of Ritual, Time, and Growing Older," in *Age and Anthropological Theory*, eds. David I. Kertzer and Jennie Keith (Ithaca and London: Cornell University Press, 1984), especially pp. 306, 328. For symbols as communication, see Hugh Dalziel Duncan, *Symbols in Society* (New York: Oxford University Press, 1968).
5. Sheldon Rothblatt, "An Oxonian 'Idea' of a University: J.H. Newman and 'Well-Being,'" in *The History of the University of Oxford*, VI, Nineteenth-Century Oxford, Part I, eds. M.G. Brock and

M.C. Curthoys (Oxford: Clarendon Press, 1997), p. 293. In one set of writings, Newman made integrity in education the special province of the Church of Rome.

6. John Thelin and James Yankovich also make the point about the American attachment to a physical place. "'Bricks and Mortar,' Architects and the Study of Higher Education," in *Higher Education: Handbook of Theory and Research*, 3 (1987), pp. 57, 61. On pp. 74–75, they report that agencies of state government in California balked at the expense involved in providing sculpture and student lounges. This is why even public universities, aware of the importance of signs, pursue private donors.

7. Paul Venable Turner, *Campus: An American Planning Tradition* (Cambridge, MA and London, 1984), pp. 6, 21. I believe this is essentially correct; but in a fascinating unpublished paper, Lyse Roy, "Le jardin du savoir: La representation de l'espace universitaire du Moyen Âge au XXIe siècle" (2003), suggests that in the charters and papal bulls of the middle ages there is an ideal conception of university space as a joint production of Athens and Paradise, of knowledge and healthy, amenable space, or a cross between Christian otherwordliness and classical temporality. As the American campus inherited these notions, she says, there is a certain continuity.

8. Robert Brentano, "The Medieval University and Space" (unpublished paper May 1996).

9. Robert Willis and John Willis Clark, *The Architectural History of the University of Cambridge*, III (Cambridge: Cambridge University Press, 1988), pp. 270–272. First published in 1886.

10. Antoine Grumbach, *La formation et l'évolution des sites et bâtiments universitaires en Europe et aux États-Unis* (April 1994), p. 39, makes the point that the contemplative (otherworldly) vision of American education required colleges to be sited "loins des 'vices et contaminations' de la vie urbaines."

11. Skidmore College in upstate New York, approached via a long landscaped road, is an exception. The campus is indeed a fine park but without the suggested courts.

12. Lisa R. Findley, "The Pedagogical Building," in *Academe* (July–August 1991), p. 32.

13. Ragnar Widegren (text) and Iva Sviestins (photographs), *Konsten på Kungl Tekniska Högskolan* (Stockholm: Byggförlaget, 1992); *Lallerstedt redivivus, Kungl Tekniska Högskolan* (Stockholm: Byggförlaget, n.d.). KTH was founded under a different name in 1827.

14. Herbert Muschamp, "Democratic Decorations at Bard College," in *The New York Times* (October 31, 1993), p. 42.

15. Thomas A. Gaines, *The Campus as a Work of Art* (New York: Praeger, 1991), pp. 11–12.

16. Thomas Hall, ed., *Frescati: Huvudstadsuniversitet och arkitekturpark* (Stockholm: Stockholm University, 1998).

17. The university was first called Howard. A two-volume history of Samford University, James E. Sulzby, Jr., *Toward a History of Samford University* (Birmingham, AL: Samford University Press, 1986), is full of references to winning football games but contains little pertaining to the lovely appearance of this southern campus.

18. Colonial leaders like Jefferson, who knew the ancient sources, picked up on the anti-urban themes. Virtue resided in the countryside and amongst the yeoman farmers. Sylvia Doughty Fries, *The Urban Idea in Colonial America* (Philadelphia: Temple University Press, 1977), pp. xiii, xvii.

19. Hyde, p. 15.

20. Peter Hall, *Cities in Civilization: Culture, Innovation, and Urban Order* (London: Orion Books, 1999), p. 3.

21. Stefan Muthesius, *The Postwar University: Utopianist Campus and College* (New Haven and London: Yale University Press, 2000).

22. William Whyte, "'Redbrick's Unlovely Quadrangles': Reinterpreting the Architecture of the Civic Universities," in *History of Universities*, XXXI/1 (2006), pp. 164–165. Muthesius, p. 59, thinks of Downing College Cambridge, founded in the early nineteenth century, as in the campus mode.

23. Whyte, pp. 154–159; Rothblatt, *Modern University*, chap. 5.

24. Sally Woodbridge, "How to Make a Place," in *Progressive Architecture* 5.74, p. 76.

25. Clark Kerr, *The Uses of the University*, 5th ed. (Cambridge, MA: Harvard University Press, 2001), pp. 5, 102. (First published in 1963.)

26. Joseph Bilello, "Deciding to Build: University Organization and the Design of Academic Buildings" (unpublished Ph.D. dissertation, University of Maryland, 1993), pp. 40–43.

27. Turner, p. 294.

28. Ian Brown, "The Irrelevance of University Architecture," in *The Shape of Higher Education*, ed. Tyrell Burgess (London: Cornmarket Press, 1972), pp. 182–185,

29. Jean Pierre Frey, "Equipements universitaires et processus d'urbanisation en Europe," in *Les annales de la recherché urbaine*, nr. 62–63 (June 1994), pp. 186ff.

30. Hastings Rashdall expressed this finding in his monumental three-volume work, *The Universities of Europe in the Middle Ages*, new edition, eds. F.M. Powicke and A.B. Emden (Oxford: Oxford University Press, 1969). (First published in 1895.)

31. T.J.N. Hilken, *Engineering at Cambridge University, 1783–1965* (Cambridge: Cambridge Unviersity Press, 1967), chap. 3.

32. See, *e.g.*, "Reflections on the Transition from Elite to Mass to Universal Access: Forms and Phases of Higher Education in Modern Societies since WWII," in *International Handbook of Higher Education*, eds. James J.F. Forest and Philip G. Altbach (New York: Springer, 2006), pp. 243–280.

33. Oliver Lowenstein, "Slap on Factor Four," in *The Times Higher Education Supplement* (March 24, 2000), p. 37.

34. E.M. Farrelly writing in *The Sydney Morning Herald* (December 9, 1997), Arts 16.

35. "A Campus Village," in *Architecture California*, 11 (January/February 1989), p. 26.

36. See for example *New Universities and Regional Context*, eds. Urban Dahllöf and Staffan Selander, *Uppsala Studies in Education*, 56 (Uppsala: Uppsala University, 1994).

37. I am grateful to Kerstin Thörn for information on Linköping and other aspects of Nordic architecture and campus planning.

38. Claes Caldenby and Catharina Dyrssen, "Building and the City: Project Description of University and Place, Universities as Interface between Global Forces of Change and Regional Ambitions" (Lund, Umeå and Chalmers Universities, n.d.), p. 33.

39. Photographs appear in the *Architectural Competition for Södertörns Högskola, the University College of South Stockholm* (Stockholm: National Premises Authority, November 1995).

40. I am indebted to the insights of Nicole Fardet Carrié as expressed in her doctoral thesis, *Les universités américaines: de la genèse à la planification des campus* (May 11, 1995).

41. Oslo University's Long Range Plan, 2000–2004, mentions the desirability of increasing "the feeling of satisfaction among staff and students" by raising "the standard of the physical environment" and developing new communal areas.

42. Brochure, 1992, and personal visit.

43. This was made apparent to me when I was a research student at King's College Cambridge and was informed by my tutors that in essence neither the University nor the College really knew how to accommodate postgraduate students to the collegiate culture. One decision was to create a "middle combination room" and a special table for dining in Hall.

44. Henry Rosovsky, *The University: An Owner's Manual* (New York and London: W.W. Norton, 1990), p. 161.

45. Richard Chait, "The 'Academic Revolution' Revisited," in *The Future of the City of Intellect*, ed. Steven Brint (Stanford: Stanford University Press, 2002), pp. 299, 294–321. Writing a decade or so earlier, Clark Kerr thought that market forces had given more influence to students and professors. *Troubled Times for American Higher Education, the 1990s and Beyond* (Albany: State University of New York Press, 1994), p. 41. Lars Engwall and Thorsten Nybom are not convinced that market discipline in Sweden has had a positive influence on Swedish academic research.

"The Visible Hand vs. the Invisible Hand," in *The Sociology of the Sciences Yearbook 2006* (Fall), eds. Richard Whithley *et al.*, 14. But on the issue of who really controls academic life and output see also Sheldon Rothblatt, "Many Masters, Many Servants," in *Whose University is It*, eds. Douwe Breimer *et al.* (Amsterdam: Amsterdam University Press, 2006), pp. 17–26.

46. A.H. Halsey, *Decline of Donnish Dominion: The British Academic Profession in the Twentieth Century* (Oxford: Clarendon Press, 1992); Ted Tapper and Brian Salter, *Oxford, Cambridge and the Changing Idea of the University: The Challenge to Donnish Domination* (Buckingham: Society for Research into Higher Education and Open University Press, 1992).

47. Tony Becher discusses inner and outer academic circles, open and closed groups over the spectrum of the academic profession as a whole in *Academic Tribes and Territories: Intellectual Inquiry and the Cultures of Disciplines* (Buckingham: The Society for Research into Higher Education and Open University Press, 1989), e.g., pp. 66–67.

48. Muthesius, p. 38.

49. Muthesius, p. 158. Another sign of serious and systematic interest was the formation by the Royal Institute of British Architects of the Higher Education Design Quality Forum in 1995 to encourage discussion between architects and university representatives.

Old Books and
E-Books

ROBERT DARNTON

*A*nyone who has admired a copy of Gutenberg's Bible might be tempted to argue that everything in the history of printing has been downhill since Gutenberg. The first important book to be printed from movable type in Europe is also the most perfect. Even its ink cannot be matched today. Yet bibliographers have demonstrated that the Bible was designed with the intention of cutting costs. By increasing the number of lines on a page—at first 40, then 41, finally 42—and by other devices such as running the books of the Bible together instead of starting each of them on a new page, Gutenberg saved a great deal on one of the most expensive items of the production process: paper and vellum. The result leaps to the eye if you compare his layout with that of the "Giant Bible of Mainz," a manuscript work completed between 1452 and 1453, which is profligate in its use of space. Gutenberg's Bible was a down-market product, appropriate for collective use in austere monasteries rather than for prelates and princes. Gutenberg set in motion the process that ultimately brought books into the hands of common people. He began the democratization of print culture.

That story, great as it is, does not necessarily lead to a happy ending. The publishing industry is now going through a crisis. I don't want to join the prophets of doom who have announced "the death of the book," because the book has been declared dead so often that it must be very much alive. Book sales in the United States have increased by 36 percent since 1997. They increased by 4.6 percent last year and brought in $23.4 billion. The number of titles published worldwide each year is now greater than ever—more than a million, or a rate of publication that comes to one book every 30 seconds. Jeremiads about the death of the book remind me of my favorite graffito in the men's room of the university library in Princeton:

God is dead.
          —[*signed*] *Nietzsche*

Nietzsche is dead.
          —[*signed*] *God*

No, the book is not dead, but university presses are dying. Faced with bankruptcy a few years ago, the Presses Universitaires de France closed its famous bookshop in the Place de la Sorbonne. A notice appeared in the shop window explaining that academic publishing had become economically unfeasible and recommended an article on the subject that I had recently published in *Le Débat*. Some polemics followed, which raised enough noise to be called "the Darnton Affair." I was flattered to be the subject of an affair in Paris, but I thought I had been misunderstood. So I would like to explain.

The cost of periodicals has skyrocketed. (The example librarians most love to cite is *Brain Research*, which now costs $21,269 a year, but subscriptions to many journals in the life sciences come to more than $2,000.) The increased expense has produced havoc in the acquisitions budgets of research libraries, which have responded by cutting back drastically in their purchase of monographs. And the drop in the orders for monographs has forced university presses to phase out whole areas of scholarship, especially dissertations on esoteric subjects. You know the cruel catchword of American academic life: "Publish or perish." I fear that young scholars are perishing everywhere—and that many university presses are doomed, too.

Although I am simplifying a complex situation, I think it fair to say that while the commercial book has adjusted to twenty-first century conditions, the scholarly book is an endangered species. My own modest attempt to cope with the danger is a project sponsored since 1999 by the American Historical Association: Gutenberg-e. Thanks to a generous grant from the Andrew W. Mellon Foundation, the AHA conducts an annual competition for the best dissertations. The winners receive a prize of $20,000, which they use to transform their doctoral theses into electronic books. Instead of merely rewriting their text, they redesign it in a fundamental way so that, with the help of computer scientists and specialized editors, they can incorporate all sorts of supplementary material: digitized manuscripts, images, recordings, films, and hyperlinks to other works. Seventeen Gutenberg-e books are now available online. They have received excellent reviews, sold well, and helped their authors to advance in their careers, despite some initial grumbling by older scholars who claimed that e-books can't be real books. The experiment has proved to be such a success that the publisher, the Columbia University Press, now plans to continue Gutenberg-e as a self-supporting operation.

I agree that older scholars should volunteer as guinea pigs instead of performing experiments with the young. So I am now attempting to write an e-book of my own. It is a daunting experience, and it may not work; but I thought I might offer a brief, preliminary report as a way to illustrate my general thesis—namely, that e-books are not inimical to old books; on the contrary, the Internet can open up new ways of understanding how the printed word became one of the most powerful forces in our civilization.

Fifteen years ago, I set out to write a general study of the book trade in prerevolutionary France. I had been through so much material—50,000 letters in the papers of an important Swiss publisher, the Société typographique de Neuchâtel, and thousands of other documents from the archives of the Book Trade Administration in Paris—that I could follow all the twists and turns of the book business in every corner of the kingdom. But that was the problem: I had too much information. I began drafting a chapter on the Loire Valley. First, Orléans, a great market for books: I had to tell the story of Couret de Villeneuve, the most eminent bookseller in town who cut quite a figure as a man of letters in the local academy but had difficulty honoring his bills of exchange. Then Blois: here the underground trade flourished, and I would have to provide a full portrait of Lair, a schoolmaster, who supplied his townsmen (and his students) with a spicy diet of pornography and philosophy. Next, Tours: not important in itself but rather as a base for the operations of the three Letourmy brothers, who peddled woodcuts and topical pamphlets throughout the area, undercutting the business of the established dealers. Then Amboise, then Saumur, then Anger, then Nantes, and detours to key distribution centers like Loudun and La Rochelle, each with its own constellation of fascinating characters. Fascinating, that is, to me. I realized that I would bore my readers to death, and so gave up.

Then along came the Internet. I saw the possibility of writing a new kind of book—in part a conventional narrative, which would be printed in the normal way; in part a collection of monographs and documents, which could be accessed through the Internet.

I now am drafting the first part, a survey of the book trade in provincial France from 1769 to 1789. Because I am trying to write for the general educated reader, not merely for other college professors, I hope to hold the reader's interest by following a "sales rep" (traveling salesman) of the STN, Jean-François Favarger, who climbed on a horse in Neuchâtel at the beginning of July 1778 and spent the next five months riding through southern and central France and inspecting every bookshop on his route. He sent regular reports to the home office, received fresh instructions in reply, and kept a diary with a detailed expense account: 10 sols for refurbishing his pistols in Marseille (the road to Beaucaire was infested with bandits), 26 livres for new breeches in La Rochelle (the old pair did not hold up against the friction in the saddle), and 8 louis for a new horse in Loudun (the old one kept collapsing on the muddy trails and finally had to be sold). It was an extraordinary, un-sentimental journey. When Favarger arrived back in Neuchâtel in December, he knew more about the French book trade than any historian can ever hope to discover.

I intend to tap Favarger's street-level knowledge in writing a chapter about each of his main stops. To avoid boring the reader, I will move fast, in picaresque fashion. Anyone who wants further information can click down through successive layers in the electronic version of the book, which will contain a vast data base linked to the original text.

For example, on July 7 Favarger crossed a pass through the Jura Mountains into France and spent the night in Pontarlier. On this leg of the journey, he concentrated on inspecting supply lines at the border, because the STN did a large business in il-legal and pirated books, which had to be smuggled. Favarger negotiated over costs and risks with a half dozen smugglers in villages on both sides of the border. His re-ports and the correspondence of the smugglers are so rich that I decided to devote a chapter to the smuggling industry.

In it, I explain that two basic kinds of smuggling had evolved in the course of the eighteenth century. The first was called "insurance." Self-styled "assureurs" guaranteed to get shipments of books from taverns on the Swiss side of the border to secret storehouses on the French side. They hired teams of "porters," who re-ceived a stiff shot of schnapps in a Swiss inn—a favorite launching site was chez Jannet in the hamlet of Les Verrières—and then set off along tortuous mountain trails with the books on their backs. A backpack weighed 60 pounds (30 kilos) under normal conditions, 50 pounds when the snow was deep. The men got 25 sols for a successful crossing. If caught, they could be branded with the letters GAL for "galérien" and sent to row for nine years as a galley slave in Marseille. Their boss, the insurer, would then have to reimburse the STN for the full value of the mer-chandise. On January 23, 1784, Ignace Faivre, an insurer in Pontarlier, signed a contract with the STN that committed him to get its shipments across the border "[. . .] at my peril and risk for 15 livres per hundredweight [. . .] In case of any un-happy event ["événement facheux"], I will reimburse the value of the merchandise within one year according to the prices on the bill of lading."

But Faivre had to compete with rivals from the second type of smuggling, what I call the paralegal system. Using various techniques such as "marrying" (larding leaves of prohibited books inside leaves of inoffensive ones), publishers hid danger-ous works in normal shipments and sent them through the legal channels of the trade. The crates were sealed by French customs officers at the border, inspected by officials of the booksellers' guilds in certain "villes d'entrée" such as Lyon, and then sent to their final destination, where they might be inspected yet again. Specialized forwarding agents ("commissionnaires") like Jacques Revol in Lyon often bribed the inspectors to do a superficial job. To be safe, however, they arranged for the wagon drivers to stop at an inn on the outskirts of the ville d'entrée—Aux Trois Flacons in the Croix Rousse suburb of Lyon, for example—where they would unpack the ille-gal books, replace them with legal works, bind the crate back up under a counterfeit seal, and send it on for inspection in the guild hall. Then they would forward the "philosophical books" (a term in the trade for everything dangerous, from pornography to atheism) along another route, disguised as domestic merchandise.

This system worked well so long as the shipping agents and the wagon drivers coordinated their activities. But things were always going wrong. The STN often dealt with Jean-François Pion, a shipper in Pontarlier, who had a great stable of horses. He could yoke up five at a time to get loads over the Juras at the height of winter. His business suffered from one drawback, however: his stupidity. He once sent a crate to Nantes instead of to Rennes. He also forwarded a barrel of sauerkraut to the wrong address in Lyon, and it was devoured before the mistake could be corrected. So despite the excellence of Pion's horses, the STN constantly tried to find a substitute for him among the brainier peasants of the border towns—François Michaut and the Meuron brothers of Les Verrières and St. Sulpice, for example. But the small farmers could not supply enough horses during spring planting and "the cheese season" in the fall. As a consequence of all these complications, the STN kept readjusting its smuggling operations. It was always playing off one smuggler against another, hoping to find cheaper rates and safer service, while calculating factors such as snow, horses, and the maturing rate of cheese.

I subscribe to the slogan: The devil is in the detail. But I do not want to smother my reader with stories about the colorful characters who got books across the Swiss-French border. I therefore plan to restrict my account of smuggling to a short chapter in the printed version of the book. The readers who desire more can log on to the electronic version, click down one level and take their pick of short monographs about Faivre, Revol, Pion, and a half-dozen other professional smugglers. If those short narratives excite more interest, the readers can click down to level three and read through selections from the smugglers' correspondence translated into English. Really serious readers can pursue the trail deeper down to level four, where I will provide transcriptions of entire dossiers in the original French. Transcriptions are invariably imperfect, however, owing to ambiguities in the manuscripts; so specialized scholars can click down to level five and study digitized versions of the originals.

This kind of book will require a new kind of reading, one that proceeds vertically as well as horizontally. It could also involve diagonal zig-zagging, because I plan to intersperse each sector with maps, contemporary engravings of mountain passes, scenes of city streets, accounts of life in country inns, and hyperlinks to related themes in other dossiers. Each reader will find his or her own path through the material. Each will print out the parts that he or she finds most interesting. Each print-out can be trimmed and bound in a matter of minutes, thanks to technological advances in what is already a major industry: Print-on-Demand. The result should be an endless supply of custom-made paperbacks, every one different from all the others.

E-books of this kind will transform the relationship between writers and readers. Readers will become collaborators, or even adversaries, of the scholars who provide the components of each book. Although the material will conform to strict, academic standards, everyone can make of it what they want. There will be no fixed text and no limits, aside from built-in guarantees against falsifying the documents, to the empowerment of the readers.

Perhaps I am succumbing to utopianism. I know there are many objections to what I have proposed. I know it has flaws, and I shudder at the prospect of becoming entangled in the World Wide Web. But whether or not I succeed in this particular task, I hope I have said enough to substantiate the argument that old books and e-books are not enemies. They are allies. We need to build a strong alliance if we are to overcome the current crisis and to collaborate in the effort to expand Gutenberg's everlasting galaxy.

Notes on Contributors

Tiziana Bascelli is a Ph.D. student in the Department of Philosophy at the University of Padua where she is working on the science of motion in the sixteenth century and the heritage of Archimedean mechanics. She recently edited (with Willliam R. Shea) Campanus of Novara's *Equatorium Planetarum.*

Jean-François Battail is professor emeritus in Scandinavian studies at Sorbonne. At present he is guest professor at the University of Tartu in Scandinavian studies. His latest publications are *Les Destinées de la Norvège moderne, 1814–2005* (Paris, 2005) and *Un Savant suédois à la conquête du monde— Carl von Linné* (Paris, 2007).

Marco Beretta teaches history of science at the University of Bologna and is Vice-Director of the Institute and Museum of History of Science in Florence. He has recently co-edited *Linneaus in Italy* (2007) and *Lucrezio: La natura e la scienza* (forthcoming).

Janet Browne is Aramont Professor of the History of Science at Harvard University. Her interests range widely over the history of the life sciences and natural history. After a first degree in zoology she studied for a PhD in the history of science at Imperial College London, published as *The Secular Ark: Studies in the History of Biogeography* (1983). Ever since then she has specialised in reassessing Charles Darwin's work, first as associate editor of the early volumes of *The Correspondence of Charles Darwin,* and more recently as author of a major biographical study that integrated Darwin's science with his life and times.

Robert Darnton is Carl H. Pforzheimer University Professor at Harvard and Director of the Harvard University Library. His latest book was *George Washington's False Teeth: An Unconventional Guide to the Eighteenth Century.* He has recently completed a large-scale study of slander and the art and politics of libel in eighteenth-century France.

Lorraine Daston is director at the Max Planck Institute for the History of Science in Berlin, Germany. She is the author of *Classical Probability in the Enlightenment,* coauthor (with Katherine Park) of *Wonders and the Order of Nature, 1150–1750* and (with Peter Galison) of *Objectivity.*

Paolo Galluzzi is Director of the Institute and Museum of History of Science in Florence and, since 1982, Professor of History of Science at the University of Florence. He is a member of the Royal Academy of Sciences in Stockholm and *Socio* of the Accademia Nazionale dei Lincei. He is the author of more than 200 publications on Renaissance science and technology, on Galileo and his school, on the history of the European scientific academies, and on the birth and history of the historiography of science.

Karl Grandin is Professor and Director of the Center for History of Science at the Royal Swedish Academy of Sciences, Stockholm. He has co-edited The *Science–Industry Nexus: History Policy Implications* (2004), and recently published "The rise and fall of the Berzelius Museum," in *The laboratorio Chimico of the Polytechnic school of Lisbon* (2008) and with A. Eriksson "Scientific instruments," in *The Linnaeus Apostels—Global Science and Adventure* (2008). In 2006, he became editor of *Les Prix Nobel.*

John L. Heilbron was a Professor of History at the University of California, Berkeley. His most recent publications, in out-of-the-way places, are centered on the polymath Francesco Bianchini. His current projects include a biography of Bianchini and a study of parallel developments in historiography and natural science in early modern times.

Daniel J. Kevles is the Stanley Woodward Professor of History at Yale University. His latest publications include: "The Poor Man's Atomic Bomb," *New York Review of Books,* April 12, 2007, pp. 60–63; "Howard Temin: Rebel of Evidence and Reason," in Oren Harman and Michael Dietrich, eds., *Rebels,*

Mavericks, and Heretics in Biology (forthcoming, Yale University Press, 2008); and "Patents, Protections, and Privileges: Intellectual Property Protection in Animals and Plants," *Isis*, June 2007, pp. 323–331. He is currently completing a book on the history of innovation and intellectual property in living matter.

SVANTE LINDQVIST is Director of the Nobel Museum. Previously he was Professor of History of Technology at the Royal Institute of Technology, Stockholm. Among his latest publications is "The R & D Production Model: A Brueg(h)elesque Alternative," in Guy Neave *et al.*, eds., *The European Research University: An Historical Parenthesis?* (New York: Palgrave Macmillan, 2006), pp. 77–90. Lindqvist has for some years been working on a cultural history of Swedish science and technology, focusing on Hannes Alfvén.

THORSTEN NYBOM is Professor of History and Deputy Vice-chancellor at Örebro University. Among his recent publications are: with G. Neave and K. Blückert, *The European Research University—An Historical Parenthesis?* (Palgrave, 2006); "A Rule-governed community of Scholars: The Humboldt Vision in the History of the European University", in P. Maassen & J.P. Olsen, eds., *University Dynamics and European Integration* (Springer, 2007); with L. Engwall, "The Visible Hand vs. the Invisible Hand: Allocation of Research Resources in Swedish Universities," in R. Whithley, ed., *The Changing Governance of the Sciences: The Advent of Research Evaluation Systems* (Springer, 2007).

MARY JO NYE is Horning Professor of the Humanities and Professor of History at Oregon State University in the USA. She recently has written *Blackett: Physics, War, and Politics in the Twentieth Century* (Harvard University Press, 2004) and an *Isis* (2006) essay on "Scientific Biography: History of Science by Another Means?" Her latest articles on Michael Polanyi in the *Journal of Computational Chemistry* (2007) and in *Historical Studies in the Physical and Biological Sciences* (2007) are part of a larger study on "Michael Polanyi: Scientific Life and the Philosophy of Science in the Twentieth Century."

SHELDON ROTHBLATT is Professor of History Emeritus at the University of California, Berkeley. His latest book is a study of the meritocratic concept in three nations. Its title is "Education's Abiding Moral Dilemma: Merit and Worth in the Cross-Atlantic Democracies, 1800–2006." He is currently preparing the Sir Douglas Robb Lectures to be given at the University of Auckland in March 2008 on the occasion of that university's Jubilee.

NICOLAAS A. RUPKE is professor of the history of science at Göttingen University. Among his latest publications is *Alexander von Humboldt: A Metabiography* (Chicago University Press, 2008). He is currently working on non-Darwinian traditions in late-modern biology.

WILLIAM R. SHEA is Galileo Professor of History of Science at the University of Padua. He is the author of several books including *Designing Experiments and Games of Chance: The Unconventional Science of Blaise Pascal* (Science History Publications, 2003), and with Mariano Antigas *Galileo in Rome* (Oxford University Press, 2003) and *Galileo Observed* (Science History Publications, 2006).

H. OTTO SIBUM is Hans Rausing Professor of History of Science and Director of the Office for History of Science at Uppsala University. He has published extensively on the history of experimentation and is co-editor of: *Instruments, Travel and Science. Itineraries of Precision from the 17th to the 20th Century*, London and New York: Routledge 2002; *Scientific Personae*. Science in Context 16 ($\frac{1}{2}$) Cambridge University Press 2003; *The Heavens on Earth. Observatory Techniques in the Nineteenth Century* forthcoming Duke University Press 2009. He is guest-editing a double volume in Studies in History and Philosophy of Science (September 2008) titled *Science and the Changing Senses of Reality circa 1900*. He is currently preparing a book on *Science and the Knowing Body*.

SVEN WIDMALM is Professor at the Department of Technology and Social Change, Linköping University. He has worked on the history of astronomy, physics, and biochemistry. At the moment he is working on the relations between policy and science in post-war Sweden.

Index

Académie des sciences of Paris, 56, 61, 66, 77
Academy of Vestiges, 52
Accademia degli Inquieti, 52, 63
Accademia del Cimento, 52, 55
Accademia della Traccia, 52, 63
Accademia delle accademie, 65
Accademia dell'esperienze filosofiche e
    naturale, 55
Accademia fisico-matematica romana (AFR),
    53, 55, 56
Accademia reale, 53, 54, 57
Adlum, John, 136
AFR. *See* Accademia fisico-matematica romana
    (AFR)
Afrodisio, Selvaggio, 57
Agassiz, Louis, 73, 78
Aggiunti, Niccolò, 32
Agnello, Andrea, 59
Alabama A&M, 286
Albani (Cardinal), 58
Albani, Allessandro, 62
Alberti, 26
Aletofili, 54, 64
Alexander, James, 136
Alexander VII, 53, 55
Alexander VIII, 56, 58
Alfvén, Hannes, 217–230
Althoff, Friedrich, 120
American Geophysical Union, 227–228
American Historical Association, 300
American Pomological Society, 133, 137, 141,
    142
Ampère, André Marie, 168

Andrew W. Mellon Foundation, 300
Apollonius of Perga, 44
Aquinas, Thomas, 53, 175
Arcadi, 57–58
Arcadia, 65
Arcadians, 57–58, 60
Archimedes, 34, 44, 169
Århus University, 289
Aristotle, 3, 6, 20, 72, 174
Arizona State University, 280
Armenian Academy of Sciences, 224
Armitage, Angus, 246
Arrhenius, Svante, 216
Association of Scientific Workers (ASW), 235,
    239, 245
Atkins, Hedley, 90, 94
Auerbach, Felix, 181, 188–190
Aurora Borealis, 184

Babbage, Charles, 103
Bacon, Francis, 64, 150
Bacchini, Benedetto, 53, 59–61, 66
Baer, Karl Ernest von, 75
Baker, John, 249
Baldwin, Colonel, 131
Balzac, Honoré de, 148
Bank Charter Act, 1844, 95
Bank of England, 95–96
Bard College, 281
Barry, Patrick, 140, 142
Bartlett, Enoch, 132
Bartram, John, 132
Bayesianism, 168, 173

Beagle, 91, 98
Bech, Ulrich, 260
Beethoven, Ludwig van, 283
Belidor, Bernard Forest de, 182
Ben-David, Joseph, 117
Bergson, Henri Louis, 158
Bergström, Lars, 218, 222, 229
Bernal, John Desmond, 121, 236–239, 242,
    244–251
Bernalism, 236, 245, 248, 251
Bianchini, Francesco, 51–66
Big-bang theory, 216, 219–222, 225, 228–230
Big Science, 259
Birkbeck College, 239
Birkbeck Physics Society, 235, 241
Bismarck, Otto Eduard Leopold von, 157
Björklund, Jan, 267
Blackett, P.M.S., 235–246, 248, 250–252
Blomberg, Stig, 281
Blumenbach, Johann, Friedrich, 72, 74–77, 80,
    83
Bohr, Niels, 119, 198, 202, 203, 206, 216
Bondi, Hermann, 225
Book Trade Administration (Paris), 301
Born, Max, 195, 196, 198
Bosch, Carl, 119
Boston Public Garden, 133
Boyle, Robert, 54
Bragg, William Lawrence, 197–200, 236
Brahe, Tycho, 56
Braithwaite, R.B., 240, 242
Brasavola, Girolamo, 53
Brenning, Nils, 226
Brentano, Robert, 279
Brillouin, Léon, 197
British Society for the History of Science, 235,
    246
Bronn, Heinrich Georg, 77
Browne, Janet, 79
Brush, Stephen G., 224
Buch, Leopold von, 77
Buckland, William, 73
Buffon, Georges Louis Leclerc, Comte de, 185
Bukharin, Nikolai, 237, 238
Bull, Ephraim, 141
Bunsen, Robert, 185
Burdach, Karl Friedrich, 75
Burke, Edmund, 280
Burmeister, Hermann, 77
Bush, Vannevar, 121, 262

Butterfield, Herbert, 247

Cambridge University, 186, 287, 293
Cambridge Wranglers, 185–186
Campani, Giuseppe, 56
Campbell, Norman R., 242
Candolle, Alphonse de, 79
Cantor, Georg, 152
Caprone, Giovanni Battista, 63
Carcavy, Pierre, 47
Carlqvist, Per, 218–220, 229
Carlyle, Thomas, 102
Caront, Sadi, 105, 151
Cartwright, Nancy, 166
Carus, Carl Gustav, 80, 81
Cassini, Gian Domenico, 61
Castelli, Benedetto, 33, 41
Castells, Manuel, 260
Catholic University of Nijmegen, 288
Celsius, Anders, 163
Chapman, John, 131
Charlottenburg, 291
Christina, queen of Sweden, 53, 54, 57, 58, 66
Churchill, Winston, 239
Ciampini, Giovanni Giusti, 53–58, 63, 64, 66
Clausius, Robert, 151
Clement IX, 53
Clement X, 53
Clement XI, 52, 58, 61, 62
Cockroft, John, 237, 239
Colbert, John Baptiste, 56
Collegio Romano, 55
Collins, Harry, 166
Columbia University Press, 300
Comte, Auguste, 149, 150, 151, 158, 162
Conant, James B., 242
Confederation of Swedish Enterprise, 264
Conferenza dei concili, 53, 54, 64
Congresso medico romano, 53, 55
Cook, Hal, 91, 104
Copernicus, Nicolaus, 56
Corbusier, 288
Council of Trent, 52
Coxe, William, 134
Crescimbeni, Giovanni, 57, 58, 65
Cronin, James W., 220
Crossman, Richard H.S., 249
Crowther, James G., 237, 238, 242, 246, 247,
    252
Czolbe, Heinrich, 72

Dampier, William, 249
Dante Alighieri, 163
Darwin, Charles Galton, 195, 197, 202
Darwin, Charles, 5, 71–72, 79, 81–83, 87–95,
    97–106, 149, 153, 283
Darwin, Emma (wife of Charles), 90, 91, 94, 97,
    99, 100
Darwin, Erasmus (brother of Charles), 94
Darwin, Francis (son of Charles), 91
Darwin, Dr. Robert Waring (father of Charles),
    90
Darwin, William (son of Charles), 94
Davis, Natalie Zemon, 167
de Beer, Gavin, 249
Debye, Peter, 195, 196, 197, 198
Democritus, 7
Descartes, René, 52, 53, 54, 150, 153, 176
Desmond, Adrian, 103
Dessler, Alex J., 227
Dickens, Charles, 97
Diderot, Denis, 182
Dietrichstein, Franz, 32
Dietzgen, Joseph, 187
Dilthey, Wilhelm, 120
Dingle, Herbert, 246, 249
Dirac, Paul A.M., 194, 200–209, 221, 239
Docklands (East London), 287
Drake, Stillman, 39
Dumas, Alexandre, 148
Du Pont de Nemours, Eleuthère-Irénée, 132
Dupré, John, 166

Edison, Thomas, 154
Edquist, Charles, 264
Ehrenfest, Paul, 202
Einstein, Albert, 119, 195, 209
Eliade, Mircea, 160
Ellwanger, George, 140, 142
Elsevier, Louis, 32, 33
Elster, 154
Emmanuel College, Cambridge, 279
Engels, Friedrich, 248
Enlightenment, the, 112
Epicureanism, 2, 3
Epicurus, 3, 4, 8, 10
Eriksson, Per, 268
Erskine, Ralph, 282
Erxleben, Johann Christian Polykarp, 184
Eschinardi, Francesco, 55, 57
Etzkowitz, Henry, 261

Euclid, 19, 35, 173, 174
Ewald, Paul Peter, 197
Experimental vorlesungen, 185

Faivre, Ignace, 302, 303
Fälthammar, Carl-Gunne, 226
Farrington, Benjamin, 246
Favarger, Jean-François, 301, 302
Faxén, Hilding, 196
Fermi, Enrico, 196
Ficino, Marcilio, 23, 26
Findlay, Alexander, 249
Firth, E.M., 198
Fitch, Val L., 220
Flamsteed, John, 63
Flexner, Abraham, 118
Fokker, Adriaan Daniël, 197
Forbes, Edward, 81
Forman, Paul, 252
Fowler, Ralph, 239
Franck, James, 196
Frängsmyr, Tore, 51, 66, 71–72, 81, 87, 91,
    148–149, 153, 155, 156, 158, 164
Franklin, Benjamin, 132
Fraunhofer, Joseph, 185
Freud, Sigmund, 158
Friedrich Wilhelm III, king of Prussia, 117
Friedrich-Wilhelms-Universität zur Berlin, 116,
    118
Frosini, Fabio, 26

Gaitskell, Hugh, 249
Galeate, Leucoto, 58
Galen, 54
Galiani, Celestino, 62, 63
Galileo Galilei, 31–38, 40, 41, 43–49, 60, 230, 242
Galison, Peter, 197
Galton, Francis, 87–88, 100–101, 105
Gamow, George, 202
Gassendi, Pierre, 53
George I of Hanover, 62
German Physical Society, 185
Gibbons, Michael, 263
Gibbs, Isabella, 136
Gilman, Daniel Coit, 118
Ginzburg, Carlo, 167
Glass, Bentley, 82
Goethe, Johann Wolfgang von, 81
Goitel, 154
Gorky, Maksim, 148

Gothenburg University, 289
Göttingen University, 112, 184
Graham, Loren, 249
Great Eastern, The, 155
Greeks, 1
Gregory, Richard, 248
Grisebach, August, 79
Guidobaldo del Monte, 45
Gustafsson, Bengt, 229
Gutenberg, Johann, 299, 304
Gutenberg-e, 300

Haber, Fritz, 119
Hacking, Ian, 166
Haldane, John Burdon Sanderson, 248
Hall, Peter, 283
Halley, Edmond, 63, 245
Handwerksgelehrte, 185, 187
Hardenberg, Fürst, 117
Harnack, Adolf von, 120
Harrach, Ernest Adalbert, 33
Hartree, Douglas R., 198, 199, 200, 203, 209
Harvard University, 279
Harvey, William, 186
Hasbrouck, Jonathan, 131
Hauksbee, Francis, 63
Hedrick, U.P., 129, 135
Hegel, Georg Wilhelm Friedrich, 149
Heisenberg, Werner, 196, 202, 208
Heitler, Walter, 202
Helmholz, Hermann von, 119, 185–187
Heraclitus, 177
Hertz, Heinrich Rudolf, 154
Herzfeld, K., 197
Hessen, Boris, 236–237, 238, 240,
    242, 243
Hetzel, 147, 157
Hevelius, Johann, 56
Hilbert, David, 196
Hodgkin, Dorothy, 236
Hoff, K.E.A. von, 77
Hoffman, E.T.A., 151
Hogben, Lancelot, 237
Holtsmark, Johan, 202
Homer, 162
Hooke, Robert, 245
Hossfeld, Uwe, 82
Hovey, Charles M., 134, 135, 139–142
Hovey, Phineas, 134, 139–140
Humanistisches Gymnasium, 114

Humboldt, Alexander von, 114
Humboldt, Wilhelm von, 112–115, 118, 120,
    122, 123
Humboldt University, 111–114, 120, 262
Huxley, Aldous, 157, 237, 251
Huxley, Julian, 237, 240
Huxley, Thomas, 103, 149

Imperial College of Technology, 291
Innocent XI, 52–53, 54, 55, 57
Innovation Society, 266–268, 272
International Summer School in the History of
    Science, 51, 66

Jacobi, Hermann Moritz, 185
James, Patricia, 101
James, Reginald W., 197, 198, 199, 200, 203, 209
James III, 62
Jefferson, Thomas, 129, 136, 282
Joffe, Abram F., 237
Johnsson, Ivar, 281
Jordan, Pascual, 202
Jörg, Christian Gottfried, 74
Joule, James, 185
Jubilee campus (Nottingham), 287
Junker, Thomas, 82

Kafka, Franz, 157
Kaiser-Wilhelm-Gellschaft, 120
Kaiser-Wilhelm Institute, 120
Kant, Immanuel, 112, 115, 151
Kapitsa, Peter, 237, 238
Karlstad University, 288, 289
Kármán, Theodor von, 195
Keill, John, 62
Kenrick, John, 133
Kenrick, William, 133
Kepler, Johannes, 171
Kerr, Clark, 284, 285
Kevles, Daniel, 229
King's College, Cambridge, 279
King's College, London, 291
Kitcher, Philip, 166
Klein, Oskar, 198, 202, 216–218, 220, 226, 227,
    230
Klein-Alfvén theory on cosmology, 216, 217,
    218, 219, 221, 221, 222, 223, 224, 225, 228,
    229, 230
Knight, Thomas Andrew, 141
Koch, Robert, 119, 176

Koerner, Lisbet, 91
Koyré, Alexandre, 247
Kragh, Helge, 221–222, 228
Krahwinkel (Berlin), 112
Krauss, Lawrence M., 225, 226
Kronig, Ralph, 202
Kuhn, Thomas, 185, 271

Lallerstedt, Eller, 281
Lamarck, Jean Baptiste Pierre Antoine de
    Monet, 153
Lamethérie, Jean-Claude de, 81
Lancisi, Giovanni Maria, 53, 58
Landau, Lev, 238
Laplace, Pierre Simon de, 151
Larsson, Axel (Nordhalt), 196
Latour, Bruno, 98, 166
Laue, Max von, 197, 198
Laurent, Antoine, 105
Leibniz, Gottfried Wilhelm von, 41, 56, 63
Leiden University, 112
Leijonborg, Lars, 267
LeMaître, Abbé, 218, 224
Lenin, Vladimir Ilyich, 248
Leonardo da Vinci, 13–26, 154
Lerner, Eric J., 229
Letourmy brothers, 301
Leuven University, 290
Leydesdorff, Loet, 261
Lichtenberg, Georg Christoph, 184
Liebing, Justus, 119, 185
Lindemann, Frederick, 239
Linköping Univeristy, 288, 290, 293
Linnaean Botanic Garden, 133
Linné (Linnaeus), Carl von, 71, 72, 81–83, 91,
    152
Lobachevski, 152
Longfellow, Henry Wadsworth, 136
Longworth, Nicholas, 135, 136, 138
Louis XIV (king of France), 56, 62
Louvain-la-Neuve, 290–291
Lowell, John, 132
Lucretius, 1–10, 72
Lund University, 263, 290
Lundvall, Bengt-Åke, 260
Lyell, Charles, 78, 81, 105
Lysenko, Trofim, 249

Mabillon, Jean, 53
Mach, Ernst, 177

Machiavelli, Niccolò, 26–27
Magliabecchi, Antonio, 53, 60
Magnus, Gustav, 185
Maitland, Frederic, 283
Mallarmé, Stephané, 148
Malpighi, Marcello, 58
Malthus, Thomas Robert, 101–102
Manfredi, Eustachio, 52, 63
Mann, Thomas, 119
Mark, H., 197
Marklund, Göran, 226
Marsili, Luigi Ferdinando, 52, 60
Martineau, Harriet, 102
Marx, Karl, 248
Massachusetts Horticultural Society, 133, 134,
    135, 141
Massachusetts Institute of Technology, 291
Massachusetts Society for the Promotion of
    Agriculture, 133
Maupertuis, Pierre Louis de, 163
Maxwell, James Clerk, 185, 186, 187
Mayer, Anna-K., 247
McKie, Douglas, 246
McOuat, Gordan, 104
Meijer, Cornelius, 55
Meitner, Lise, 207
Memmius, 2
Merton, Robert K., 116, 242, 250, 252
Merton College, Oxford, 36
Merton rule, 36
Meuron brothers, 303
Meyen, Franz, 79, 81
Micanzio, Fulgenzio, 32
Michaut, François, 303
Milles, Carl, 281
Mittelstrauss, Jürgen, 116
Moltke, Helmuth von, 170
Montanari, Geminiano, 51, 52, 55–56, 63, 64
Moore, James R., 103
Mount Hope Botanical and Pomological
    Garden, 140
Mozart, Wolfgang Amadeus, 161
Müller, Johannes, 76
Munch, Edvard, 281
Muratori, Ludovico Antonio, 58–61, 65, 66
Museum of the History of Science (Oxford),
    235
Muthesius, Stefan, 283, 294

Needham, Joseph, 247–249

Newman, John Henry, 123, 277–278, 280, 292
Newton, Isaac, 43, 48, 49, 62, 63, 163, 169, 242, 245
New York University, 292, 293
Nietzsche, Friedrich, 118, 158, 165, 166, 300
Nilsson, Peter, 229
Nishina, Yoshio, 202
Nobel, Alfred, 152
Nollet, Jean Antoine, 183, 185
Nordenskiöld, Adolf Erik, 153
Nordström, Thomas, 263, 265
Nowotny, Helga, 260, 263
Nozick, Robert, 172

Occhialini, Giuseppe, 238
Odense University, 282
Offenbach, Jacques, 156
Oken, Lorenz, 77, 81
Ørsted, Hans Christian, 168
Orwell, George, 157
Oseen, C.W., 195–197, 202–203, 209
Ott, H., 197
Oxford University, 287, 293

Pagel, Walter, 247
Paris Academy of Sciences. *See* Académie des sciences of Paris
Parkland College, 288
Parkman, Francis, 133
Pascal, Blaise, 31, 230
Pasteur, Louis, 73, 119, 152, 153
Pauli, Wolfgang, 195, 197, 200, 201, 203, 204, 207, 208, 216
Pease, Sebastian, 222
Peebles, P. James E., 221
Pell, Robert L., 139
Penn, William, 136
Penzias, Arno, 218
Perkin, Harold, 88, 103
Perthes, Jacques Boucher de, 152
Pestalozzi, Johann Heinrich, 118
Piazza di San Marco (Venice), 284
Piccolomini, Ascanio, 31
Pieroni, Giovanni, 32
Pinch, Trevor, 166
Pion, Jean-François, 303
Pirie, Bill, 238
Pirie, Tony, 238
Pitt, William the younger, 101
Planck, Max, 119, 216
Plato, 7, 48, 161, 172, 244, 288

Poe, Edgar Allan, 161, 162
Polanyi, Michael, 248–252
Poovey, Mary, 105
Pope, Alexander, 242
Porter, Theodore M. (Ted), 87, 91, 104
Postan, Michael, 247, 249
Pouchet, Félix Archimède, 81, 152
Presses Universitaires de France, 300
Prince, Robert, 131, 132
Prince, William, 131, 132
Prince, William (junior), 133
Prince Nursery, 131
Pritanio, Lamindo, 58–61, 65, 66
Pritchard, James Cowles, 80
Prochaska, Frank K., 100

Ranke, Leopold, 165
Rathke, Martin Heinrich, 76
Raven, Charles, 247
Réaumur, René Antonio Ferchault de, 185
Repubblica letterata d'Italia (RLI), 58–60, 65, 66
Republic of Letters, 51, 58, 60, 65, 183
Research Grant University, 284, 285
Revol, Jacques, 302, 303
Rickert, Heinrich, 120
Rimbaud, Jean Nicholas Arthur, 148
Ritgen, Ferdinand August, 74
Ritter, Carl, 78
RLI. *See* Repubblica letterata d'Italia (RLI)
Roman Catholic Church, 65
Romano, Elisa, 1–2
Romans, 1, 2
Röntgen, Wilhelm, 119, 181
Rosovsky, Henry, 292
Rosseland, Svein, 202
Rostow, W.W., 252
Rothblatt, Sheldon, 112
Royal Holloway College, 291
Royal Institute of Technology (KTH), 217, 226, 281, 282, 291
Royal Observatory, 61
Royal Society of London, 56, 63, 64, 66, 88, 185
Royal Swedish Academy of Engineering Sciences, 264, 268
Royal Swedish Academy of Sciences, 218
Rudolphi, Karl Asmund, 80
Rutherford, Ernest, 119, 236, 239

Saarinen, Eero, 285
Samford University, 282

Sand, George, 148
Sarpi, Paolo, 34
Sarton, George, 242–246, 249
Schabas, Margaret, 91
Scheiner, Christopher, 32
Schelstrate, Emmanuel, 54, 63
Scherrer, Paul, 197
Schinkel, 281
Schmarda, Ludwig, 79
Schouw, Joakim Frederik, 79
Schrödinger, Erwin, 197, 203
Scientific Revolution, 111
Seckel, Mr., 131
Sedgwick, Adam, 73
Shakespeare, William, 283
Shapin, Steven, 99
Siegbahn, Manne, 194, 196, 209, 217
Siemens, Werner von, 120
Silk, Joseph, 221
Simpson, Celia, 237
Singer, Charles, 237, 242, 246, 247, 249
Singer, Dorothy Waley, 247
Singh, Jagjit, 221
Smiles, Samuel, 99, 102
Smith, Adam, 91
Snow, C.P., 239 240, 249, 251, 152
Société typographique de Neuchâtel (STN), 301
Society for Freedom in Science, 249
Society for Social Studies of Science, 251
Södertörn University, 289
Spary, Emma, 91
Spencer, Herbert, 94, 105, 149
Stahl, P.J., 147
Stanford University, 284
Steady-state theory, 216
Stehr, Nico, 260
Steigman, Gary, 220
Stockholm University, 282, 288, 290
Strindberg, August, 148, 154
Sunshine Coast University College, 282, 287, 288
Sussex University, 287
Svanberg, Jöns, 163
Swarthmore College, 284
Swedenborg, Emanuel, 154
Swedish Research Council, 261
Syme, Ronald, 1

Tansley, George, 249
Tawney, R.H., 242
Taylor, Frank Sherwood, 235, 242, 246, 249

Technische Hochschulen, 120
Technische Universitäten, 286
Temkin, Owsei, 82
Tham, Carl, 260
Theophrastus, 6
Thomas, Hugh Hamshaw, 247, 249
Thomson, William, 105
Tizard, Henry T., 236
Torell, Lena Treschow, 264, 268, 270
Torre di Mangia (Siena), 283
Torricelli, Evangelista, 31, 35, 39, 41–44, 48–49
Treschow, Michael, 270, 271
Treviranus, Gottfried Reinhold, 75
Trevisan, Bernardo, 58, 59
Triple Helix model, 261, 263, 266
Trollope, Anthony, 99
Trow, Martin, 286
Turner, Paul Venable, 278
Tyson, Jonathan, 131

Umeå University, 289–290
University College London, 291
University of
    Berlin, 114, 116, 123, 281
    Birmingham, 283
    Bologna, 52
    California, Berkeley, 278, 283, 285, 289
    California, Los Angeles, 287
    California, San Diego, 287
    California, Santa Cruz, 284, 287
    Canterbury, 287
    Essex, 293
    Kent, 288
    London, 284, 291
    Montréal, 290
    Norwich, 287
    Oslo, 281, 283, 290
    Padua, 51
    South Denmark, 282
    Twente, 281
    Virginia, 282
    York, 287
Uppsala University, 280–281, 283, 288, 290
U.S. National Academy of Science, 66

Valentin, Gabriel Gustav, 74
Vanderbilt University, 287
Van Mons, Jean Baptiste, 141
Vanni, Francesco, 55
Vassar, Matthew, 291
Vassar College, 291

Vavilov, Nikolai I., 237, 248
Veblen, Thorsten, 285
Venturi, Robert, 281
Verne, Jules, 147–164
Vico, Giambattista, 66
Villeneuve, Couret de, 301
Vinnova, 261, 263, 266–270
Virchow, Rudolf, 72
Virgil, 283
Viviani, Vincenzo, 38–41
Vogt, Carl, 78
Volta, Alessandro, 184
vom Stein, Freiherr, 117
von Hartmann, 158

Waddington, C.H., 240
Wallace, Alfred Russel, 105
Waller, Ivar, 193–200, 202–209
Warwick, Andrew, 194
Washington, George, 132
Watson, James, 251
Weber, Max, 116, 120
Weber, Wilhelm, 185
Wedgwood, Emma, 91

Wedgwood, Josiah, 94
Wedgwood, Susanna, 88
Werskey, Gary, 250, 251
Whewell, WIlliam, 246
Whitehead, Alfred North, 242
Whittaker, E.T., 249
Wilder, Marshall P., 133, 137–138, 139, 142,
     143
Wilkes, Commander Charles, 142
Willey, Basil, 247
Wilson, Harold, 249
Wilson, Robert, 218
Winkler, Johann Heinrich, 184
Wolf, Abraham, 242, 246
Wolff, Christian, 182
Woolgar, Steve, 98
Wren, Christopher, 245
Wyckoff, R.W.G., 197

Yale University, 285
Young, Robert, 82, 88, 103

Zuckerman, Solly, 239

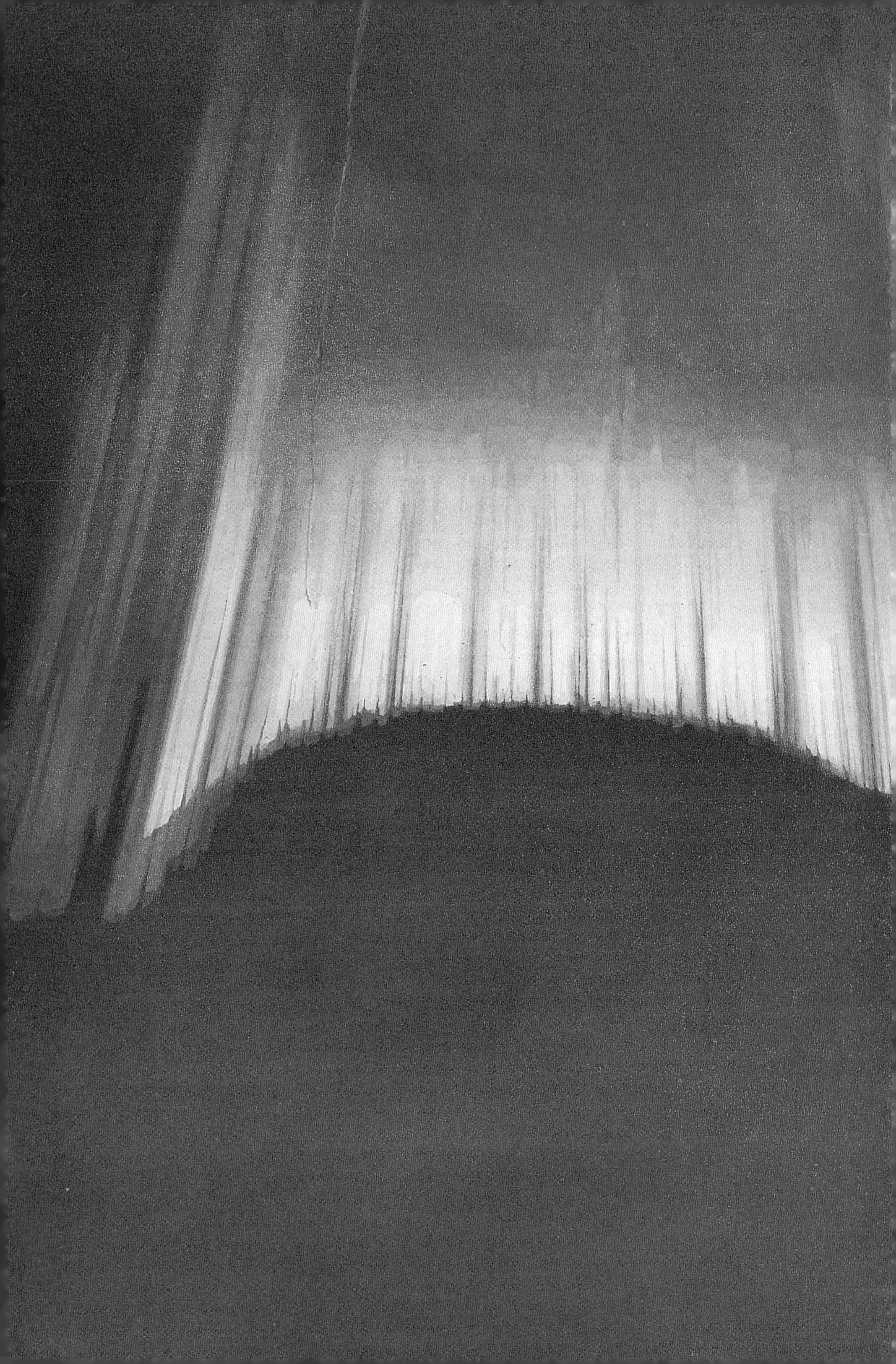